AF887190

Cooperation in the Law of Transboundary Water Resources

Climate change, population growth and the increasing demand for water are all capable of leading to disputes over transboundary water systems. Addressing these challenges will require enhancement of adaptive capacity, improvement of the quality of water resources management and reduction in the risk of conflict among riparian states. Such changes can be brought about only through significant international cooperation.

Christina Leb's analysis of the duty to cooperate, as well as the related rights and obligations, highlights the interlinkages between this duty and the principles of equitable and reasonable utilisation and the prevention of transboundary harm. In her analysis, she considers the law to be applicable to both international watercourses and transboundary aquifers. She explores the complementarities and interaction between the rules of international water law and the related obligations of climate change and human rights law.

CHRISTINA LEB is an associate member of the Platform for International Water Law at the University of Geneva and a consultant for the World Bank. Her primary focus is on transboundary water resources management and cooperation.

CAMBRIDGE STUDIES IN INTERNATIONAL AND COMPARATIVE LAW

Established in 1946, this series produces high quality scholarship in the fields of public and private international law and comparative law. Although these are distinct legal sub-disciplines, developments since 1946 confirm their interrelations.

Comparative law is increasingly used as a tool in the making of law at national, regional and international levels. Private international law is now often affected by international conventions, and the issues faced by classical conflicts rules are frequently dealt with by substantive harmonisation of law under international auspices. Mixed international arbitrations, especially those involving state economic activity, raise mixed questions of public and private international law, while in many fields (such as the protection of human rights and democratic standards, investment guarantees and international criminal law) international and national systems interact. National constitutional arrangements relating to 'foreign affairs', and to the implementation of international norms, are a focus of attention.

The Series welcomes works of a theoretical or interdisciplinary character, and those focusing on the new approaches to international or comparative law or conflicts of law. Studies of particular institutions or problems are equally welcome, as are translations of the best work published in other languages.

General Editors James Crawford SC FBA
Whewell Professor of International Law, Faculty of Law, University of Cambridge
John S. Bell FBA
Professor of Law, Faculty of Law, University of Cambridge

A list of books in the series can be found at the end of this volume.

Cooperation in the Law of Transboundary Water Resources

Christina Leb

CAMBRIDGE
UNIVERSITY PRESS

CAMBRIDGE
UNIVERSITY PRESS

University Printing House, Cambridge CB2 8BS, United Kingdom

Published in the United States of America by Cambridge University Press, New York

Cambridge University Press is part of the University of Cambridge.

It furthers the University's mission by disseminating knowledge in the pursuit of education, learning and research at the highest international levels of excellence.

www.cambridge.org
Information on this title: www.cambridge.org/9781107035973

© Christina Leb 2013

This publication is in copyright. Subject to statutory exception and to the provisions of relevant collective licensing agreements, no reproduction of any part may take place without the written permission of Cambridge University Press.

First published 2013

A catalogue record for this publication is available from the British Library

Library of Congress Cataloguing in Publication data
Leb, Christina.
Cooperation in the law of transboundary water resources / Christina Leb.
pages cm. – (Cambridge studies in international and comparative law)
Includes bibliographical references and index.
ISBN 978-1-107-03597-3 (hardback)
1. Water rights (International law) 2. Water resources development – Law and legislation. 3. Watershed management – Law and legislation. 4. Riparian rights. 5. Watershed management – International cooperation. 6. Water resources development – International cooperation. 7. Water-supply – Management. I. Title.
K3498.L43 2013
341.4′4 – dc23 2013013120

ISBN 978-1-107-03597-3 Hardback

Cambridge University Press has no responsibility for the persistence or accuracy of URLs for external or third-party internet websites referred to in this publication, and does not guarantee that any content on such websites is, or will remain, accurate or appropriate.

In memoriam Professor Enzo Grilli

Contents

Acknowledgements

This book is the result of a long journey that started at the School of Advanced International Studies of Johns Hopkins University, where I was introduced to the complexities related to the management of transboundary resources by Professor Ariel Dinar. It was Professor Enzo Grilli's advice to first work in the field of interest before choosing a topic for doctoral research that eventually led to this book. I am immensely grateful to him for this advice. This book benefits from the experience I was privileged to gain when working for many years with the Nile Team at the World Bank. It is based on my doctorate thesis written at the University of Geneva under the skillful guidance of my thesis advisor, Professor Laurence Boisson de Chazournes.

Many other people directly or indirectly contributed to this book by providing guidance and advice, sharing their knowledge, reviewing and commenting on earlier drafts, making incidental comments or simply by offering encouragement as colleagues, friends and parents. Special thanks go to Undala Alam, Don Blackmore, Christian Bovet, Stefano Burchi, Lindsey Cameron, Marwa Daoudy, Edouard Fromageau, Jakob Granit, David Grey, Julia Grignon, Birgit Kainz-Labbé, Elizabeth Kistin, Anne-Marie Leb, Georg Leb, Gerhard Loibl, Andrew Macoun, Stephen McCaffrey, Owen McIntyre, Barbara Miller, Franz Xaver Perrez, Anne Petitpiere, Claudia Sadoff, Salman M. A. Salman, Komlan Sangbana, Simon Schropp, Ashok Subramanian, the Nile Team, Mara Tignino, Elisabeth Türk, the W8+ CAV, Emma Wyatt, Julian Wyatt, William Zartman, Mark Zeitoun, the anonymous reviewers and many others who may remain unnamed but are not forgotten.

I am also immensely grateful to Constance Burt and Eugenia Lamont for copy-editing the text and to the editorial team at Cambridge University Press. Special thanks also go to the Swiss National Science Foundation and the Boninchi Foundation for providing generous financial support for this research endeavour.

Table of treaties

Table of cases

Central American Court of Justice

European Court of Human Rights

Human Rights Committee

Inter-American Court of Human Rights

International Arbitration Tribunals

International Court of Justice

Neutral Expert (Indus Waters Treaty 1960)

Permanent Court of International Justice

List of abbreviations

ACTO	Amazon Cooperation Treaty Organization
ADB	Asian Development Bank
AfDB	African Development Bank
AJIL	*American Journal of International Law*
ASEAN	Association of Southeast Asian Nations
AU	African Union
AWG-LCA	Ad-Hoc Working Group on Long-Term Cooperative Action
BBl	*Bundesblatt* (Switzerland)
BYIL	*British Yearbook of International Law*
CARU	Administrative Commission of the River Uruguay
CBDR	common but differentiated responsibilities
CESCR	Committee on Economic, Social and Cultural Rights
CJIELP	*Colorado Journal of International Environmental Law and Policy*
Consol. TS	Parry's Consolidated Treaty Series
COP	Conference of the Parties
cusecs	cubic feet per second
EC	European Community
ECAFE	Economic Commission for Asia and the Far East
EEC	European Economic Community
EFAS	European Flood Alert System
EIA	environmental impact assessment
ENVSEC	Environment and Security Initiative
EU	European Union
EU WFD	European Union Water Framework Directive
EUTSer	EU Treaty Series
FAO	Food and Agriculture Organization (UN)
GA	General Assembly
GEF	Global Environmental Facility

GHG	greenhouse gas
GoS	Government of Sudan
HRC	Human Rights Council
ICCPR	International Covenant on Civil and Political Rights
ICESCR	International Covenant on Economic, Social and Cultural Rights
ICJ	International Court of Justice
ICPDR	International Commission for the Protection of the Danube River
ICPR	International Commission for the Protection of the Rhine
IDA	International Development Agency
IIL	Institute of International Law
IJC	International Joint Commission (United States, Canada)
IJGEnvI	*International Journal of Global Environmental Issues*
ILA	International Law Association
ILC	United Nations International Law Commission
ILM	International Legal Materials
INIWEH	Institute for Water, Environment and Health
IPCC	Intergovernmental Panel on Climate Change
IUCN	International Union for the Conservation of Nature
IWRM	integrated water resources management
JAWRA	*Journal of the American Water Resources Association*
KBO	Kagera Basin Organization
LCBC	Lake Chad Basin Commission
LDCF	Least Developed Countries Fund
LN	League of Nations
LNTS	*League of Nations Treaty Series*
LVFO	Lake Victoria Fisheries Organization
MDG	Millennium Development Goal
MEA	multilateral environmental agreement
MOP	Meeting of the Parties
MRC	Mekong River Commission
NAPAs	National Adaptation Programs of Action
NBI	Nile Basin Initiative
OAU	Organization of African Unity
OAU Doc.	Official Document of the Organization of African Unity
OECD	Organisation for Economic Cooperation and Development
OJ	*Official Journal of the European Union* (unless otherwise indicated)
OMVS	*Organisation pour la Mise en Valeur du Fleuve Sénégal*

OSCE	Organization for Security and Cooperation in Europe
PCIJ	Permanent Court of International Justice
ppm	parts per million
RCADI	*Recueil de Cours de l'Académie de Droit International*
RECIEL	*Review of European Community International Environmental Law*
RGDIP	*Revue Générale de Droit International Public*
RIAA	*Reports of International Arbitral Awards*
SADC	Southern African Development Community
SBSTA	Subsidiary Body for Scientific and Technological Advice
SCCF	Special Climate Change Fund
SDD	USSR Treaty Collection (*Sbornik desitvuiushchikh dogovorov, soglashenii I konventsii zakliuchenyuhk SSR inostranymi gosudarstvami*)
SICJ	Statute of the International Court of Justice
SOGED	*Agence de Gestion et d'Exploitation du barrage de Diama*
SOGEM	*Société de Gestion de l'Energie de Manantali*
SPA	Strategic Priority for Adaptation
SPLM/A	Sudan People's Liberation Movement/Army
UN	United Nations
UNC	Charter of the United Nations
UNCED	United Nations Conference on Environment and Development
UNCLOS	United Nations Convention on the Law of the Seas
UN Doc.	Official Document of the United Nations
UNDP	United Nations Development Program
UNECE	United Nations Economic Commission for Europe
UNEP	United Nations Environment Programme
UNESCO	United Nations Educational, Scientific and Cultural Organization
UNESCWA	United Nations Economic and Social Commission for Western Asia
UNFCCC	United Nations Framework Convention on Climate Change
UNFPA	United Nations Population Fund
UNITAR	United Nations Institute for Training and Research
UNTS	*United Nations Treaty Series*
UNU	United Nations University
US	United States of America
USTS	*United States Treaty Series*

WFD	Water Framework Directive
WHO	World Health Organization
WMO	World Meteorological Organization
WTO	World Trade Organization
YBILC	*Yearbook of the International Law Commission*

Introduction

Water permeates everything; it provides a link among living things, land and the atmosphere. Water sustains and can destroy life; it shapes the landscapes that it travels through, and it is part and parcel of the air we breathe. Water also connects States: with the exception of island nations, virtually all countries share one or more transboundary freshwater systems with their neighbours. These international rivers, lakes and aquifers create inevitable linkages and interdependencies among States. The use of transboundary water systems by one State almost invariably impacts other States sharing the same system. Such impacts can be of a factual or an immaterial nature; they can affect a State's physical territory or its existing or future rights; they may cause harm; and they can be beneficial. States are therefore well advised to coordinate their activities regarding the utilisation of their shared water resources to prevent mutual damage and to enhance beneficial effects; in fact, the hydrologic interconnectedness almost necessitates coordination. Cooperation is the process by which States take coordination to a level at which they work together to achieve a common purpose that produces additional mutual benefits that otherwise would be unavailable with unilateral action.[1] Experience in transboundary basins shows that cooperation among riparian States more often leads to greater benefits for all basin States than non-cooperation.[2]

The present analysis explores the principles and norms of international water law through the prism of cooperation; by examining international

[1] I. W. Zartman, 'Concept: Cooperation' (2008) 30 *PIN Points* 5–7.

[2] World Bank, 'Creating Benefits from Cooperation in Shared Water Resources: Lesotho's Highlands Water Project Phases IA and IB' (2008) 12 *Water Feature Stories*; D. Phillips *et al.*, *Trans-boundary Water Co-operation as a Tool for Conflict Prevention and Broader Benefit Sharing*, Global Development Studies (Stockholm: Ministry of Foreign Affairs, 2006).

water law from this angle, the analysis intends to illustrate that the legal principles and rules that promote cooperation can help to navigate the challenges of transboundary water resources management. At this point, it is worth clarifying that the term 'transboundary' is applied here, primarily, relative to water resources that cross political boundaries between sovereign States, thus only referring to freshwater resources that are shared internationally. Other definitions of the concept 'transboundary' – which may refer to administrative boundaries delimiting communities or electoral districts or to boundaries between provinces or states in federal systems or others – are not applied here.

In the course of history, riparian States on many international watercourses engaged in and enhanced their cooperative relationships in managing the utilisation, protection and development of their shared waters. They frequently sealed their mutual agreements with international treaties that described how the resources should be managed and the specific rules of conduct; this contributed to the crystallisation of a law of transboundary freshwater resources (hereinafter also referred to as 'international water law'), including customary principles and rules. The scope of application of the principles and rules expanded over time and continues to widen. Earlier treaty practice mostly concerned surface water bodies; later, as knowledge and utilisation of groundwater resources grew, these resources were also considered in water treaties, first separately in agreements managing boundary waters and then as resources that are connected to and form integrative parts of surface water bodies. More recently, the focus has turned to non-renewable, or fossil, transboundary groundwater resources and to aquifers not connected to watercourses. In the future, it is likely that the ever-increasing water demand of the continuously growing world population and the effects that climate change will have on the global distribution of water resources will lead to further expansion in the scope of the law of transboundary freshwater resources.

In parallel to the developments in international law more generally, a trend away from a law of coexistence towards a law of cooperation can also be observed in international water law.[3] Early treaties generally focused on water and its uses in relation to individual States' territories: this

[3] L. Boisson de Chazournes, '*Eaux internationales et droit international: vers l'idée de gestion commune*', in L. Boisson de Chazournes and S. Salman (eds.), *Les ressources en eau et le droit international – Water Resources and International Law*, Académie de droit international, (Leiden/Boston: Martinus Nijhoff Publishers, 2005), pp. 10–30. For a conceptualisation of this evolution, see W. Friedman, 'General Course in Public International Law' (1969) 127 *RCADI* 39–224.

included the drawing of political boundaries along rivers or across lakes, as well as regulating rights related to the passage over one State's territory by ships navigating under the flag of another. Agreements on managing shared water resources in a coordinated and cooperative manner became increasingly more common once the negative impacts of water uses by one riparian State on other riparian States grew, particularly with respect to the increase in industrial and agricultural water uses and their negative effects on water quantity and quality. These treaties and coherent State practices led to the emergence of a general duty to cooperate on transboundary freshwater resources. This general duty is now widely recognised as one of the so-called general principles of international water law, together with the principle of equitable and reasonable utilisation, the obligation not to cause significant harm and other principles.[4]

Given the importance of cooperation in the management of transboundary water resources that interlink State territories, it is interesting that the recognition of the general duty to cooperate as a 'general principle' has been open to question. Legal instruments concerning transboundary water systems refer more often to cooperation than they do, for example, to the principle of equitable use.[5] However, the question of recognition was subject to debate, for example, during the United Nations (UN) International Law Commission's (ILC) discussions and elaboration of the 1994 *Draft Articles on the Law of Non-Navigational Uses of International Watercourses* (hereinafter ILC 1994 *Draft Articles*),[6] which form the basis of the 1997 *UN Watercourses Convention*. The general duty to cooperate also received comparatively less scholarly attention than other general principles of international water law.[7] Yet, cooperation is essential not only for the management and development of shared resources; it is also

[4] Part II of the *United Nations Convention on the Law of Non-Navigational Uses of International Watercourses* (hereinafter '1997 *UN Watercourses Convention*'), New York, 21 May 1997, not yet in force, UN Doc. A/RES/51/299.

[5] See Annex: about 56 per cent of treaties mention cooperation whereas only half as many refer to equitable utilisation.

[6] ILC, 'Summary Records of the Meetings of the Thirty-Ninth Session 4 May–17 July 1987' (1987) I *YBILC*, UN Doc. A/CN.4/SER.A/1987, pp. 70–95.

[7] For example, A. D. Tarlock and P. Wouters, 'Are Shared Benefits of International Waters an Equitable Apportionment?' (2007) 18 *CJIELP* 523–536; K. Hayward, 'Supporting Basin-Wide Reforms with an Independent Assessment Applying International Water Law: Case Study of the Dnieper River' (2007) *CJIELP* 633–664; I. Kaya, *Equitable Utilization: The Law of Non-navigational Uses of International Watercourses* (Aldershot: Ashgate 2003); J. W. Dellapenna, 'The Customary International Law of Transboundary Fresh Waters' (2001) 1 *IJGEnvI* 282–284, J. G. Lammers, *Pollution of International Watercourses: A Search for Substantive Rules and Principles of Law* (Boston/The Hague: Martinus Nijhoff, 1984).

necessary for bringing other principles and rules of international water law to fruition. The principle of equitable and reasonable utilisation, for example, cannot be adequately implemented in the absence of cooperation; notification obligations and consultations yield information that is necessary to assess whether a respective use is equitable and reasonable and can be used to enhance joint management of the shared waters.

It is for these reasons that this book intends to add to the existing literature with insight into recent developments of treaty practice, highlighting noteworthy trends. The book is divided into three parts, which examine the evolution, current state and future trends of the rules of cooperation that apply to transboundary water resources management and development. Part I explores factors that motivate States to engage in cooperative behaviour. The global hydrologic cycle creates interlinkages among territories, people and the environment. Interdependence contributes to States' decisions to engage in international cooperation. At the same time, it is not the only determining factor of their behaviour; other factors, such as concerns for security and economic development as well as historic relationships, also intervene. Chapter 1 analyses why States choose to cooperate and why they enter into legal agreements as part of the cooperative process. It illustrates that the development of international relations and growing interdependence among States have changed the nature of sovereignty and international law, reflecting the increasing importance of international cooperation. Chapter 2 traces these developments with respect to international water law. Similar to what is happening in other domains, the development of this body of norms is characterised by an expansion of its normative reach, both geographically and in terms of the different subject areas that it has come to address.

Principles and norms that form the current legal basis of cooperation among riparian States of transboundary water bodies comprise the subject of analysis in Part II. Chapter 3 illustrates that legal principles reinforce one another and that individual principles in the area of international water law would not be complete without the others; that is, they cannot achieve their intrinsic value and objectives on their own. This is one reason for the argument made in Chapter 4 – namely, that it is not sensible to conceive of the general duty to cooperate and the specific obligations that derive from it as purely procedural norms. These rights and duties contain procedural as well as substantive normative elements. Chapter 5 emphasises this point by illustrating the large variety of measures that have been adopted by States with a view to achieving specific water management objectives, including water quality control, protection of ecosystems and regulation of flow through cooperation. In the majority of cases, States have established and assigned water management responsibilities to joint mechanisms, which serve as a platform for cooperation and dialogue and have led, at times, to a widening and deepening of collaborative action among States. Joint mechanisms play an important role in the process of State cooperation: as structures that are designed to ensure regular interaction among States, they can contribute to preventing and settling disputes as well as to ensuring compliance with legal obligations.

The final part of the analysis, Part III, examines where the evolution of cooperation among States might lead in the future. Water availability is characterised by high regional and temporal variability, resulting in uneven access for individuals and communities – across not only basins but also regions and the planet. Climate change impacts on the global water cycle compound these effects and particularly harm poor States and regions already experiencing water stress, thereby causing increasing inequity regarding access to and availability of freshwater resources at the global level. Recent trends in legal development indicate that different areas of international law are responding to these challenges: the rights and interests of individuals and State cooperation relative to water are also increasingly considered at the regional and universal levels.[8] Issues concerning the existence of international cooperation obligations to satisfy vital human water needs are explored in Chapter 6. The focus of this chapter is to respond to the question of whether a prioritisation of vital human needs among different water uses, combined with the crystallisation of a human right to water, can be the source for additional or new international cooperation obligations. Chapter 7 addresses the emerging trends of international cooperation relative to climate change and the effects of its phenomenon on hydrologic variability. Climate change highlights the fact that States are connected not only through transboundary watercourses and aquifers but also through the atmosphere and the much larger global water cycle. Economic activity in one international basin can affect water availability in basins elsewhere. The increasing severity of water events, such as floods and droughts, requires preparedness and cooperation among riparian States for the design of adequate

[8] M. Tignino, '*Les contours du principe de la participation publique et la protection des ressources en eau transfrontières*' (2010) 7 *Vertigo*, http://vertigo.revues.org/9750; K. Bourquain, *Freshwater Access from a Human Rights Perspective* (Leiden: Martinus Nijhoff, 2008), p. 47; Boisson de Chazournes, '*Eaux internationales et droit international*', pp. 32ff.

adaptation measures. Climate change law promotes international cooperation for adaptation. This final chapter explores whether international water law, together with the legal framework for climate change, can facilitate advancing the regulation of equitable water utilisation beyond the basin level while also considering atmospheric water streams and global hydrologic interdependence.

The method applied in the research is a combination of the analysis of primary resources, the decisions of international courts and tribunals and the secondary literature complemented by discussions with practitioners. The core of the analysis rests on the examination of international water treaties concluded among States since 1900: 219 agreements were analysed relative to their role in the overall context of cooperative basin management, and the principles and rules of cooperation included in their text. For the latter, the assessment of the treaties' text followed the question of to what extent these instruments establish cooperation duties for States. Cooperation obligations were grouped into different categories and their occurrence was quantitatively assessed. A detailed description of the methodology of this analysis and a table of the treaties that were evaluated are provided in the Annex. Some of the treaties analysed were superseded by later instruments; nevertheless, reference is made to them to demonstrate the wealth of possibilities and ways in which States can regulate the various aspects of water resources management. Regarding earlier treaties, the research drew heavily on the writings of international scientific and expert bodies – the ILC, the International Law Association (ILA) and the Institute of International Law (IIL) – for their interpretation. These entities conduct comprehensive analyses of treaties and State practice in the process of elaborating codifications of universal norms of international water law.

The treaty analysis primarily focused on agreements with the explicit objective of protection, management or development of transboundary waters and thus on water treaties in a strict sense. Yet, because the law of transboundary water resources is neither a separate nor a self-contained system of rules and because it interacts with other domains of public international law, rules from these other areas that are relevant to the cooperative management of shared waters were also analysed. This was done specifically in regard to human rights law and the international climate change regime within the context of Part III. International environmental law rules and multilateral environmental agreements (MEAs) were also considered in the analysis relative to how they apply to shared water resources. Decisions of international courts and tribunals and the writings of legal scholars were examined to complete the analysis of the current state of international water law and with respect to interpretation of norms.

The research analyses international water law from the perspective of natural resource utilisation. Navigational uses are therefore considered herein only as they impact the condition of water resources, such as through pollution or changes in flow characteristics. Navigational uses and rights related solely to the right of passage over the territory of another State by taking advantage of the specific surface texture of water are not analysed. Another area of law related to transboundary waters that is not considered herein is boundary law. International water bodies had interconnected land and communities long before the demarcation of State territories. However, rivers in particular are frequently perceived as natural geographic lines of separation and were often used to mark political boundaries. Although the focus of the present research is the analysis of principles and rules that regulate the legal relationships among States, this divider function of watercourses as political boundaries is not analysed; on the contrary, it is rather the contribution of international water law to the dilution of boundaries that constitutes the primary focus. For similar reasons, the role of international water law in dispute settlement is considered only as a tool of cooperation that assists in overcoming, managing or preventing disputes among States and the differences concerning their interests and positions.

It is recognised that cooperation among States also takes place outside of the formal setting of international agreements and international law; the role of treaties – as one of many instruments or simply one step in the process of State cooperation – has been taken into account. Secondary literature on international waters cooperation was reviewed and discussions with practitioners were conducted to complement the analysis using insights from current State practice.

Finally, certain clarifications are required regarding the use of words and definitions applied in this volume. The term 'water resources' is used in a wide sense; it refers to freshwater primarily in its liquid but also in its solid (i.e., ice) and gaseous (i.e., atmospheric water) states. Bodies of water that cross or form international boundaries and are therefore situated on or traverse the territories of more than one State are referred to as 'transboundary water bodies'.[9] The historical distinction among

[9] Art. 2 (b), 1997 *UN Watercourses Convention*.

'international rivers', navigable waterways of international concern and other rivers is not used here;[10] the definitions denoting different types of transboundary water bodies follow the legal definitions developed by the ILC. A 'watercourse' is a 'system of surface waters and groundwaters constituting by virtue of their physical relationship a unitary whole and normally flowing into a common terminus'.[11] This definition includes rivers, lakes, aquifers, glaciers and canals as long as they are interrelated with one another.[12] An 'aquifer' is defined as 'a permeable water-bearing geological formation underlain by a less permeable layer and the water contained in the saturated zone of the formation', which may or may not be linked to watercourses.[13] When two or more aquifers are hydraulically connected, they form an 'aquifer system'.[14]

The ILC discussed the use of the terms 'system' and 'basin' when it elaborated the legal definitions applicable to transboundary water bodies. The term 'watercourse system' was rejected by States as not being 'distinguishable to any appreciable extent from the "drainage basin" concept'.[15] The 'drainage basin' concept recognises land–water linkages and the importance of terrestrial plants, soils and minerals for aquatic ecosystems. An 'international basin' is the 'geographical area extending over two or more States determined by the watershed limits of the system of waters, including surface and underground waters flowing into a common

[10] Within the context of regulating navigation on Europe's great rivers, navigable rivers with access to the sea were declared 'international rivers'. This definition, which can be traced back to Article 108, 1815 Final Act of Vienna, does not include connected aquifers and lakes. See *Territorial Jurisdiction of the International Commission on the River Oder*, judgment, PCIJ (1929), Ser. A.23 No. 16, p. 25; L. Caflisch, '*Règles Générales du Droit des Cours d'Eau Internationaux*' (1989), 219 *RCADI* 30f.; S. McCaffrey, *The Law of International Watercourses – Non-Navigational Uses*, 2nd edition (New York: Oxford University Press, 2007), p. 41.

[11] Art. 2 (a), 1997 *UN Watercourses Convention*.

[12] ILC, 'Draft Articles on the Law of the Non-navigational Uses of International Watercourses and Commentaries Thereto and Resolution on Transboundary Confined Groundwater' (1994), II(2) *YBILC* 90.

[13] Art. 2 (a), *Draft Articles on the Law of Transboundary Aquifers* (hereinafter 2008 ILC *Draft Articles*) UN Doc. A/RES/63/124. This definition, which was adopted by the ILC, is interesting from a legal perspective because, as a consequence, it unites divisible resources – the geological rock formation – and indivisible resources – freshwater – under one legal regime.

[14] Art. 2 (b), 2008 ILC *Draft Articles*.

[15] ILC, 'Second Report on the Law of the Non-navigational Uses of International Watercourses, by Mr. Jens Evensen, Special Rapporteur' (1984) II(1) *YBILC* 105.

terminus'.[16] This concept acknowledges that land and water, as well as terrestrial and aquatic fauna and flora of the catchment area, constitute an interdependent 'ecosystem'.[17] Although 'basin' is the preferred notion of the ILA, State practice is not uniform regarding its use in water treaties. It has therefore not become a concept of customary law and applies only when States determine its applicability by treaty.[18]

In this book, the term 'water system' or 'water body' is used as the general term when referring to the interconnected system of water resources flowing in a transboundary basin, including the glaciers that might be their source. The two terms are applied equally to surface water bodies, watercourses, and aquifers. When analysis is more specifically concerned with one or another of these systems, the ILC terms and definitions of 'watercourse' 'aquifer' and 'aquifer system' are used. When the more complex land–water linkage is considered, the term 'ecosystem' or 'basin' is used. The more precise terms are employed particularly when it is necessary to refer to legal differences that result from applying the different definitions. Depending on whether the definition of 'watercourse' is used for cooperative management of an international river or the 'river basin' concept, the substance of obligations related to resource protection may differ. For example, whereas the first concept would focus on the quality of the shared waters, the second would apply also to the quality of the ecosystem interacting with the waters. Regarding participants in water resources management, in many cases, it would not make a significant difference whether the basin concept or the system concept is applied; the geographic space covered by these two concepts is congruent in most cases.[19] Therefore, for purposes of simplification – and despite legal nuances that might apply – the terms 'riparian', 'system' and 'basin' States[20] are used here without distinguishing among them.

[16] Art. II, 1966 *Helsinki Rules*, reproduced in ILA, *Report of the Fifty-Second Conference Held at Helsinki, August 14th to August 20th, 1966* (London: ILA, 1967), pp. 484f. For aquifer systems, the recharge zone is equivalent to the drainage basin.

[17] *Ibid*., pp. 472 and 485. [18] Caflisch, '*Règles Générales*', p. 29.

[19] McCaffrey, *The Law of International Watercourses*, p. 37.

[20] Art. III, 1966 *Helsinki Rules*: 'A "basin State" is a State the territory of which includes a portion of an international drainage basin'.

PART I · ACKNOWLEDGING INTERDEPENDENCE

> "In the desert of life, the wise person travels in caravan while the fool prefers to travel alone".
>
> Arab Proverb

Water cycles through the oceans, the atmosphere, the land and the living matter of the biosphere changing among gaseous, liquid and solid states. This cycle creates links and interdependencies among fauna, flora and human beings, as well as among sovereign States. A change in the cycle in one place, regardless of whether it occurs due to natural or human causes, impacts the cycle elsewhere. Transboundary water bodies form part of the global cycle. At the same time, they constitute hydrologic units, in which changes that occur in one place can frequently be attributed directly to natural events or human activities that occurred in another area of the basin. Chapter 1 describes the dynamics of the global water cycle and the links established among sovereign States. It analyses how States address this interconnectedness and how they discharge the responsibilities that they assume as sovereign States in an interdependent world. Chapter 2 traces the results of State cooperation in transboundary basins by examining the development and expansion of the scope of the law of transboundary water resources.

1 Cooperation between sovereign States

States interact with respect to transboundary waters because they are inevitably linked through these resources to one another's territorial space. Due to this interlinkage, water utilisation by one riparian State can have beneficial or harmful effects for other States sharing the same water system; these effects may be the result of intentional State action or they may be unintended. When these hydrologic links and the coordination of water uses of different riparian States can result in mutually beneficial outcomes to those involved, the nature of the resource may become a source of international cooperation.

1.1 The nature of water

Water is a unique resource. It is life-giving and has tremendous destructive force. As a dynamic resource, it evades complete control by human beings.

1.1.1 The water cycle

Water that falls on landmass in the form of rain, snow or dew is captured by soil, plants and other living organisms. Water not tapped by life forms or frozen into ice caps continues its journey as runoff into rivers, lakes and oceans through infiltration in the ground and directly as surface flow. Unless captured and stored in sealed tanks, it will eventually escape again into the atmosphere. A certain amount of the water stored in oceans, lakes and reservoirs evaporates. With respect to reservoirs, evaporation is often referred to as 'water loss'; however, rather than being lost, this water is reintroduced as vapour in the atmosphere and continues the cycle.

The water cycle is kept in motion by energy. The heat of the sun causes evaporation; its light allows plants to transform water through the process of photosynthesis into organic matter, thereby permitting the production of food for other living organisms.[1] Gravity and the geography of the Earth's surface generate kinetic energy by which water moves through the ground and on the surface towards the oceans and seas. The shape of the Earth's surface transports water to places where humans and other living beings can have easy access for drinking and where they can immediately utilise water and its kinetic energy or store it for later use. Water becomes the source of life for a multitude of terrestrial, avian and aquatic organisms.

At the same time, water can become a source of destruction and death: in its absence; in situations in which too much accumulates too quickly; and when it becomes a vector or breeding ground for disease.[2] Floods, mudslides and avalanches are among the most devastating natural catastrophes. Land-use change has increased the occurrence of floods in recent decades.[3] Between 50 and 300 inland floods causing an average of 25,000 deaths and affecting about 520 million people occur worldwide each year.[4] Floodwaters are often contaminated with raw sewage and other pollutants. As a consequence, these events are often followed by a shortage of drinking water and outbreaks of waterborne diseases.[5] Deforestation of watersheds, urban development in floodplains and the covering of otherwise permeable surfaces with asphalt and concrete have all reduced the water retention capacity of the soil and increased surface runoff volumes and, along with this, the risk of floods. Identification of these causal linkages has more recently led to significant re-naturalisation initiatives on a number of rivers, such as in the Rhine Basin.[6]

The 'global water cycle' is composed of many circuits, or sub-cycles; each has a different geographic and temporal range.[7] Depending on the

[1] R. Kandel, *Water from Heaven* (New York: Columbia University Press, 1998), p. 44.

[2] For example, puddles and stagnant water in hot, humid climates promote the spread of malaria and bilharzia.

[3] Around the world, there were twice as many catastrophic inland floods per decade between 1996 and 2005 as there were between 1950 and 1980; property damage was five times as great. P. J. W. Gore, 'Floods and Flood Control', in C. W. Allin (ed.), *Encyclopedia of Global Resources* (Pasadena, CA: Salem Press, 2010), p. 445.

[4] *Ibid.*, p. 441.

[5] World Meteorological Organization (WMO), *Legal and Institutional Aspects of Integrated Flood Management* (Geneva: WMO, 2006), p. 1.

[6] International Commission for the Protection of the Rhine (ICPR), *Conference of Rhine Ministers 2001: Rhine 2020 – Program on the Sustainable Development of the Rhine* (Koblenz: ICPR, 2001), p. 16.

[7] Kandel, *Water from Heaven*, p. 6.

Table 1. *The Earth's Water*[8]

Type of Water	Share of Total Water (% or ppm)	Typical Renewal or Residence Time
Saltwater (oceans, inland seas, salt lakes)	96.3%	3,700 years
Present-day 'permanent' snow/ice	2.0%	30 to 600,000 years
Present-day permafrost	0.21%	8,000 years
Soil moisture	0.005%	1 month
Underground freshwater	0.7%	50 to 5 million years
Underground saltwater	0.9%	very long
Freshwater lakes	0.008%	10 to 10,000 years
Rivers and streams	1.4 ppm	18 days
Biosphere	0.7 ppm	hours
Atmospheric water vapor and cloud water/ice	91 ppm	8 to 10 days

circuit, water travels at various speeds and over different distances. Only relatively small quantities of all global water resources cycle continuously through sub-cycles. As shown in Table 1, 2 per cent of the Earth's total water is captured as permanent snow or ice, with a typical renewal time of about 30 to 600,000 years; 1.5 per cent of total freshwater and saltwater is stored underground, with a residence time of about 50 to 5 million years; and about 96 per cent of global water resources is saltwater flowing in the oceans, inland seas and salt lakes. Thus, only a small portion of global water circulates in the biosphere, rivers and lakes.

The amount of water flowing in the global water cycle is finite; that is, the total does not change over time. What does change is the availability of freshwater on the local as well as the global levels; broadly speaking, two mechanisms are responsible. The first factor is human intervention: new pumping techniques that facilitate access to groundwater at great depths increase the availability and accessibility of freshwater, as do technologies that transform saltwater into freshwater (e.g., desalination) and methods of water recycling that restore water to a quality suitable for human use. Therefore, the 'same drop' of freshwater can be used multiple times on its way from the clouds to the oceans. The second factor that impacts freshwater availability over space and time is hydrologic

8 Data reproduced from Kandel, *Water from Heaven*, p. 86; ppm = parts per million = 0.0001 per cent.

variability. Water is unevenly distributed throughout our planet. In addition, the speed at which water travels through the cycle varies, and global and local precipitation patterns are subject to change. Seasonal changes and the El Niño and La Niña phenomena impact regional rainfall and drought conditions during the course of multiple years. Global climate change has long-term effects on freshwater availability: seasonal increase in rainfall and long-term increase of glacier melt are expected to temporarily augment surface runoff volumes and local freshwater availability. The continuous changes in freshwater availability over space and time present a challenge to human water resources management.

1.1.2 Human intervention and interdependence

Humans form part of and thus share the global water cycle with and within terrestrial and aquatic ecosystems. For millennia, human society has devised mechanisms to harness the positive qualities of water for its own benefit and to control and mitigate the destructive forces of water. Humans have appropriated increasing amounts of freshwater for their own social, economic and cultural uses and are currently using 50 per cent of all renewable and accessible freshwater flows. Of these, about 14 per cent is used for domestic purposes (e.g., drinking, cooking and washing); 70 per cent is used to grow food and fibre; and 16 per cent is used for industrial and energy production.[9] Humans interfere with atmospheric water. Technologies that artificially modify weather patterns have been developed and applied since the 1960s. Cloud-seeding techniques allow for some control over rainfall to benefit agriculture. Through the injection of chemical agents, rain can be produced from clouds over desired areas, thereby preventing crop loss. As a consequence, however, other regions are deprived of rainwater that otherwise would have fallen there. Extraction and utilisation of resources other than water also have an impact on the water cycle. Most climate scientists argue that the exploitation and burning of fossil fuels are changing precipitation patterns worldwide, which causes rainfall distribution and the availability of local and regional water to change. Mineral dust from emissions that floats in the atmosphere impacts local – and sometimes not so local – availability and

[9] M. Palaniappan and P. Gleick, 'Peak Water', in P. Gleick *et al.*, *World's Water 2008–2009* (Washington, DC: Island Press, 2009), p. 11; T. J. Clausen and C. Bjerg, *The Blue Revolution: Adapting to Climate Change*, Thought Leadership Series No. 6 (Copenhagen: Climate Council, 2009), p. 7.

quality of water.[10] Large-scale heavy-metal industries in Central Europe have caused acidification of Nordic freshwater lakes; airstreams pushed emission pollution northwards, where it eventually fell over the Scandinavian lakes region in the form of acid rain.[11]

Human intervention relative to surface waters equally impacts the ecosystem. Reservoirs established for agricultural use or energy production alter flow volumes and velocity in the immediate area as well as downstream. Storage dams often destroy or cut off access to spawning grounds for fish. Depending on whether water is released from a reservoir's surface or from the colder water layers at the bottom, downstream water temperature will change accordingly. All of these interventions are not without an impact on biodiversity. Changes in flow velocity affect the temperature of water and the amount of oxygen contained in it, but certain plants and species can thrive only at specific water temperatures.[12] The amount of oxygen in the water determines the existence of microorganisms and plants that can filter toxins found in rivers, lakes and aquifers.[13] Flow availability determines the extent to which human domestic and industrial water use – and the pollution it causes – can be tolerated by the environment before serious or irreversible ecological damage occurs. These environmental water requirements must be balanced against water needs for human use to maintain healthy ecosystems and to preserve the life-giving quality of water.[14]

Interference in the water cycle affects the environment as well as water availability for use by others. In transboundary (or international)[15] water systems, utilisation of water by one State might be in conflict with the

[10] R. J. Moses and J. N. Corbridge, *Legal Structures for International Supervision of Weather Modification Techniques* (Washington, DC: Water for Peace Conference, Doc. P/635, 1967), p. 380.

[11] F. B. Pyatt, 'Acid Rain in Scandinavia – Some Current Data' (1987) 7 *The Environmentalist* 197.

[12] For detailed information on instream flow requirements and environmental flows, see C. Brown and J. King, *Environmental Flows: Concepts and Methods*, Water Resources and Environment: Technical Note C.1 (Washington, DC: World Bank, 2003); and R. Hirji and R. Davis, *Environmental Flows in Water Resources Policies, Plans and Projects* (Washington, DC: World Bank, 2009).

[13] United Nations Development Program (UNDP), *Human Development Report 2006 – Beyond Scarcity: Power, Poverty and the Global Water Crisis* (New York: Palgrave McMillan, 2006), p. 308.

[14] V. Smakhtin, C. Revenga and P. Doell, 'A Pilot Global Assessment of Environmental Water Requirements and Scarcity' (2004) 29 *Water International* 307–317.

[15] The two terms, 'transboundary' and 'international', are used interchangeably to denote water systems that are shared by two or more States.

interests of another State, particularly when uses are consumptive. Many uses are non-consumptive; in a transboundary context, this means that they do not impact the flow volume that naturally continues onto another State's territory. These uses also do not significantly change water quality; therefore, usage by other States remains largely unaffected.[16] Navigation, for instance, is a non-consumptive use as long as pollution emanating from engine-powered vessels – or from accidental spills or leakage of the substances they transport – is maintained at a level that is sustainable for the aquatic ecosystem. Other non-consumptive uses include fisheries, most recreational uses and hydroelectricity production. In specific circumstances, hydroelectricity production may constitute a consumptive use.[17] In cases in which hydropower stations are installed in dams that retain large reservoirs and the reservoirs elicit high evaporation rates, water can be considered 'consumed' because of evaporation 'losses'. In addition, large reservoirs cause temporary consumptive use relative to potential use by other States, which means that flow to another State's territory is temporarily reduced during the reservoir-filling period. For example, filling of the reservoir behind the Ataturk Dam, located on the Turkish part of the Euphrates River, took more than four years.[18] Reduced flow and temporary stoppage of the river's flow led to failed harvests and interrupted water services in Syria.[19]

1.1.3 Addressing hydrologic interdependence through cooperation

In the case of transboundary bodies of water, one State cannot control how other riparian States utilise the resource. Conflicting usage interests can be the source of disputes among States, whereas cooperation is most likely when usage interests align. At the same time, competing interests can also result in cooperation.[20] In transboundary water system

[16] In contrast to 'consumptive' water uses, which cause a diminishment of the source at the point of its appropriation, 'non-consumptive' water uses are those that do not cause a diminishment of the source or return the same quantity of previously diverted water to the source and at a quality that meets the standards of the source. See Department of Ecology, State of Washington, *POL 1020 – Consumptive and Nonconsumptive Water Use*, www.ecy.wa.gov/programs/wr/rules/images/pdf/pol1020.pdf.

[17] Other consumptive uses include agricultural irrigation, drinking water supply and industrial uses in which water is an input factor or is polluted to an extent that it cannot be applied to other uses afterwards.

[18] N. Kliot, *Water Resources and Conflict in the Middle East* (New York: Routledge, 1994), p. 106.

[19] J. Warner, 'Contested Hydrohegemony: Hydraulic Control and Security in Turkey' (2008) 1 *Water Alternatives* 282.

[20] Competing interests coexist with cooperation. M. Zeitoun and N. Mirumachi, 'Transboundary Water Interaction I: Reconsidering Conflict and Cooperation' (2008) 8

settings, it is not unusual for cooperation to exist when States have competing interests relative to resource use. Differences in seasonal water requirements of Kazakhstan, Uzbekistan and the Kyrgyz Republic are a case in point. Before the breakup of the Soviet Union, Kyrgyzstan was provided with electricity in the winter so that water in the Toktogul reservoir located in its territory could be stored for summer irrigation use downstream. These electricity subsidies disappeared with the Soviet Union; therefore, Kyrgyzstan began meeting its electricity demands with releases from the reservoir during winter, leaving insufficient reserves for the summer irrigation needs downstream.[21] The conflicting demands eventually became a catalyst for a cooperation agreement among the three riparian States of the Syr-Darya Basin. In the *Agreement on Joint and Complex Use of Water and Energy Resources of the Naryn Syr Darya Cascade Reservoirs*, downstream States agreed to provide electricity and coal to the Kyrgyz Republic during the winter months in exchange for water releases during the summer growing season.[22]

Hydrologic interdependence can be a driver for cooperation. However, other than water use, a State's decision to cooperate is driven by a variety of concerns including political, economic and social concerns. Regarding cooperation on transboundary water resources, decisions are not made based solely on environmental and hydrologic concerns; other factors, such as the economy and national security, also have a significant role in decision making. The remainder of this chapter highlights factors that intervene in a State's choice to cooperate and analyses the role of cooperation in international law more generally.

1.2 State motivation for cooperation

A brief exploration of political thought on the drivers of State action provides useful insights about why States may (or may not) use international law to facilitate cooperation. International law is not immune to politics

International Environmental Agreements, pp. 297–316; Warner, 'Contested Hydrohegemony', p. 276.

21 L. Moller, 'Transboundary Water Conflicts over Hydropower and Irrigation: Can Multilateral Development Banks Help?', CREDIT Research Paper No. 05/09, University of Nottingham (2009), p. 23.

22 *Agreement on Joint and Complex Use of Water and Energy Resources of the Naryn Syr Darya Cascade Reservoirs, Bishkek, Kyrgyzstan*, 17 March 1998, http://ocid.nacse.org/tfdd/tfdddocs/592ENG.pdf. On the viability and effectiveness of the agreement, see Moller, 'Transboundary Water Conflicts'.

but rather is inextricably intertwined with it. Factors such as distribution of power, national security concerns and international solidarity all have a role in the genesis, implementation and effectiveness of international law. Both international law and international relations theory study the 'laws' of State behaviour.[23] The former is concerned with rules that were established by States with the aim of peacefully managing their coexistence; the latter helps to understand political phenomena and the variety of factors that motivate State cooperation and implementation of rules.

1.2.1 Elements intervening in the choice to cooperate

Transboundary water resources interlink States. Functionalist theories explain why States choose to cooperate in such a situation of interdependence. These theories argue that 'social actors may strive to cooperate in order to combat problems that... require a joint effort'.[24] State cooperation thus is seen as the logical consequence of interdependence. As obvious as this might seem, this approach does not provide the full picture; by disregarding other important drivers of State action, it cannot explain the lack or failure of cooperation in some transboundary basins. Other approaches that trace cooperation – and the absence of it – to factors such as the concern for national security, historic relationships and intervention of third parties, as well as the quest for the optimisation of gains (i.e., benefits), are better suited to shed light on these phenomena. Finally, hydrologic conditions of basins also influence the nature of State interaction.

State concern for security and power

The point of departure for the analysis of State behaviour is the architecture of the international system. In this respect, the structure of the international legal system comes rather close to the underlying assumptions of political 'realism'.[25] In the international legal system, States interact as individual players in the absence of a higher centralised enforcement mechanism. According to realist thought, States are perceived as acting rationally and as self-interested, unitary actors within a system of

[23] A.-M. Slaughter Burley, 'International Law and International Relations Theory: A Dual Agenda' (2003) 87 *AJIL* 205.

[24] J. Klabbers, *An Introduction to International Institutional Law* (Cambridge: Cambridge University Press, 2002), p. 30.

[25] The realist school has evolved over time; more recent strands of this school are generally referred to as 'neo-realism'. Because here we refer only to the main tenets of the theory, the term 'realism' is used for simplicity.

anarchy.[26] 'Anarchy' is described as the state that a society finds itself in when it lacks a higher enforcement authority. The existence of a State is never guaranteed in this state of anarchy. States survive by gaining power and by ensuring that, at the very least, they maintain a balance of power in the system that maximises their own security.[27] State behaviour is determined by security considerations and concerns with the distribution of power. As long as cooperation serves to maximise national security and relative power gains,[28] States may cooperate. The moment that it puts them at risk of a change in their relative power position, States are likely to abandon cooperation. Hence, cooperation, according to realism, is ephemeral. States will defy systems that lock them into long-term cooperation (e.g., international law or international organisations) the moment they consider their national security concerns to be threatened.[29]

Therefore, the nature and lifespan of cooperation, to a large extent, depends on the concept and definition of 'national security' that prevails for each State individually.[30] Traditional realism is based on a military security concept. However, with intensification of industrialisation, increasing trade linkages and increasing pollution and destruction of ecosystems, the national security concept evolved in many States to increasingly include economic, environmental and water dimensions.[31]

[26] It is recognised that every State act is the outcome of a complex process, which involves multiple actors and is influenced by a variety of domestic forces, interest groups, values and ideas. From the perspective of international law, however, the State is conceived as a single unit; each State act – for example, the ratification of a treaty – binds the State as a whole and not only the domestic actors who were in favour of it.

[27] T. Dunne and B. C. Schmidt, 'Realism', in J. Baylis and S. Smith (eds.), *The Globalization of World Politics*, 3rd edition (New York: Oxford University Press, 2005), pp. 169f.

[28] 'Relative gains' as opposed to 'absolute gains' focus on the importance that one's own gains are higher than gains of others. When States are concerned with 'absolute' rather than 'relative' gains, the emphasis is on the fact that there are benefits that accrue to oneself and not on how much this amounts to relative to gains made by others.

[29] Friedman, 'General Course', p. 77; Slaughter Burley, 'International Law and International Relations Theory', p. 207.

[30] Two sides of the security concept can be distinguished: (1) in a subjective sense, 'security' is defined as perception – that is, the absence of fear and the feeling of being safe from harm by others; and (2) in an objective sense, 'security' means the factual absence of threats. A. Wolfers, 'National Security as an Ambitious Symbol', in A. Wolfers (ed.), *Discord and Collaboration: Essays on International Politics* (Baltimore, MD: Johns Hopkins University Press, 1962), pp. 147–165.

[31] T. F. Homer-Dixon, *Environmental Scarcity and Global Security* Headline Series No. 300 (Ephrata: Foreign Policy Association, 1993); World Commission on Environment and Development, *Report of the World Commission on Environment and Development: Our Common*

The 'economic security' definition has changed in time from one that focuses on immunity against economic manipulation by other States to a new concept that is more focused on socioeconomic security. In addition to economic aspects, this definition stresses the vulnerability of individuals and societies as well as the stability of internal social peace that accompanies it.[32]

Availability of sufficient and safe drinking water is an important factor in maintaining socioeconomic security. 'Water security' itself would be achieved by a State when it can secure a long-term national water supply in adequate quantity and quality to satisfy national needs.[33] When water security forms an integral part of the national security goal, realist thought predicts that this national security concept will not necessarily lead to continued cooperation. The quest for evermore water can hamper cooperation and might even result in conflict – for example, in transboundary water basins where the overall flow volume cannot satisfy the national demand of all riparian States; or situations of pronounced water scarcity in which the restricted water supply is not matched by reasonable demand management by the riparian States. According to realism, cooperation would not survive over time in these situations, or it would take the shape of submission or coercion of relatively weaker States, in which large power asymmetries exist among basin States, thus leading to highly inequitable distribution of cooperation benefits.[34]

In contrast to the political theories, the architecture of the international legal system per se does not provide for relative power differences of States. The influence of power asymmetries on cooperation and its outcomes is mitigated in international law through the underlying assumption of the equality of sovereign States.[35] In the international legal order, States have equal rights and obligations and thus equal possibilities (or 'power'). Sovereign States are initially endowed with the same rights. The legal system is based on the premise that each State decides for itself

Future, 4 August 1987, UN Doc. A/42/427, pp. 286–299; and 'Ministerial Declaration of The Hague on Water Security in the 21st Century', adopted at the Second World Water Forum in 2000, www.waternunc.com/gb/secwwf12.htm.

32 M. Kahler, 'Economic Security in an Era of Globalization: Definition and Provision' (2004) 17 *The Pacific Review* 486, 488.

33 U. Alam, 'Questioning the Water Wars Rationale: A Case Study of the Indus Water Treaty' (2002) 168 *The Geographical Journal* 347.

34 Klabbers, *An Introduction to International Institutional Law*, p. 30.

35 Cited, for example, by Zeitoun and Mirumachi, 'Transboundary Water Interaction I'; M. Zeitoun and J. Warner, 'Hydro-hegemony – A Framework for Analysis of Trans-boundary Water Conflicts' (2008) 8 *Water Policy* 435–460; T. Bernauer, 'Explaining Success and Failure in International River Management' (2002) 64 *Aquatic Science* 1–19.

whether to give up or limit some or part of its sovereign rights by entering into agreements with other States. At the same time, the rules of international law that emerged within this architecture address the actual power differences that exist among States. International law, for example, has devised norms and mechanisms that allow less developed or relatively 'weaker' States to comply with international obligations according to their capacities.[36]

Historical theories and political interest

History is another factor that determines the present and the future of State relations. It can be cited as a catalyst for cooperation among member States of the European Union (EU) as well as among the States of the Senegal Basin, albeit for two very different reasons. Regarding the European continent, it was the experience of (at least) two consecutive and devastating wars that led to cooperation and to the establishment of a process of economic cooperation, which ultimately led to the emergence of the EU.[37] In the Senegal Basin, French colonial influence and subsequent development aid are often cited as contributing to close cooperation among riparian States.[38] Other historians ascribe the success of cooperation in the Senegal Basin to the close ties that the region experienced under the Empire of Mali. It has been argued that the fame and reputation of the imperial system of Mali lingered in people's minds and in the form of a loose customary allegiance long after the Empire's political decline.[39] Senegal, Mali and Mauretania (as well as Guinea)[40] established

[36] These include, for example, rules of progressive implementation of economic, social and cultural rights and the concept of due diligence.

[37] In 1951, the six founding members – Belgium, France, Germany, Italy, Luxembourg and the Netherlands – signed the *Treaty Establishing the European Coal and Steel Community* to organise free movement and access to sources of production. Cooperation was expanded in 1957 with the signature of the Treaty of Rome, establishing the European Economic Community (EEC). Since then, membership has expanded to twenty-seven States, and regional economic integration and political cooperation was enhanced further, with the Treaty of Lisbon – in force since 1 December 2009 – as the most recent instrument of consolidation.

[38] S. M. Seck, A. Lericollais and G. Magrin, '*L'aménagement de la vallée du Sénégal: Logiques nationales, crises et coopération entre les Etats riverains*', in J.-P. Raison and G. Magrin (eds.), *Des fleuves entre conflits et compromis* (Paris: Karthala, 2009), pp. 47ff.

[39] The Empire of Mali dates to the early thirteenth century with its heyday in the fourteenth century. It started disintegrating around the time of the arrival of the first Europeans on West Africa's coast and had completely disappeared around 1600. For details, see B. Davidson, *The History of West Africa to the Nineteenth Century* (New York: Anchor Books, 1966), pp. 116 and 209.

[40] Guinea participated in the *Comité inter-etats des pays riverains du fleuve Sénégal* established by the Convention of Bamako in 1963. Guinea abandoned cooperation due to lack of

an extraordinary framework of cooperation, including joint ownership of key river-regulation infrastructure[41] and a river basin organisation[42] with extensive decision-making power.

History and the political interest of third parties have resulted in cooperation and the establishment of coordination mechanisms including, for example, in the Mekong River Basin. The first basin development studies of the Mekong River were carried out in the early 1950s by both the US Bureau of Reclamation and the UN Economic Commission for Asia and the Far East (ECAFE). Whereas the United States became involved due to an interest in containing the spread of communism in regions under its own sphere of influence, the role of ECAFE was to promote postwar economic development in the region.[43]

External actors have had a role in bringing about cooperation not only in the Lower Mekong Basin but also in many other transboundary rivers and aquifers.[44] It is not unusual for third parties to act as facilitators in establishing cooperation among basin States. Development agencies occasionally serve as catalysts for cooperation by providing the prospect of project financing or other funding;[45] their involvement can be especially beneficial when joint management initiatives require specialised technical know-how for their implementation.[46] The effectiveness of third party

incentives because it gained relatively few benefits from it, as well as because of political differences with Senegal in 1971. The country rejoined the cooperative framework in 2005. Seck, Lericollais and Magrin, '*L'aménagement de la vallée du Sénégal*', pp. 50f.

41 *Convention Concerning the Legal Statute of Common Works*, Bamako, 21 December 1978, in FAO, *Treaties Concerning the Non-Navigational Uses of International Watercourses: Africa – Traités concernant l'utilisation des cours d'eau internationaux à des fins autres que la navigation: Afrique*, FAO Legislative Study – 61 (Rome: FAO, 1997).

42 The character of the *Organisation pour la Mise en Valeur du Fleuve Sénégal* (OMVS) – which was established by the *Convention Establishing the Senegal River Basin Organization*, Nouakchott, 11 March 1972 (OJ of the Republic of Senegal, 26 July 1972, Law 72–71, 72–73) – is the subject of more detailed analysis in subsequent chapters.

43 F. Molle, T. Foran and P. Flach, 'Introduction: Changing Waterscapes in the Mekong Region – Historical Background and Context', in F. Molle, T. Foran and M. Käkönen (eds.), *Contested Waterscapes in the Mekong Region – Hydropower, Livelihoods and Governance* (London: Earthscan, 2009), p. 4. The Chinese and Burmese parts of the basin were not considered in these studies.

44 B. Hettne and F. Söderbaum, *Regional Cooperation* (Göteborg: International Task Force on Public Goods, 2004), p. 34.

45 See, in detail, U. Alam, 'Water Rationality: Mediating the Indus Water Treaty', unpublished Ph.D. Thesis, University of Durham, England (1998).

46 F. Biermann, *Internationale Meeresumweltpolitik – Auf dem Weg zu einem Umweltregime für die Ozeane* (Berlin: Peter Land, 1994), p. 224. The intervention of third parties, however, is not always without controversy, having been accused of undermining local ownership of cooperation or imposing cooperation from the outside.

involvement in cooperation among basin States – although not always uncontroversial – has been demonstrated in the Mekong, Indus and Senegal River Basins, among others.

Benefit optimisation

The increase in international cooperation for the management of international problems has also been explained by political theories that emphasise that cooperation is rational, not only because it increases one's own benefits but also because it increases overall benefits for all.[47] According to 'regime theory', States will be satisfied with additional absolute gains that may be lower than the gains of others participating in a regime: they know that the pursuit of relative gains (i.e., gains that are higher than those of others) will ultimately yield relatively lower net benefits for all.[48] The underlying assumption of regime theory is that cooperation increases the overall benefits and absolute gains for every participant. It becomes the self-interest of States to be concerned with absolute rather than relative gains because rewards from cooperation are typically higher than those from unilateral action, even if they are distributed asymmetrically among basin States.[49]

In the context of transboundary water systems the increase of overall benefits due to cooperation was demonstrated by Sadoff and Grey.[50] They distinguish four types of benefits that can be derived from the cooperative management of transboundary watercourses. The first type concerns ecological benefits to a river if riparian States join together to maintain a

[47] International cooperation with respect to economic development of former colonies and to environmental degradation has been manifested by the foundation of international organisations and programs with the objective of addressing these issues, including the foundation of the International Development Agency (IDA) in 1960, the African Development Bank (AfDB) in 1963, the UNDP in 1965, the Asian Development Bank (ADB) in 1966 and the United Nations Environment Programme (UNEP) in 1972, as well as by the establishment of other international and regional organisations active in this field.

[48] Regime theory was developed in the 1970s as a response to realism and was later subsumed under the more general rubric of institutionalism. See Slaughter Burley, 'International Law and International Relations Theory', pp. 206, 218–226. The leading work on regime theory is by R. O. Keohane, *After Hegemony – Cooperation and Discord in the World Political Economy*, 1st edition (Princeton, NJ: Princeton University Press, 1984/2005).

[49] J. Waterbury, 'Between Unilateralism and Comprehensive Accords: Modest Steps toward Cooperation in International River Basins' (1997) 13 *Water Resources Development* 280.

[50] C. Sadoff and D. Grey, 'Beyond the River: The Benefits of Cooperation on International Rivers' (2002) 4 *Water Policy* 389–403; C. Sadoff, T. Greiber, M. Smith and G. Berkamp, *SHARE – Managing Water across Boundaries* (Gland, Switzerland: International Union for the Conservation of Nature [IUCN], 2008), pp. 23–25.

healthy aquatic environment in the river basin. To a certain extent, these benefits *to* the river are a prerequisite for the second type of benefits: cooperation will increase economic benefits that can be reaped *from* the river. The third type of benefit focuses on the reduction of costs that arise *because of the river*. This type of benefit alludes to the costs of political conflict over the river. Cooperation can reduce these tensions, thereby avoiding costs that would occur in the case of conflict. Finally, benefits *beyond* the river will occur as a follow-on effect of cooperation in the transboundary water system. The argument is that cooperation about water would also foster cooperation in sectors of the economy that are related only indirectly to water management, such as trade in commodities and goods or exchange of technological and scientific know-how.

Hydrologic conditions as a source of aligning and competing interests

Characteristics that are endogenous to hydrologic systems can also be determinants of the nature of State cooperation. Endogenous system conditions influence State interests in the resources of the system and the benefits that can be derived from them. Transboundary systems that are characterised by relatively low water volumes, for instance, are more likely to give rise to competing demands for allocation of water than hydrologic systems, which can satisfy the volumetric demands of all basin States.

At the same time, it is true that even in cases in which water quantity is not a specific issue, State interests may still be of a conflicting nature, and they may compete. The uses of water resources, particularly within the context of surface water bodies, are manifold. Different non-consumptive water uses can be in conflict with one another. Long-term development scenarios for the Lower Mekong Basin, for example, put interests of hydropower production in competition with fisheries, even though both are primarily non-consumptive water uses.[51] Hydropower revenue from the construction of additional main-stem dams on the Mekong River would benefit Laos because the country could sell the electricity generated; however, the impacts on fisheries could threaten the livelihoods of millions of people, especially in Cambodia.[52]

As discussed previously, it is generally believed that rational and self-interested actors who share common interests will act collectively to

[51] Large reservoirs cause some evaporation loss of surface water volume.
[52] 'Reshaping the Mekong', *Bangkok Post*, 23 August 2010.

achieve them.[53] The majority of authors would predict that cooperation is most easily achieved when State interests align – that is, when States have a common perspective on a given issue and similar value systems on how to solve problems.[54] The development of rules for cooperation in environmental law, for example, has been traced to such shared and coinciding interests.[55] Yet, the causality of aligning interests and cooperation is not automatic. A number of authors point to the limits of the catalytic function of coinciding interests.[56] Cooperation itself has costs; transaction costs, for example, are higher when more basin States are involved,[57] and the larger the group of States involved, the more difficult it is to achieve collective action.[58] When agreement of several actors must be sought or when States have the possibility of free-riding relative to actions taken by others, cooperation becomes more difficult to achieve. This is particularly the case when the gains to be made through cooperation are not known or appreciated by all States in the same way.

1.2.2 Factors contributing to the use of law in State cooperation

Political theories describe the reasons that States decide to engage in certain behaviour; in this case, cooperation. They can also explain the States' choice of instrument to structure their cooperative relationships.

Larger gains through transparency

The reasons that States formalise cooperation through international agreements and the establishment of international organisations are clarified by the theory of institutionalism. This approach analyses the dynamics of how institutions facilitate cooperation among States that have conflicting interests because they enable the flow of information. By

53 P. G. Harris, 'Collective Action on Climate Change: The Logic of Regime Failure' (2007) 47 *Natural Resources Journal* 200.

54 J. Brunnée, '"Common Interest" – Echoes from an Empty Shell' (1989) 49 *Zeitschrift für ausländisches öffentliches Recht und Völkerrecht* 791–808; D. G. Le Marquand, *International Rivers: The Politics of Cooperation* (Vancouver: University of British Columbia, Westwater Research Centre, 1977); Bernauer, 'Explaining the Success and Failure', pp. 1–19; Biermann, *Internationale Meeresumweltpolitik*, p. 224.

55 Brunnée, 'Common Interest', pp. 791–808.

56 Harris, 'Collective Action on Climate Change'; Waterbury, 'Between Unilateralism and Comprehensive Accords', pp. 279–289; M. Olson, *The Logic of Collective Action: Public Goods and the Theory of Groups*, 2nd edition (New York: Schocken Books, 1968).

57 Waterbury, 'Between Unilateralism and Comprehensive Accords', p. 280.

58 The stalling of the Doha Round in international trade negotiations epitomises this argument. Harris, 'Collective Action on Climate Change', p. 195.

applying the 'iterated prisoner's dilemma', scholars have demonstrated that information deficits and asymmetries lead to non-cooperative results that might be rational in the short term for individual actors but that prevent optimisation of benefits for all in the long term.[59] The prisoner's dilemma is generally used in game theory. It is a symmetric two-person game in which non-cooperation pays for both players individually regardless of what the other does. Yet, together, they are both better off if they cooperate. In the format of an iterated prisoner's dilemma, multiple participants are locked into an interactive process over a long period. They learn over time that cooperative behaviour pays off and yields higher benefits, even if there are short-term incentives for defection.[60]

Also of interest are findings that show that when gains of cooperation are known, cooperation will not only happen; it also will happen faster than in the absence of information.[61] The more that participants know about the possibility of achieving higher net benefits from cooperation, the faster cooperation evolves and the higher the net benefits.[62] International organisations and many international treaties tie States together into systems of continuous and structured interaction; this generally leads to increases in the mutual availability of information. International organisations and treaties have their source in the willingness of States to cooperate, and they enhance this cooperation with the increase in transparency that they generate. 'Efficient cooperation is more likely when information is more symmetrically distributed and institutional structures provide a better guarantee that agreed measures will be implemented'.[63] In addition, institutions facilitate cooperation by decreasing the fear of being cheated because non-compliance is more easily detected through structures that ensure regular interaction.

The situation of the iterated prisoner's dilemma is similar to the one faced in the management of transboundary water systems. Basin States cannot escape their hydrologic interdependence. In the absence of information exchange and joint institutions, States may find it difficult to acquire information that is necessary to estimate or be aware of additional gains that can be achieved through the joint management of shared water resources. Without transparency and joint institutions, it is difficult for basin States to quantify and to realise the four types of

59 R. Axelrod and W. D. Hamilton, 'The Evolution of Cooperation' (1981) 211 *Science* 1390–1396.

60 This game yields the same results for two-player and multi-player settings.

61 R. Axelrod, *The Evolution of Cooperation*, revised edition (New York: Basic Books, 2006), p. 189.

62 *Ibid.*, p. 24.

63 Bernauer, 'Explaining Success and Failure', p. 9.

benefits – 'to, from, because of, and beyond the river' – that were described by Sadoff and Grey. Cooperation may be slow to develop; however, because of their hydrologic interconnectedness, basin States will most likely interact. According to findings of the iterated prisoner's dilemma game, these States would eventually discover that unilateral development of shared water resources leads to sub-optimal net benefits in the basin. With the additional knowledge gained through iterative interaction, they learn that management of the system can be optimised through increasing cooperation. The establishment of international institutions and of structured information exchange facilitates this process and is therefore often a preferred choice of procedure for basin States.

Reciprocity

Experiments on the evolution of cooperation in the situation of an iterated prisoner's dilemma demonstrated that cooperation takes place in the framework of a 'quid pro quo psychology'.[64] To elicit cooperation, a first concession must be made; this takes place in good faith that it will be reciprocated. Hence, the underlying condition for cooperation to occur is reciprocity in its manifestation as the assurance of similar, beneficial return behaviour in the future.[65] If reciprocity does not take place as a response to concessions, good faith is frustrated and cooperation is unlikely to continue or to occur at all.

Reciprocity[66] can effectively enhance good faith and cooperation in State relations,[67] and it may become a catalyst for the emergence of customary law rules.[68] As Simma observed, 'the anticipation of reciprocal

64 Through the process of iterated interaction, players learn that 'tit for tat' applied in a positive 'give-to-get' manner reaped higher benefits than non-cooperation. A player who knows that defection will be reciprocated will make the rational choice of not starting a process of defections. At the same time, he will defect after someone else defects to show that he will not be exploited. The key for doing well lies in eliciting the cooperation of others. Axelrod, *The Evolution of Cooperation*, pp. 189f.

65 Zartman, 'Concept: Cooperation', p. 5.

66 Reciprocity as social behaviour must be distinguished from reciprocity in the legal context; that is, 'the maxim that a State basing a claim on a particular norm of international law must accept that rule also as binding upon itself'. B. Simma, 'Reciprocity', in R. Bernhardt (ed.), *Encyclopedia of Public International Law, Volume Four* (Amsterdam: North-Holland/Elsevier Science, 2000), p. 30.

67 M. Fleishman, 'Reciprocity Unmasked: The Role of the Mexican Government in Defense of its Foreign Nationals in United States Death Penalty Cases' (2003) 20 *Arizona Journal of International and Comparative Law* 362.

68 '*La réciprocité s'impose, en effet comme la base sur laquelle peuvent être édifiés les premières relations juridiques . . .*', M. Virally, '*La Réciprocité dans le droit international*' (1967) 122 *RCADI* 6.

concession and reciprocal restraint may stimulate auto-limitation and thus account for the growth of general and concordant practice eventually accepted as law'.[69] Customary law emerges through widespread and consistent international practice, which is accompanied by the belief that the respective behaviour is required by law.[70] Through this process, international law becomes the legal result of the continuous collective interest of States in the benefits of cooperation.[71] Once international rules have crystallised, a primary reason for their effectiveness is the reciprocal interest of States to adhere to the rules and in the stability of established relationships.[72]

Law is not a necessary component of cooperation but it can be a useful one. Rules and legal agreements provide predictability and stability to relationships among States.[73] Although the effectiveness of international law remains subject to politics, the normative structure of international law contributes to the cooperation process.[74] It uncouples effectiveness from being solely dependent on the political will of States. Effectiveness of cooperation partially becomes a function of the quality of the design of its legal framework, which establishes the substance and the procedural structure for the process of cooperation. The legal framework also becomes the reference document in the case of a lack of political will by one of the parties to implement the agreement. In the following section, the general nature and evolution of cooperation in international law is analysed in more detail.

1.3 The nature of cooperation in international law

Reciprocity, the interest in stable and predictable relationships, and ultimately cooperation itself all contributed to the emergence of

[69] Simma, 'Reciprocity', p. 30.

[70] The ICJ confirmed this doctrine; for example, when it declared on the *North Sea Continental Shelf Case* that 'Not only must the acts concerned amount to a settled practice, but they must also be such, or be carried out in such a way, as to be evidence of a belief that this practice is rendered obligatory by the existence of a rule of law requiring it'. *North Sea Continental Shelf* (1969) ICJ Reports 3, para. 77. See also *Military and Paramilitary Activities in and against Nicaragua (Nicaragua v. United States of America)* (1986) ICJ Reports 14, paras. 183 and 207.

[71] Brunnée, 'Common Interest', p. 794.

[72] Simma, 'Reciprocity', p. 31.

[73] L. Boisson de Chazournes, 'The Role of Diplomatic Means of Solving Water Disputes: A Special Emphasis on Institutional Mechanisms', in Permanent Court of Arbitration, *Resolution of International Water Disputes* (Leiden: Kluwer Law International, 2003), p. 91.

[74] As Friedman reminded his audience at the Academy of International Law in 1969, 'history has provided abundant evidence that States, big or small, will defy international law where they consider vital "national interests" as threatened'. Friedman, 'General Course', pp. 47f., 77–80.

international law. States enter into legal relationships and accept restraints on their sovereignty based on the expectation of similar concessions by other States.[75] Actual reciprocation of these good faith expectations has led to the emergence of an international legal order under which States coexist and cooperate. The concept of good faith has a specific role in this respect.

1.3.1 Good faith as an essential assumption

Good faith, or *bona fide*, is one of the fundamental principles of international law.[76] As discussed herein, good faith is essential for cooperation and fundamental to the development of international law: if a State cannot trust that its concessions will be reciprocated, neither will happen. Grotius observed that 'good faith . . . is not only the principal hold by which all governments are bound together, but [it] is the key-stone by which the larger society of nations is united'.[77] Although some authors consider good faith to be the basis of an international morality and the 'source and origin of all law', others are more cautious.[78] Schwarzenberger seems little inclined to bestow moral values with too preeminent a role in international law; he does not consider good faith as an all-pervading principle of international law. He agrees to the fundamental character of the principle only because of its paramount role in treaty law, which in turn holds a key role as a source of international law.[79] The ICJ confirmed that 'the principle of good faith is one of the basic principles governing the creation and performance of legal obligations, whatever their source; [yet] it is not in itself a source of obligation where none would otherwise exist'.[80]

Good faith facilitates the identification of a reasonable balance in the exercise of rights in which the rights of one State conflict with the rights of another. As the ICJ confirmed, if States are in conflict with one another when exercising their respective rights, they are under the mutual

75 Simma, 'Reciprocity', p. 31.

76 See *Case Concerning Land and Maritime Boundary between Cameroon and Nigeria, Preliminary Objections* (1998) ICJ Reports 275, para. 38; G. Schwarzenberger, 'The Fundamental Principles of International Law' (1955) 87 *RCADI* 193–383; and Advisory Committee of Jurists, *Procés-Verbaux of the Proceedings of the Committee June 16th–July 24th 1920* (The Hague: Van Langenhuysen Brothers, 1920), p. 335.

77 H. Grotius, *The Rights of War and Peace*, reprint (New York: Cosimo Classics, 2007), p. 417.

78 J. B. Scott, 'Introduction', in *Selections from Three Works of Francisco Suarez* (Oxford: Clarendon Press, 1944), p. 143, cited in O'Connor, *Good Faith in International Law*, p. 28.

79 Schwarzenberger, 'The Fundamental Principles', p. 325.

80 *Nuclear Tests Case (Australia v. France)* (1974) ICJ Reports 269, paras. 46 and 49; and *Border and Transborder Armed Actions (Nicaragua v. Honduras)* (1988) ICJ Reports 105, para. 94.

obligation to negotiate in good faith in order to arrive at a reasonable and equitable solution. In the *Fisheries Jurisdiction Case*, the ICJ decided that in the negotiations, which were to take place on the basis of the judgment, the parties must 'in good faith pay reasonable regard to the legal rights of the other in the waters around Iceland . . . '.[81] The ICJ continued its dictum on the role of good faith in governing the exercise of rights and obligations: 'trust and confidence are inherent in international cooperation, in particular in an age when this cooperation in many fields is becoming increasingly essential'.[82] Good faith is the basis of trust and is thus the essence of cooperation within a society of States. Therefore, the ubiquity of good faith as the principle that guides the creation and exercise of rights is also the basis of the reverse paradigm that bad faith cannot be presumed. This conclusion was recognised for the sphere of cooperation on international watercourses in the decision of the *Lac Lanoux* arbitration: ' . . . *car il est un principe général de droit bien établi selon lequel la mauvaise foi ne se présume pas*'.[83]

The absence of a hierarchically superior giver and enforcer of the law in the international legal system places a legal duty on States to act in good faith. This duty exists to prevent the arbitrary or abusive exercise of rights.[84] In a community of States that is characterised by the absence of a hierarchical order, the societal norms – and thus the norms of international law – can achieve effectiveness only if States comply with the obligations that they assume in good faith.[85] In the presence of good faith, States have established an ever more tightly knit network of legal relationships over time.

1.3.2 Evolution of cooperative coexistence of States

Throughout history, States have generally met expectations of *bona fide* and reciprocity in their mutual relationships, and they continue to do so. In hindsight, it can be argued that this led to ever-closer economic, political and legal ties among them. States are formalising their relationships in a constantly growing body of treaties and norms. These form the basis

[81] *Fisheries Jurisdiction (Federal Republic of Germany v. Iceland)* (1974) ICJ Reports 202, paras. 69 and 70 and ICJ Reports 205, para. 77xs.

[82] *Nuclear Tests Case (Australia v. France)*, 269, para. 46.

[83] *Sentence du Tribunal arbitral franco-espagnol dans l'affaire de l'utilisation des eaux du lac Lanoux* (1958) XXIX *RGDIP* 79–119, para. 9.

[84] G. Fitzmaurice, 'The General Principles of International Law' (1957) 92 *RCADI* 55.

[85] A. Verdross and B. Simma, *Universelles Völkerrecht*, 3rd edition (Berlin: Duncker & Humblot, 1984), p. 46.

of a continuous process of development of international law, which has also been described as the evolution from a law of coexistence to a law of cooperation.[86]

First developed was the law of coexistence, which was defined by Verdross and Simma as the law underlying the not-yet organised society of States.[87] Its main role was to identify spatial, temporal and personal limits of State sovereignty; this then was linked to the obligation to omit interference in the sphere of sovereignty of others. Friedman characterised the law of coexistence as composed of rules of abstention.[88] It governs relationships and interactions among States and establishes standards of mutual treatment of State agents and civilians during times of war and peace.[89] Its norms are found primarily in treaties determining political boundaries, in diplomatic law and in the *jus ad bellum*.

In time, particularly with the increase in economic relations among States and the transport and exchange of goods, the law of coexistence was no longer sufficient. Positive obligations of assistance and obligations of *facere* (i.e., to do rather than to abstain) increasingly formed part of international agreements.[90] The establishment of both the League of Nations (LN) in 1919[91] and its successor, the UN,[92] must be considered milestones in this development. The LN was the first attempt to create an international governance structure for lasting world peace.[93] The principle objectives – to maintain the political independence and the territorial integrity of its member states – were still marked by the spirit of the

[86] Friedman, 'General Course', pp. 92–99; M. Sahovič, 'Codification of the Law of Friendly Relations between States' (1972) 137 *RCADI* 243–310; Brunnée, 'Common Interest', 795; 'Discussion', in J. Delbrück (ed.), *International Law of Cooperation and State Sovereignty – Proceedings of an International Symposium of the Kiel Walther-Schücking-Institute of International Law, May 23–26, 2001* (Berlin: Duncker & Humblot, 2002), p. 124.

[87] Verdross and Simma, *Universelles Völkerrecht*, p. 41.

[88] Friedmann, 'General Course', p. 92.

[89] Verdross and Simma, *Universelles Völkerrecht*, p. 41.

[90] *Ibid.*

[91] The Pact of the League of Nations was incorporated into the Peace Treaties of Versailles, 28 June 1919 (1981), 225 Consol. TS 188–406. It also forms the first twenty-six articles of the subsequent Treaties of Saint-Germain and Neuilly (1981), 226 Consol. TS 8–169 and 332–433, respectively.

[92] Charter of the United Nations, 24 October 1945, 1 *UNTS* XVI.

[93] President Woodrow Wilson enunciated his 'Fourteen Points' for world peace in a speech before the U.S. Congress in January 1918. The first five points outline general principles for peace. Points six through thirteen address territorial questions, and the fourteenth point envisages the establishment of an 'association of nations . . . for the purpose of affording mutual guarantees of political independence and territorial integrity to great and small States alike'. See D. Thomson, *Europe since Napoleon* (London: Penguin Books, 1990), pp. 571f.

law of coexistence. In further developing the governance structure, the Charter of the UN is generally considered to have laid the foundation of the law of cooperation.[94] One objective of the Charter is to 'achieve international cooperation in solving international problems of an economic, social, cultural, or humanitarian character . . . '.[95] In addition to the principles and rules for the proactive and cooperative engagement of States in the implementation of the Charter outlined in Article 2, Articles 55 and 56[96] are commonly considered the primary treaty source from which a general principle of cooperation can be derived.[97]

Recognition of cooperation as a customary principle of international law was later manifested in UN General Assembly (GA) resolutions, such as the 1970 *Declaration on Principles of International Law Concerning Friendly Relations and Cooperation among States*[98] and the 1974 *Declaration on the Establishment of a New International Economic Order*.[99] Both *Declarations* endorse the principle of cooperation; they also emphasise sovereign equality as a confirmation and reminder of the principle that States interact and cooperate as equals in the international legal system.[100]

Later, the law of cooperation received a particular stimulus with the development of international environmental law.[101] The common interest of States in the protection of the natural environment and the realisation that a number of related issues can be resolved only at the universal level led to the addition of a large body of norms of cooperation. The

94 Verdross and Simma, *Universelles Völkerrecht*, p. 310.

95 Art. 1 (3), 1945 UN Charter.

96 Art. 55: 'With a view to the creation of conditions of stability and well-being which are necessary for peaceful and friendly relations among nations based on respect for the principle of equal rights and self-determination of peoples, the United Nations shall promote: (a) higher standards of living, full employment, and conditions of economic and social progress and development; (b) solutions of international, economic, social, health, and related problems; and international cultural and educational cooperation; and (c) universal respect for, and observance of, human rights and fundamental freedoms for all . . . '.

Art. 56: 'All Members pledge themselves to take joint and separate action in cooperation with the Organization for the achievement of the purposes set forth in Article 55'.

97 Sahovič, 'Codification of the Law of Friendly Relations', p. 287; Verdross and Simma, *Universelles Völkerrecht*, p. 310.

98 GA, *Res. 2625 (XXV)*, 24 October 1970, UN Doc. A/8082.

99 GA, *Res. 3201 (S-VI)*, 1 May 1974, UN Doc. A/RES/S-6/3201.

100 Judge Higgins suggested that GA resolutions, which repeat similar content and are adopted by overwhelming majorities, give rise to or are evidence of a general *opinio juris* and can be taken as an indication of the emergence of a customary law. R. Higgins, 'The United Nations and Lawmaking: The Political Organs' (1970) 64 *AJIL*, 43.

101 P. Sands, *Principles of International Environmental Law*, 2nd edition (Cambridge: Cambridge University Press, 2003), pp. 249ff.

general obligation to cooperate was affirmed in multilateral conventions governing global commons, including the oceans, biodiversity and the atmosphere.[102]

It is necessary to point out that the development from a law of coexistence to a law of cooperation is not a linear transformation. Historic antecedents[103] of the law of cooperation existed before the creation of the LN and the establishment of the UN System.[104] Today, States do not always interact in cooperation with one another. Coexistence continues to play a role as an 'umbrella' for cooperation and peaceful relations among States,[105] and the law of coexistence continues to exist today in parallel with the law of cooperation.[106] State relations are more frequently in a state of cooperative coexistence than in a state of cooperation. Sovereignty remains a primordial concept, and States are careful to remind one another of their respective sovereign rights.[107] At the same time, the legal development from coexistence to cooperation left its mark, and it contributes to a new or, rather, a better understanding of sovereignty itself.

1.3.3 *Changing paradigms*

'A State is sovereign, but has to respect his obligations and the rights of other States, in particular the right to integrity and inviolability'.[108] Sovereignty, like good faith, is a fundamental building block of the international legal system. It confers legal personality and the capacity to

102 See, for example, Convention on Biological Diversity, 5 June 1992, 1760 *UNTS* 79; *United Nations Framework Convention on Climate Change* (*UNFCCC*), 9 May 1992, 1771 *UNTS* 165; *United Nations Convention on the Law of the Seas* (UNCLOS), 10 December 1982, 1833 *UNTS* 397.

103 For example, the European Commission of the Danube, which was established based on the Peace Treaty of Paris 1856 that ended the Crimean War, and the International Postal Bureau, which was created in 1874; see B. Vitànyi, *The International Regime of River Navigation* (Alphen aan den Rijn: Sijthoff & Noordhoff, 1979), p. 83.

104 Friedmann, 'General Course', p. 95.

105 Sahovič, 'Codification of the Law of Friendly Relations', p. 306.

106 Verdross and Simma, *Universelles Völkerrecht*, p. 41.

107 Treaties on shared resources often include references to sovereignty over the same – for example, *Revised Protocol on Shared Watercourses in the Southern African Development Community* (2000 *SADC Revised Protocol*), 7 August 2000, www.sadc.int, ECOLEX TRE/001360; *Convention and Statute Relating to the Development of the Chad Basin*, 22 May 1964, OJ of Cameroun, 15 September 1964, p. 1003; and *Convention and Statute on the Regime of Navigable Waterways*, 20 April 1921, 7 *LNTS* 37. The evolution of how States conceive of sovereignty over transboundary waters is analysed in more detail in Chapter 2.

108 *Isla de Las Palmas Arbitration* (1928) II *Recueil de sentences arbitrales* 839.

participate in the process of the creation of international law. The nature and scope of sovereignty is defined and developed by States through consistent international practice and the establishment of international rights and obligations. As Lauterpacht summarised in the introduction to his book, *Private Law Sources and Analogies of International Law*, the nature of the principle of sovereignty is not determined by the prevailing theories but instead by state practice and is thus an evolving concept:

> The international lawyer, if he wishes to assist in the searching investigation which is now taking place on the subject of external sovereignty, will be well employed if he confines himself to the examination of the question whether the doctrine of sovereignty is in accordance both with the practice of States as expressed in customary and conventional international law, and with a rational and consistent system of international law as developed by international publicists, positivists and non-positivists. If he succeeds in showing that certain widely spread notions, directly and demonstrably resulting from the doctrine of sovereignty, run counter to the practice of States and to the conception of international law as taught by the general body of writers, he has already done his share. The doctrine of sovereignty as a legal conception cannot be shattered by well-meaning publicists and pacifists who denounce it as mischievous and as out of date. It will only disappear when notions usually associated with it are shown to be inconsistent both with the practice of States and with the science of international law in its individual branches.[109]

The character of sovereignty in law is dynamic and constantly evolving.

Concepts of sovereignty

The different conceptualisations of sovereignty identified in this chapter illustrate that its character is subject to change in time. These concepts include personal and national sovereignty, internal and external sovereignty, political sovereignty and legal sovereignty.

Jean Bodin is considered the first to have developed the concept of sovereignty in the sixteenth century, at the time of monarchic rulers.[110] His concept described the absolute power – that is, personal sovereignty – of princes and kings over their territory and their people; they were not

109 H. Lauterpacht, *Private Law Sources and Analogies of International Law* (London: Longmann, Green and Co. Ltd., 1927), p. x.

110 Jean Bodin, a French jurist and philosopher, is considered the first to have developed the theory of sovereignty in his *De Republica* in 1576, followed by Thomas Hobbes in 1651 with his *Leviathan*. See R. P. Anand, 'Sovereign Equality of States in International Law' (1986) 197 *RCADI* 23.

accountable to anyone, except perhaps to God or custom.[111] Development of political thought and the birth of 'nation-States', with the peace of Westphalia in 1648 and the American and French revolutions, led to the emergence of the concept of national sovereignty. This concept testifies to the fact that the bearer of sovereignty has changed in democratic systems from one individual to a group, from the monarchic ruler to the government, to the people.[112]

'Internal sovereignty' refers to the supremacy of the sovereign over the territory, its resources within and its people, where the sovereign has the liberty to adopt any constitution and arrange the internal administrative and legal systems according to his or her discretion. 'External sovereignty' is the capacity to assert independence from external interference by third parties in practice and in law. Today, both internal and external sovereignty are limited by fact and by law. States are de facto limited by their relative political power and both internal and external sovereignty are subject to international law. In the case of conflict between national and international law, States cannot invoke their constitution or other national legal norms as an excuse for non-compliance with international obligations. This has been consistently confirmed by international courts and tribunals, and it was set out by Article 13 of the *Draft Declaration on Rights and Duties of States*, prepared by the ILC and adopted by the GA in 1949: 'Every state has the duty to carry out in good faith its obligations arising from treaties and other sources of international law, and it may not invoke provisions in its constitutions or its laws as an excuse for the failure to perform this duty'.[113]

'Political sovereignty' refers to a country's freedom of action and independence from interference by other countries based on its relative military, economic and political power. In today's world, one could argue, a single State that is endowed with full political sovereignty, entirely free from external interference, no longer exists. Political independence is restricted, for instance, by economic and military ties that have created a network of interdependencies impacting the scope of freedom of political

111 M. R. Fowler and J. M. Bunck, *Law, Power, and the Sovereign State – The Evolution and Application of the Concept of Sovereignty* (University Park: Pennsylvania State University Press, 1995), p. 4.

112 Jean Bodin's concept was further developed by Thomas Hobbes in 1651 with his *Leviathan* and many others, such as John Locke and Jean-Jacques Rousseau who, like Hobbes, advocated the idea that not only an individual but also a group or people as a whole could be the bearer of sovereignty. Anand, *Sovereign Equality of States*, pp. 23–25.

113 GA, *Res. 375 (IV)*, 6 December 1949, UN Doc. A/RES/375(IV).

action. It is important to understand that interdependence and independence are not mutually exclusive. Interdependencies merely restrict the freedom of choice due to political, economic, military or other considerations. Only when a State is fully dependent on an outside State for its action is the first State's political sovereignty in doubt.

'Legal sovereignty' describes the scope of rights and duties of a State as a participant in the international society of States. This concept is based on a set of fictitious assumptions that underlie the international legal order. These legal fictions, which define the code of conduct within international society, are the legal equality of States, their independence in the creation of internal (i.e., national) and external (i.e., international) law, and the supremacy of international law over legal sovereignty. The third attribute gives rise to the changing character of legal sovereignty: it leads to a situation in which sovereignty is subordinate to the law to which sovereign States decide to submit. That is, States can willfully restrict their independence through the creation of international law; however, they never fully relinquish their original freedom of action.[114] Theoretically, they could return to their original state of full sovereignty. However, with the progressing development of international law and the vast network of bilateral and multilateral treaties to which they have consented, States have maneuvered into a position from which it is practically impossible to return to the original state of full legal independence.

The evolving character of sovereignty

Legal interdependence has resulted in the current character of legal sovereignty. The original concept of sovereignty – derived from the *souverain*, who was not accountable to anyone – translated in legal terms into sovereignty of States that are independent in their action, limited only by the principles and rules of natural law that originate in spheres outside of law. Sources of the natural law can be found in morals, ethics or religion; they preexist the willful consent of States; and they preexist rules of international law. For those scholars who accept these pre-legal

[114] The Permanent Court of International Justice (PCIJ) pronounced in its judgment of 17 August 1923 in the *S. S. Wimbledon* case: 'The Court declines to see in the conclusion of any Treaty by which a State undertakes to perform or refrain from performing a particular act an abandonment of its sovereignty. No doubt any convention creating an obligation of this kind places a restriction upon the exercise of the sovereign rights of the State .... But the right of entering into an international engagement is an attribute of State sovereignty'. PCIJ (1923), Ser. A No.1, p. 25.

concepts as a source of international law once transformed into legal principles, sovereignty has never been completely unrestricted.[115]

Stated differently, a sovereign State cannot be seen as a structure that is independent from its environment. Sovereign States form part of the wider international society of States. As members of this society, States have not only rights but also certain duties and responsibilities that come with this membership. The creation of rules of international law is a manifestation of State participation in this international society. Over time, States developed a complex code of conduct with a vast set of international rules to regulate their coexistence, which in turn further restricted their sovereignty. They have limited the scope of their freedom of action through the development of international law and the conclusion of treaties. With these legal instruments, the duty to perform certain obligations, to agree to refrain from acting in a certain way or even to transfer specific functions of their sovereignty to international organisations is imposed.[116] For example, by becoming members of the UN, States transfer the primary responsibility to maintain international peace and security to the Security Council.[117] Within the EU framework, member States have even transferred their sovereign rights to conduct important aspects of their commercial policy to the EU.[118] On many occasions, States established these rules limiting the scope of their sovereignty because cooperation creates new opportunities and freedoms. The existence of commonly accepted rules of conduct and/or the action through joint mechanisms can be mutually beneficial. In this sense, the creation of international law is an act of cooperation in the expression of sovereignty.

Cooperation is also increasingly becoming an objective of international law in and of itself, comprising the content of new rules such

[115] See M. Kohen, 'Is the Notion of Territorial Sovereignty Obsolete?', in M. Pratt and J. A. Brown (eds.), *Borderlands under Stress* (London: Kluwer Law International, 2000), p. 36.

[116] At this point, it is important to remember that the transfer of specific functions of sovereignty does not mean that these functions have been relinquished by the States. Theoretically, States maintain the legal capacity to recover these functions if they wish to do so, even if practically this becomes increasingly more difficult with the progressive development of international law and the ever-increasing network of interdependencies it creates.

[117] Art. 24 (1), 1945 Charter of the United Nations.

[118] On the EU role in representing the interests of the EU and of its members States within the World Trade Organization (WTO), see F. Hoffmeister, 'Outsider or Frontrunner? Recent Developments under International and European Law on the Status of the European Union in International Organizations and Treaty Bodies' (2007) 44 *Common Market Law Review* 41–68.

as in the fields of environmental law, trade law and international criminal law. The questions arise as to whether the character of sovereignty has changed and whether sovereignty, which was originally conceived of as a concept of autonomy, has been transformed into a notion that considers States' responsibilities and duties as members of an interdependent international community. It seems to be widely accepted that legal sovereignty is accompanied by the responsibility to participate and engage in good faith in matters of common concern. Perrez went one step further by proposing that a concept of 'cooperative sovereignty' is emerging.[119] This conceptualisation inextricably links sovereignty to the principle of cooperation; it sees cooperation as an integral part of sovereignty. According to Perrez, 'cooperation is not an independent principle supplementing State sovereignty or even a counterweight to State sovereignty, but . . . cooperation has to be conceived as an element of the very notion of sovereignty itself'.[120] Arguably, this progressive understanding of sovereignty already informs the process of integration and cooperation among EU member States; however, it is debatable whether this interpretation is also accepted outside that region.

1.4 Conclusion

The global water cycle establishes intricate links among the territories of all States: human activities affecting freshwater resources in one State frequently impact the quantity or quality of freshwater available in other States' territories. Coordination is required to manage these transboundary effects of water utilisation. This chapter's analysis highlights the many reasons why States may decide to cooperate or not in their pursuit to maximise benefits. Political theory identifies various factors that impact a State's decision-making process; probably most preeminent is the factor of interdependence. This may be interdependence established by history, through economic ties or by inescapable facts of nature such as the hydrologic interlinkages in transboundary water systems. Cooperation and conflict emerge as a consequence of interdependence.[121] By the same

[119] F. X. Perrez, *Cooperative Sovereignty – From Independence to Interdependence in the Structure of International Environmental Law* (The Hague: Kluwer Law International, 2000), pp. 331–333.

[120] *Ibid.*, Preface.

[121] M. Bernad y Alvarez de Eulate, '*La coopération transfrontalière régionale et locale*' (1993) 243 *RCADI* 347.

process, interdependence pervades international law, which is a factor contributing to and a result of interdependence.

The evolution of international law seems to confirm the findings of the iterated prisoner's dilemma game. As many authors have pointed out, States developed a system of international law that evolved from a system of coexistence in which 'good fences make good neighbours' to a system that is characterised by the coexistence of those fences, with norms that perforate the fences and establish bridges – and, indeed, even increase interdependence between the spheres of influence of sovereign States. States agree to limit their sovereignty as a concession and in expectation of additional benefits to be derived from cooperative behaviour. Sovereignty becomes increasingly more closely intertwined with cooperation; however, it still seems premature to speak of universal acceptance of the concept of 'cooperative sovereignty'. It must be recognised that cooperation is not omnipresent, either in State behaviour or in the international legal order. With full knowledge of the potential benefits of cooperation, States still choose coexistence without cooperation regarding a number of issue areas, as they did in the early stages of development of international law. However, as illustrated in subsequent chapters, mere coexistence is not a viable choice in situations in which States manage and utilise shared natural resources. The development of international water law is the legal response to interdependence and the ensuing requirements of coordination in the management and development of shared waters.

2 Development of international water law

The historic evolution of the law applicable to transboundary water resources is characterised by the transformation of the ways in which States address the inevitability of hydrologic interconnectedness. The recognition that interdependence requires some type of coordination was increasingly reflected in legal concepts, normative development and legal instruments. With the development of an ever-growing body of norms and treaties, cooperation among States about shared water resources became progressively more formalised. This process can be retraced by examining two distinct aspects: (1) the mutability of the notion of territorial sovereignty and the variations in its different conceptions, often depending on whether sovereignty is pleaded in negotiations or is, in fact, applied to the management of shared resources; and (2) the expansion of normative reach, both geographically and relative to issue areas. The regulation of sustainable water management broadened from the local to the regional and the universal levels; at the same time, an increasing number of water uses became integrated under its umbrella.

2.1 Sovereignty over shared water resources

States exert sovereignty over their territory, which is the geographic space that is delimited by a State's international boundaries; it consists of surface and subsurface areas as well as the overhead airspace.[1] States also exert sovereignty over natural resources within this geographic space. This sovereignty comprises specific rights, including the right to dispose

1 J. Verzijl, *International Law in Historical Perspective: Part III State Territory* (Leyden: A. W. Swijthoff, 1970), p. 14. Verzijl cites these elements, which also include the territorial sea, as the undisputed elements of State territory.

of, explore and exploit the resources freely; the right to use them for national development; the right to manage them according to national environmental policy; and the right to regulate resource-related foreign investment.[2] This was confirmed in GA *Resolution 626 (VII)* of 1952, which recalls 'that the right of peoples freely to use and exploit their natural wealth and resources is inherent in their sovereignty'.[3] At the same time, sovereignty over natural resources is subject to limitations by international law. With the continuous development of this body of norms, the character of sovereignty changes as well.

Resources interact with one another, and not all natural resources are solidly linked to or form an integral part of a State's territory on a permanent basis. Hence, legal issues concerning the rights and limits of control over natural resources are closely intertwined with their territorial nature. Dynamic resources including gaseous substances (e.g., air and natural gas), some liquids (e.g., water and oil) and migrating species do not linger permanently in the physical space of a given territory but instead eventually transcend its international boundaries. A State's territorial sovereignty over these transboundary resources ends where the territorial sovereignty of another State begins. When these resources can be used consumptively, questions are raised regarding the scope of territorial sovereignty: Is it absolute and can a State therefore extract, for example, the total number of migrating fish in its territory, disregarding potential interests in the catch by a neighbouring State, to which the fish would have migrated otherwise? Or is territorial sovereignty limited and does a State therefore bear certain responsibility for the impacts of activities carried out in its own territory – for instance, when it pollutes the airspace so heavily that the result is acid rain and forest decline in the territory of another State?[4]

2.1.1 Absolute territorial theories

With respect to transboundary waters, some States initially promoted an absolute approach to territorial sovereignty. Two theories have been proposed: the 'theory of absolute territorial sovereignty' and the 'theory

[2] N. Shrijver, *Sovereignty over Natural Resources: Balancing Rights and Duties* (Cambridge: Cambridge University Press, 1997), pp. 258–297.

[3] General Assembly, *Res. 626 (VII) on the Right to Exploit Freely Natural Wealth and Resources*, 21 December 1952, UN Doc. A/RES/626(VII).

[4] F. B. Pyatt, 'Acid Rain in Scandinavia – Some Current Data' (1987) 7 *The Environmentalist* 197; *Trail Smelter Case (United States, Canada)*, 18 April 1938, 11 March 1941 (1941) 3 *RIAA* 1905–1982.

of absolute territorial integrity'. Both theories are closely related to the original understanding of State sovereignty as the absolute power of a State to dispose of its territory and resources within it as it pleases, as well as the idea of absolute inviolability of a State's territory.

The theory of absolute territorial sovereignty

According to the theory of absolute territorial sovereignty, all water resources in a State's territory, including those flowing in transboundary watercourse systems, are viewed as forming an integral part of the territory and as being at full disposal to the State.[5] The incident that is most commonly cited in connection with the formulation of this theory is the disagreement that arose between Mexico and the United States towards the end of the nineteenth century, which concerned upstream diversions of waters from the Rio Grande.[6] According to the note of protest sent by Mexico to the United States, diversions by farmers in Colorado and New Mexico had led to a sharp decline in the river's flow volume and in water supply to communities downstream of the international border. The US Secretary of State referred the case to Attorney-General Judson Harmon for his legal opinion, and he responded as follows:

> The jurisdiction of the nation, within its own territory, is necessarily exclusive and absolute; it is susceptible of no limitation, not imposed by itself. Any restrictions upon it, deriving validity from an external source, would imply a diminution of its sovereignty, to the extent of the restriction, and an investment of that sovereignty, to the same extent, in that power which could impose such restriction. All exceptions, therefore, to the full and complete power of a nation, within its own territories, must be traced up to the consent of the nation itself.... The case presented is a novel one. ... [i]n my opinion, the rules, principles and precedents of international law impose no liability or obligation upon the United States.[7]

As a consequence of this statement, the theory that a State may freely dispose of the waters of an international watercourse that flow

[5] On this theory, see McCaffrey, *The Law of International Watercourses*, pp. 112–126; Caflisch, '*Règles Générales*', pp. 48–50; Lammers, *Pollution of International Watercourses*, pp. 557f.; J. Lipper, 'Equitable Utilization', in A. H. Garretson, R. D. Hayton and C. J. Olmstead (eds.), *The Law of International Drainage Basins* (Dobbs Ferry, NY: Oceana Publications, 1967), pp. 20–23.

[6] For a detailed account, see McCaffrey, *ibid.*, pp. 78–83.

[7] E. C. Brandenburg (ed.), *Official Opinions of the Attorney-General of the United States Advising the President and Heads of Departments in Relation to their Official Duties*, Vol. 21 (Washington, DC: Government Printing Office, 1898), pp. 281–283; also cited in McCaffrey, *ibid.*, pp. 113f.; and Caflisch, '*Règles Générales*', pp. 48f.

in its territory without requiring due regard to the harmful effects this might cause beyond its territory became known as 'the Harmon Doctrine'.

The legal opinion and the development of theory, however, must be put into historical perspective. Lipper, for example, observed that when the Harmon Doctrine was formulated, States rarely had to consider the mutual exclusiveness of water uses in transboundary basins, and customary rules restricting sovereignty rights were yet to be defined in the context of international law.[8] The statement by the Attorney-General therefore also can be viewed as the conclusion that – in the absence of international rules to the contrary and in the absence of State consent to any rules limiting its power in this respect – States can exert absolute sovereignty over resources in their territory.[9]

Alternatively, the Harmon Doctrine can be considered an assertion of State sovereignty that at the same time, however, has not become a rule of public international law on this scale. On several occasions, riparian States upstream of transboundary watercourses have referred to this doctrine as a negotiating position and in diplomatic notes, but the concept did not prevail in practice. In the Rio Grande case, this position eventually remained only one argument among others during the negotiation process. In 1906, the two States signed a treaty that apportions the waters of the river between them in a way that was considered equitable and acceptable to both.[10] Similar approaches were also followed in other transboundary river basins, in which upstream States initially pleaded the theory of absolute territorial sovereignty during negotiations but ultimately agreed to more equitable solutions in the international agreements that followed. These include the cases of Austria in discussion with Bavaria on the development of shared waters and subsequent treaty praxis of the former State; India in its dispute with Pakistan over the waters of the Indus; and Ethiopia with respect to downstream riparians on the Nile.[11]

The theory of absolute territorial sovereignty also did not resonate with international courts and tribunals. The arbitral tribunal in the *Trail Smelter*

[8] Lipper, 'Equitable Utilization', p. 22.

[9] Lipper, *ibid.*; McCaffrey, *The Law of International Watercourses*, p. 125.

[10] *Convention for the Equitable Distribution of the Waters of the Rio Grande for Irrigation Purposes*, 21 May 1906, 201 Consol. TS (1980), 225–227.

[11] See B. A. Godana, *Africa's Shared Water Resources – Legal and Institutional Aspects of the Nile, Niger and Senegal River Systems* (London: Frances Pinter [Publishers] Limited, 1985), pp. 35f.; McCaffrey, *The Law of International Watercourses*, pp. 115–122; Caflisch, '*Règles Générales*', p. 49.

Case[12] found that 'under the principles of international law . . . no State has the right to use or permit the use of its territory in such a manner as to cause injury . . . in or to the territory of another'.[13] The issue at question in the *Trail Smelter Case* was the damage caused in the United States by toxic fumes that resulted from operations of a smelter located in the city of Trail in British Columbia. The ICJ endorsed this view in its judgment on the *Corfu Channel Case*, stating that '[it is] every State's obligation not to allow knowingly its territory to be used for acts contrary to the rights of other States'.[14] This case arose from a dispute concerning State responsibility for damage caused by mines to two British vessels passing through Albanian territorial waters. The ICJ concluded that the laying of the minefield could not have taken place without the knowledge of Albania.[15] The principle that territorial sovereignty encompasses not only rights but also duties was recognised, furthermore, in the arbitration award in the *Island de las Palmas* case. The arbitrator found that '[t]erritorial sovereignty . . . involves the exclusive right to display the activities of a State. This right has as corollary a duty: the obligation to protect within the territory the rights of other States . . . '.[16]

Although neither case directly concerns transboundary freshwater resources, the findings nevertheless apply to this context. This was confirmed in the landmark decision of the *Lac Lanoux* arbitration that concerned a dispute between France and Spain over the diversion of waters of the Carol River by the upstream riparian France, for the purpose of hydroelectricity production. In its decision, the arbitral tribunal corroborated the fact that territorial sovereignty over shared resources is not absolute but instead is a presumption and, as such, must give way to international rights and obligations.[17]

The theory of absolute territorial integrity

As the downstream riparian of the Carol River, Spain tried to counter France's river diversion project in the *Lac Lanoux* arbitration with

12 *Trail Smelter Case (United States, Canada)*, 18 April 1938, 11 March 1941 (1941) 3 *RIAA* 1905–1982.

13 *Ibid.*, p. 1965. 14 *Corfu Channel Case* (1949) ICJ Reports 22. 15 *Ibid.*

16 *Island de las Palmas (Netherlands v. United States)*, 23 January 1925, (1928) 2 *RIAA* 829, cited in Caflisch, '*Règles Générales*', p. 54.

17 '*La souveraineté territoriale joue à la manière d'une présomption. Elle doit fléchir devant toutes les obligations internationales, quelle qu'en soit la source, mais elle ne fléchit que devant elles'. Sentence du tribunal franco-espagnol du 16 Novembre 1957 dans l'affaire de l'utilisation des eaux du lac Lanoux* (1958) 29 *RGDIP* 99.

arguments that were based on another absolute theory of territorial sovereignty. Spain claimed that diversion of the waters of Lake Lanoux away from the shared Carol River into the Ariège River – which flows exclusively on French territory and into the Atlantic Ocean, and even if compensated by full restitution of the diverted volume from another source – would violate Spain's rights to independence.[18] Spain claimed a right to absolute inviolability of its territory: a concept that is generally referred to as the 'theory of absolute territorial integrity'.[19] According to this theory, pollution, change in flow and reduction of the natural transboundary flow volume of a watercourse or aquifer by one basin State leading to physical changes in the territory of another are all considered interference in the territorial integrity of the affected State.[20]

Similar to the Harmon Doctrine, which generally strengthens the position of upstream States, the theory of absolute territorial integrity primarily supports the interests of downstream States in successive watercourses. The theory was invoked not only by Spain in the *Lac Lanoux* arbitration but also by other downstream riparians in negotiations and international disputes.[21] As the State affected downstream of infrastructure development carried out by Slovakia, Hungary claimed its right to territorial integrity before the ICJ in the *Gabčíkovo–Nagymaros Project Case*.[22] It was originally planned as a joint project between Czechoslovakia and Hungary to exploit the hydropower potential and to improve navigability of the Danube River where it forms their common border.[23] Due to internal political criticism of the project, Hungary eventually suspended work and Czechoslovakia proceeded with an alternative design. 'Variant C' allowed hydroelectricity production through a unilateral diversion of the river on Slovak territory. This diversion, however, drained surface water and groundwater from the alluvial plain at Szigetköz, which is located in

18 *Ibid.*, pp. 92f.

19 McCaffrey, *The Law of International Watercourses*, pp. 126–135; Caflisch, '*Règles Générales*', pp. 51–54; Lammers, *Pollution of International Watercourses*, p. 562.

20 In the given case, Spain did not plead the risk of alteration of water quality through the planned restitution of the diverted flow volume from a different watercourse.

21 McCaffrey and Caflisch cite the cases of Pakistan in the Indus Basin, Egypt in the Nile Basin, and Bolivia relative to Chile as a downstream riparian on the Rivers Mauri and Lauca. McCaffrey, *The Law of International Watercourses*, pp. 127–131; Caflisch, '*Règles Générales*', p. 51.

22 Slovakia became the successor State after Czechoslovakia split in 1993. *Gabčíkovo–Nagymaros Project (Hungary/Slovakia)* (1997) ICJ Reports, paras. 64 and 91.

23 *Treaty Concerning the Construction and Operation of the Gabčíkovo–Nagymaros System of Locks*, 16 September 1977, 1109 *UNTS* 236.

Hungarian territory. Hungary therefore protested the implementation of 'Variant C' because it believed that its right to territorial integrity was violated. In the *Trail Smelter Case*, the United States equally invoked the theory of territorial integrity as the 'downstream' recipient of air pollution and argued that 'it is a fundamental principle of law of nations that a sovereign State is supreme within its own territorial domain and that it and its nationals are entitled to use and enjoy their territory and property without interference from an outside source'.[24] This theory of absolute inviolability of a State's territory did not become a generally accepted concept in international law; similar to territorial sovereignty, territorial inviolability is subject to limitations by international law.

Mutual exclusiveness of absolute theories

The fact that the United States availed itself of the theory of territorial integrity in the *Trail Smelter Case* when, in the case of the Rio Grande, Attorney-General Harmon formulated a doctrine that claims largely the opposite – namely, that a State could dispose of its territory regardless of the impact on the territory of other States – might appear curious. Yet, it is evidence of the fact that both theories are primarily valuable to substantiate a specific negotiating position; the use of one or the other depends on the specific situation of a State. The theory of absolute territorial sovereignty serves the position of States carrying out projects with transboundary impact, whereas the theory of absolute territorial integrity serves the affected State.[25] From a practical as well as a legal perspective, these two theories are incompatible.

Incompatibility exists with respect to the nature of water. Transboundary water resources are not conducive to an interpretation that considers them to be an integral part of the territory of a State. The notion of 'territory' is a static concept defined as delimited space, yet water is a dynamic resource that transcends boundaries and limits. Water that is taken out of a transboundary water system and used in the territory of one State will neither reach nor become an integral part of the territory of the

[24] *Memorandum in Relation to the Arbitration of the Trail Smelter Case, United States and Canada*, 10 August 1937, prepared by Green H. Hackworth, Legal Advisor, for Swagar Sherley, Agent of the United States, cited in McCaffrey, *The Law of International Watercourses*, p. 129.

[25] Even on rivers, it is not so much a question of upstream and downstream positions of States; both can equally harm one another on contiguous systems as well as in the long term. Downstream consumptive uses may harm upstream uses over time because they can preclude future uses of upstream States by establishing acquired rights for downstream States. This is discussed in more detail in Part II of this volume.

downstream State. Water flowing in the international water system should therefore not be considered an integral part of State territory; it is rather the riverbed, rock layers and space that contains the waters during the course of their passage that belong to the territory of a riparian State.[26] A simultaneous application of the two theories to a successive watercourse is not possible. For example, the consumptive use of the waters by one riparian State is prohibited by the theory of absolute territorial integrity because it would result in changes in the territory of another State. At the same time, this restriction of the right to use shared waters when they pass through a State's territory is diametrically opposed to the notion of absolute territorial sovereignty.

Furthermore, from a legal perspective, the two theories are contrary to a core assumption of the fundamental principle of sovereignty: namely, the sovereign equality of States. Applied to the field of transboundary water systems, sovereign equality implies that all riparian States enjoy within their territory equality of rights in the utilisation of the transboundary water resources. In the exercise of their rights, States therefore must respect the correlative rights of their co-riparian States.[27] Yet, both theories – when applied in absolute terms – put restrictions on the exercise and enjoyment of territorial rights of others.[28] The theory of absolute territorial integrity, for instance, would give veto power to the State that will be affected by a measure that is planned by another basin State.[29] The existence of such a veto right was argued by Spain in the *Lac Lanoux* arbitration when it claimed that France could not proceed with the planned diversion works without the former's prior approval.[30] However, as was observed by the tribunal, granting a veto right would mean abrogating the sovereign rights of one State at the discretion of another.[31] The two theories thus overlook the fact that territorial sovereignty attributes not only rights but also duties to States, one of which is the duty to respect the rights of other States.[32]

26 A. J. Hirsch, 'From the Indus to the Jordan: Characteristics of Middle East International River Disputes' (1956) 71 *Political Science Quarterly* 207.

27 A. Tanzi and M. Arcari, *The United Nations Convention on the Law of International Watercourses* (The Hague: Kluwer Law International, 2001), pp. 13 and 15.

28 Lammers, *Pollution of International Watercourses*, pp. 557f. and p. 563.

29 Caflisch, '*Règles Générales*', pp. 51 and 53.

30 *Sentence du tribunal franco-espagnol du 16 Novembre 1957 dans l'affaire de l'utilisation des eaux du lac Lanoux*, p. 93.

31 *Ibid.*, p. 106.

32 *Ibid.*, p. 107; Caflisch, '*Règles Générales*', p. 54; J. Arechaga, 'International Law in the Past Third of a Century' (1978) 159 *RCADI* 192.

2.1.2 Limited territorial sovereignty over shared resources

The incompatibility of the two absolute territorial theories had an impact on State practice and legal development. In general, the two theories remained mere negotiation positions, and a third theory that builds a bridge between the two was crystallised as the commonly accepted approach. The theory of limited territorial sovereignty, also known as the 'theory of limited (or restricted) territorial integrity', more adequately reflects the reality that transboundary resources are not permanently attributable to any one State's territory.[33] The evolution of this theory, which also applies to other shared resources,[34] was primarily based on State practice on transboundary watercourses.[35] The theory implies that each riparian State can use the waters situated in its territory as long as this use does not cause significant harm to the rights of other States.[36] The theory is based on the notion that territorial sovereignty entails rights as well as duties. Today, it is widely considered to be the prevailing theory in international water law.[37]

Treaties that define respective rights and duties of riparian States over shared water resources reflect the idea of restricting sovereignty relative to shared waters. An earlier agreement that explicitly recognises that territorial sovereignty is subject to limitations is the 1923 *Geneva Convention Relating to the Development of Hydraulic Power Affecting More Than One State*.[38] The *Convention* was negotiated under the auspices of the LN and was one of the first multilateral water conventions with intended universal scope. The first Article states: 'The present Convention in no way affects the right belonging to each State, *within the limits of international law*, to carry out on its own territory . . . ' [emphasis added]. The Article confirms that the sovereignty of each State is restricted by international rules, including the rights that are held by other States. A similar stance is taken in Article 3 of the 2008 *Draft Articles on the Law of Transboundary Aquifers* elaborated by the ILC, which states: 'Each aquifer State has sovereignty over

33 Godana, *Africa's Shared Water Resources*, p. 40.

34 J. Barberis, *Los Recursos Naturales Compartidos entre Estados y el Derecho Internacional* (Madrid: Editorial Tecnos, 1979).

35 For an analysis of illustrative examples of state practice, see McCaffrey, *The Law of International Watercourses*, pp. 136–142.

36 *Ibid.*, p. 135; Caflisch, '*Règles Générales*', p. 55.

37 *Ibid.*; O. McIntyre, *Environmental Protection of International Watercourses under International Law* (Aldershot: Ashgate, 2007), p. 23.

38 9 December 1923, 36 *LNTS* 76. The *Convention* has been ratified by seventeen States as of December 2012.

the portion of a transboundary aquifer . . . located within its territory.[39] It shall exercise its sovereignty in accordance with international law and the present draft articles'. It is interesting that the 2008 *Draft Articles* apply the restrictions to territorial sovereignty resulting from the law and nature of shared resources not only to the transboundary waters but also to the permeable water-bearing rock formations – that is, to static resources that are permanently attributable to a State's territory. This occurs because the water-bearing rock forms part of the legal definition of the 'aquifer'. Article 2 (a) defines the term as 'a permeable water-bearing geological formation underlain by a less permeable layer and the water contained in the saturated zone of the formation'.

The theory of limited territorial sovereignty has been referred to by international courts and tribunals[40] and, furthermore, is embodied in the principle of equitable and reasonable utilisation and the obligation not to cause significant harm.[41] These two rules, which are described in more detail in Part II of this book, simultaneously acknowledge and restrict territorial use rights.[42] An interesting decision in this context is the judgment of the PCIJ in the *Case Relating to the Territorial Jurisdiction of the International Commission of the River Oder*.[43] In this case, the PCIJ confirmed the equality of rights of States that have a legal interest in the navigable use of a watercourse and denied preferential privilege to any one State. In doing so, the PCIJ put forward the notion of 'community of interest' as the basis for this equality of rights:

This community of interest in a navigable river becomes the basis of a common legal right, the essential features of which are the perfect equality of all riparian States in the use of the whole course of the river and the exclusion of any preferential privilege of any one riparian State in relation to the others.[44]

[39] Annex to GA, *Res. 63/124*, 11 December 2008, UN Doc. A/RES/64/124.

[40] See descriptions of decisions in the *Corfu Channel Case* and the *Trail Smelter* and *Lac Lanoux* arbitrations highlighted previously.

[41] McIntyre, *Environmental Protection*, p. 23.

[42] See, for example, their codification in the *United Nations Convention on the Law of the Non-Navigational Uses of International Watercourses* (hereinafter, *UN Watercourses Convention*): Art. 5 (1), '. . . In particular, an international watercourse shall be used and developed by watercourse States with a view to attaining optimal and sustainable utilization thereof and benefits therefrom, taking into account the interests of the watercourse States concerned, consistent with adequate protection of the watercourse'; and Art. 7 (1), 'Watercourse States shall, in utilizing an international watercourse in their territories, take all appropriate measures to prevent the causing of significant harm to other watercourse States'.

[43] *No. 16 – Case Relating to the Territorial Jurisdiction of the International Commission of the River Oder*, PCIJ (1929) Ser. A.23.

[44] *Ibid.*, p. 27.

The dictum of the PCIJ was later used by its successor, the ICJ, in its decision on the *Gabčíkovo–Nagymaros Project* as the legal basis for the principle of proportionality of countermeasures and links it to the principle of equitable and reasonable utilisation.[45] The ICJ perceives the concept as also having strengthened over time for non-navigational uses and cites as evidence the adoption of the *UN Watercourses Convention*, which stresses the principle of equitable and reasonable utilisation as a core rule of international water law.[46] The ICJ confirmed that there are limitations to territorial sovereignty; in this case, it declared that the diversion of the waters away from the alluvial plain at Szigetköz by the upstream riparian was disproportionate and that it deprived 'Hungary of its right to an equitable and reasonable share of the natural resources of the Danube'.[47] More recently, authors analysed the notion of 'community of interest' as a legal concept that is distinct from the theory of limited territorial sovereignty; some consider it an emerging fourth theory of sovereignty in the field of international water law.[48]

2.1.3 A community of interest?

The concept of 'community of interest' can be considered an alternative to the three territorial sovereignty theories discussed previously, as one that is based more on territorial interdependence than other theories and that goes beyond the traditional concept of sovereignty. Rather than emphasising the autonomy of sovereign entities, it highlights the interdependence of States and underlines the fact that States cannot escape their membership in the community of riparian States because it is established by the nature of their territories, which are interconnected through a shared hydrologic system. The interests of riparian States in the use of transboundary waters can be of an economic, social or cultural nature; they can be aligned or coinciding, but they also can be competing up to the point where they become diametrically opposed. The notion of

[45] *Gabčíkovo–Nagymaros Project (Hungary/Slovakia)*, para. 85.

[46] Adopted by the GA on 21 May 1997, UN Doc. A/RES/51/229.

[47] *Gabčíkovo–Nagymaros Project (Hungary/Slovakia)*, para. 85.

[48] L. Boisson de Chazournes and M. Tignino, '*Droit international et eau douce*' (2007) *Jurisclasseur – Environnement, Fasc. 2900*, para. 45; McCaffrey, *The Law of International Watercourses*, pp. 147–167; United Nations Economic and Social Commission for Western Asia (UNESCWA), *Assessment of Legal Aspects of the Management of Shared Water Resources in the ESCWA Region*, UN Doc. GENERAL E/ESCWA/ENR/2001/3, pp. 29 and 31f.; Tanzi and Arcari, *The United Nations Convention*, pp. 21–24; Lammers, *Pollution of International Watercourses*, pp. 506f.; B. Vitányi, *The International Regime of River Navigation* (Alphen aan den Rijn: Sijthoff & Noordhoff, 1979), pp. 56f.

'community of interest' evokes the necessity of cooperation to ensure that competing interests do not stand in the way of one another.[49] In its decision in the *River Oder Case*, the ICJ recognised this idea in stating that:

... [w]hen consideration is given to the manner in which States have regarded the concrete situations arising out of the fact that a single waterway traverses or separates the territory of more than one State, and the possibility of fulfilling the requirements of justice and the considerations of utility which this fact places in relief, it is at once seen that a solution of the problem has been sought not in the idea of a right of passage in favor of upstream States, but in that of a community of interest of riparian States'.[50]

The *River Oder Case* concerned the question of whether the administrative regime of a river commission that is tasked with the regulation of navigational uses geographically coincides with the regime of internationalisation of a river – a regime that is functionally linked to navigation. In this case, the community of interest was not limited merely to members connected by a geographic link but rather extended to non-riparian States that had an interest in using the particular characteristics of the internationally navigable river. This joint interest in navigation and access to upstream territory was manifested by the conclusion of a treaty: the Commission was provided for in the 1919 *Treaty* of Versailles.

A question worth analysing more closely is whether the community of interest concept crystallised as a distinct theory and/or as a legal concept that exists independently of treaties. Statements of the ICJ on this issue do not provide unequivocal evidence that the community of interest might have gained a foothold as a distinct normative concept in international practice. The ICJ cited relevant passages from the *River Oder Case* decision to support a basis for the principle of equitable and reasonable utilisation in its decision in the *Gabčíkovo–Nagymaros Case*; however, it must be pointed out that the Court merely repeated the relevant passage.[51] As in

[49] On this, the ILA, with respect to navigation, states: 'The community of interest of the riparian States in such a waterway becomes the legal basis for requiring each riparian State to give notice, consult, negotiation and cooperate in order to enable, by common agreement, a settlement of all relevant questions related to the improvement of the navigability, such as technical and administrative implication of the works needed, financial assistance and sharing of costs'. ILA, *Report of the Fifty-Sixth Conference Held at New Delhi, December 29th, 1974, to January 4th, 1975* (London: ILA, 1976), p. 128.

[50] *Case Relating to the Territorial Jurisdiction of the International Commission of the River Oder*, p. 27.

[51] *Gabčíkovo–Nagymaros Project (Hungary/Slovakia)*, para. 85.

the PCIJ decision, the idea of community of interest was not unequivocally identified as a normative concept. In the *Pulp Mills on the River Uruguay Case*, in which Argentina argued that the community of interest and law were objectively established by the 1975 *Statute of the River Uruguay*[52] and obligated the two riparian States to cooperate,[53] the Court equally limited itself to confirming the factual existence of such a community based on a treaty. It states that '[b]y acting jointly through CARU,[54] the Parties have established a real community of interests and rights in the management of the River Uruguay and in the protection of its environment'.[55]

Literature interprets these court decisions in different ways. Vitànyi and Lammers proposed that the *River Oder* judgment firmly established the community of interest concept as a theory in international river law.[56] It has been argued that the fact that the PCIJ considered the notion in parallel with the 'idea of a right to passage' is an indication that it should be equally considered as a normative concept. Furthermore, it should apply not only to problems of navigation but also to those connected with other water uses.[57] This appears to be a minority position in legal doctrine; the community of interest concept is still addressed in literature primarily as an emerging theory.[58] Caflisch supports the argument that the recognition of a community that carries legal consequences remains based on treaty arrangements by States.[59] This argument apparently was supported by the ICJ in its decision in the *Pulp Mills Case*, in which it considers the community of interest to be established through cooperation in a treaty-based river commission.[60] Godana argued that although the community of interest is the 'legal principle most appropriate for a fully developed

[52] 26 February 1975, 1295 *UNTS* 339.

[53] *Pulp Mills on the River Uruguay (Argentina v. Uruguay), Provisional Measures, Pleadings by L. Boisson de Chazournes*, Transcript of Public Sitting, 8 June 2006, CR 2006/46, pp. 45f.

[54] Abbreviation for 'Administrative Commission of the River Uruguay'.

[55] *Case Concerning Pulp Mills on the River Uruguay (Argentina v. Uruguay)* (2010) ICJ Reports, para. 281.

[56] Vitányi, *The International Regime of River Navigation*, pp. 56f.; Lammers, *Pollution of International Watercourses*, pp. 506f.

[57] Lammers, *ibid*.

[58] McCaffrey, *The Law of International Watercourses*, pp. 167–170; Tanzi and Arcari, *The United Nations Convention*, p. 21; Caflisch, '*Règles Générales*', pp. 59–61.

[59] Caflisch, *ibid*.

[60] See also *Arbitration in the Application of the Convention of December 3, 1976, on the Protection of the Rhine against Pollution by Chloride and the Additional Protocol of September 25, 1991* (2004) Permanent Court of Arbitration, para. 97.

legal community . . . the international community is far from being fully developed'.[61]

The question is whether recent treaty practice can be seen as an indication that the international community is in the process of achieving the stage of development about which Godana was writing. One initially singular treaty reference to the 'principle of community of interest', which was included in the 1995 *Protocol on Shared Watercourse Systems in the SADC Region*, was later retracted.[62] Article 2 (2) stated that 'Member States undertake to respect and apply the existing rules of general or customary international law relating to the utilization and management of the resources of shared watercourse systems and, in particular, to respect and abide by the principles of community of interests in the equitable utilization of those systems and related resources'. The *Revised Protocol on Shared Watercourses in the SADC* (hereinafter, 2000 *SADC Revised Protocol*), which replaces the earlier document, no longer includes this reference to the community of interest.[63] More recently, however, the concept reappeared in the 2002 *Water Charter of the Senegal River*[64] and the 2003 *Protocol for Sustainable Development of Lake Victoria Basin*.[65] This indicates that a small but growing number of States put into practice the argument that 'the management of international [waters] should be determined less by the traditional notion of "restricted sovereignty" than by a positive spirit of cooperation and effective interdependence'[66] – a notion that is more adequately reflected in the idea of a community of interest of riparian States than in the theory of limited territorial sovereignty.

Territorial theories developed in recognition of the special character of transboundary freshwater resources as shared resources. Consequently, the concept of territorial sovereignty is widely understood not as attributing absolute sovereignty over all resources within a State's territory but rather is perceived as a theory of limited territorial sovereignty

[61] Godana, *Africa's Shared Water Resources*, p. 49.

[62] *Protocol on Shared Watercourse Systems in the Southern African Development Community (SADC) Region*, 28 August 1995, FAOLEX, No. LEX-FAOC035640.

[63] 7 August 2000, www.sadc.int, ECOLEX TRE-001360.

[64] 'Preamble of the Senegal Water Charter', 18 May 2002, in L. Boisson de Chazournes, R. Desgagné, M. Mbengue and C. Romano, *Protection internationale de l'environnement*, 2nd edition (Paris: Pedone, 2005), pp. 301–314.

[65] Art. 4 (2) (k), *Protocol for Sustainable Development of Lake Victoria Basin*, 29 November 2003, www.lvbcom.org, FAOLEX, No. LEX-FAOC041042.

[66] P. Wouters, S. Salman, and P. Jones; 'The Legal Response to the World's Water Crisis: What Legacy from The Hague? What Future in Kyoto?' (2001) 4 *University of Denver Water Law Review* 420; McCaffrey, *The Law of International Watercourses*, pp. 167–170.

regarding transboundary waters. Whether the evolution of international water law will lead to the adoption of the more cooperative community of interest concept as a distinct legal theory remains to be seen. This issue is closely related to another question that remains to be resolved: Will international law one day give due recognition to the global dynamics of the hydrologic cycle?[67] States have moved from a national management approach to one that includes neighbourly interdependence and the concept of shared waters; it is therefore possible that they will eventually acknowledge the much wider interdependence that is created by atmospheric water streams at the universal level. Trends of legal development in this direction are discussed in Chapter 7.

2.2 Expansion of normative reach

An issue that was elevated to the universal level is the geographic applicability of a number of key principles and norms of international water law. Through coherent practice, uniform interaction and coordinated State behaviour relative to shared water resources, a number of principles and rules emerged as universally (or regionally) applicable customary law. Bilateral treaties have been augmented and complemented by multilateral instruments over time, and increasingly more water uses have been subjected to international regulation. This development has not taken place steadily in space and time, and there have been occasional reversals in this trend. Overall, however, there has been a continuous geographic expansion of normative reach as well as with respect to issue areas.

2.2.1 Expansion via navigation

Efforts to extend the normative reach of legal principles and rules concerning the geographic use of transboundary waters first occurred relative to navigation on international rivers. In the seventeenth and eighteenth centuries, questions about the use of shared water resources were settled primarily through bilateral treaties, and the majority of these agreements

[67] McCaffrey, *ibid.*, p. 54; J. Sohnle, *Le droit international des ressources en eau douce: solidarité contre souveraineté* (Paris: La Documentation Française 2002), p. 16; P. Gleick, 'Water Resources: A Long-Range Global Evaluation' (1993) 20 *Ecology Law Quarterly* 141; S. McCaffrey, 'International Organizations and the Holistic Approach to Water Problems' (1991) 31 *Natural Resources Journal* 139.

involved navigation rights.[68] In the nineteenth century, the first multilateral navigation regimes were established, on the European as well as the African continent. These were, so to speak, by-products of international agreements concluded with the objective of ending or preventing conflicts through the settlement of territorial claims, such as those adopted at the Congress of Vienna (1814–1815), the Berlin Conference (1884–1885),[69] and the Versailles Peace Conference after World War I. The core objective of these agreements was to define political boundaries and spheres of influence; hence, they also established rights over natural resources and their use. With respect to transboundary water resources, this meant primarily the regulation of navigation on rivers that provided access to products and markets. In Africa, this also meant opening up the continent to the colonising powers for further territorial expansion and exploration.

A peculiarity of these multilateral settlements was that the treaty regimes were not limited to riparian States as parties but also accorded navigation rights to non-riparian States. Before the Napoleonic Wars, States tended to monopolise navigation on their 'river territory', often denying upstream riparians access to the sea. After the wars, the European powers sought to end this situation, which had been detrimental not only to commerce but also 'prevented the growth of international relations appropriate to a state of peace'.[70] With the *Final Act of the Congress of Vienna*, freedom of navigation was introduced as a new standard. Article 109 stipulates that '[t]he navigation of the rivers, along their whole course... from the point where each of them becomes navigable, to its mouth, shall be entirely free, and shall not, in respect to commerce, be prohibited to any one... '[71] Rivers were declared 'international' and were made accessible not only to riparian States but also to all States interested in navigating the waterways. This 'regime of internationalisation'[72] was later expanded by other treaties, including those adopted at the

[68] L. Teclaff, *The River Basin in History and Law* (The Hague: Martinus Nijhoff, 1967), pp. 59f.; A. Baxter (ed.), *International Navigable Waterways – Financial and Legal Aspects of their Improvement and Maintenance* (New York: United Nations Institute for Training and Research, 1975), UN Doc. UNITAR/74/ST/6, pp. 218–220.

[69] The Berlin Conference was convened with the objective of avoiding armed conflicts among European powers over their colonies. For an account of the 'carving up' of the African continent by European powers, see T. Pakenham, *The Scramble for Africa – White Man's Conquest of the Dark Continent from 1876 to 1912* (New York: Avon Books, 1991).

[70] *Jurisdiction of the European Commission of the Danube between Galatz and Braila*, PCIJ (1927) Ser. B. No. 14, p. 38.

[71] Art. 109, *Final Act of the Congress of Vienna*, 9 June 1815 (1969), 64 Consol. TS 453–493.

[72] *Territorial Jurisdiction of the International Commission on the River Oder*, p. 23.

Berlin[73] and the Versailles[74] Conferences. Freedom of navigation became a regional customary norm in Europe and was influential on the African continent, where European States exerted colonial powers until the mid twentieth century.[75] However, with the change in political conditions after World War II and with decolonisation, the normative scope again became more limited; freedom of navigation for non-riparian States was no longer the general rule and was reserved only for riparians in most cases.[76]

An important contribution that resulted from the European peace agreements of the nineteenth century warrants mention here: that is, the foundation of the concept of river commissions. The Central Commission for Navigation on the Rhine, which has its legal basis in the *Final Act of the Congress of Vienna*, is considered the first-ever international organisation founded in the history of modern international law.[77] The European Commission of the Danube was later founded based on the 1856 *Peace Treaty of Paris* that ended the Crimean War.[78] The river commissions were established to oversee the implementation of navigation rights and of works to keep rivers free from obstacles to navigation. At the same time, they also served as neutral forums of diplomacy concerning matters of strategic importance to their members: navigation and river commerce.

Regional custom developed differently on other continents. In the Americas, the granting of freedom of navigation to foreign vessels was considered far more as an act in exercise of exclusive sovereign rights of the territorial state. Bilateral treaty practice in the eighteenth and

[73] Arts. 13 and 26 of the *Act of Berlin* declared navigation on the Congo and Niger Rivers free to merchant ships of all nations. The liberal regimes on the Congo and Niger Rivers were curtailed after World War I when freedom of navigation was limited to signatory states of the 1919 *Treaty of Saint-Germain* and to members of the LN. Vitányi, *The International Regime of River Navigation*, p. 99.

[74] The peace treaties negotiated at the end of World War I extended the geographical scope of multilateral river regulation by adding a number of Eastern European rivers to the list of previously internationalised rivers; namely, 'the Elbe (*Labe*) from its confluence with the Vlatava (*Moldau*), and the Vlatava (*Moldau*) from Prague; the Oder (*Odra*) from its confluence with the Oppa; the Niemen (*Russtrom-Memel-Niemen*) from Grodno; [and] the Danube from Ulm'. Art. 331, 1919 *Treaty of Versailles*.

[75] Caflisch, '*Règles Générales*', p. 40. [76] *Ibid.*, pp. 42 and 46.

[77] D. Akanda, 'International Organizations', in M. D. Evans, *International Law* (New York: Oxford University Press, 2003), p. 270.

[78] The European Commission of the Danube consisted of Austria, France, Great Britain, Prussia, Russia, Sardinia and the Ottoman Empire. International commissions consisting of riparian and non-riparian states were also established through the peace treaties after World War I for the rivers that were internationalised under this multilateral regime. Vitànyi, *The International Regime of River Navigation*, pp. 83 and 100.

nineteenth centuries between the United States and its neighbours was based on the assumption that 'there is no obligation in positive law on the part of the territorial sovereign to tolerate the passage of foreign vessels, riparian or non-riparian, on its rivers, but that these rights are conditioned on being awarded through treaty provisions'.[79] Similarly, in Latin America, the granting of the right to freedom of navigation was considered a sovereign act of a country.[80] A number of Latin American States awarded free navigation on their territorial rivers to foreign vessels through unilateral decrees that could be revoked at will.[81] In Asia, where most of the great navigable rivers form (or formed) part of the territory of a single country, the European rules concerning navigation likewise had little influence. As a consequence, navigation regimes did not generally emerge in the form of multilateral regimes on those continents. Navigation rights of foreign vessels on the major national rivers were granted on the basis of bilateral treaties or unilateral decrees.[82]

2.2.2 Continued efforts in multilateral codification

The principles and rules of international water law crystallised over time. Their development and continuing refinement is captured in the various codifications that occurred at the multilateral level.

Beginning of 'universalisation' by the League of Nations

The peace treaties that were concluded to end World War I initiated a gradual transformation of the international system; they formed the first step towards a nascent global governance structure. The LN was established as an organisation of collective security to maintain international peace; it also allowed for the establishment of the first permanent judiciary body with general competence: the PCIJ.[83] The organisation also made several contributions to further the development of international water law. Two multilateral conventions concerning the use of transboundary watercourses were negotiated – that is, (1) the 1921 *Convention and Statute on the Regime of Navigable Waterways of International Concern* (hereinafter, *Barcelona Convention*),[84] and (2) the 1923 *Convention Relating to the Development of Hydraulic Power Affecting More Than One State*[85] – and some of the PCIJ

[79] *Ibid.*, pp. 88–93.
[80] *Declaration of the Foreign Minister of Brazil* in 1968, cited in Vitányi, *ibid.*, p. 94.
[81] *Ibid.*, p. 95. [82] *Ibid.*, pp. 99f.
[83] The Covenant of the League of Nations was incorporated into the 1919 *Treaty of Peace of Versailles* and in the subsequent 1919 *Peace Treaty* signed in Neuilly-sur-Seine.
[84] Barcelona, 20 April 1921, 7 *LNTS* 35. [85] Geneva, 9 December 1923, 36 *LNTS* 76.

decisions and advisory opinions still serve as reference for the clarification of concepts and legal principles in this domain.[86]

Codification of multilateral instruments for the purpose of harmonisation of international norms on issues including navigation on international waterways was envisaged in the Treaty of Versailles as a future task of the LN.[87] The rationale for these efforts was the harmonisation of rules at the universal level. The objective of the *Barcelona Convention*, for example, was to find a compromise among the different regimes governing navigation;[88] as a consequence, the principle of reciprocity in granting navigation rights was developed.[89] Neither the *Barcelona Convention* nor the *Convention on Hydraulic Power* was ratified by a group of States large enough to provide these instruments with legal significance at the universal level. Yet, the codification work of the LN prepared the foundation for later projects of comprehensive codification of international water law at the universal level.

The work of the International Law Commission

Water became a prominent topic on the global policy agenda from the 1970s onwards. A series of international conferences resulted in ministerial declarations as well as in the establishment of additional international bodies with water mandates, such as the World Water Council and the Global Water Partnership.[90] The former was established in 1996 as a 'think tank' to promote awareness on critical water issues at all levels

[86] *No. 16 – Case Relating to the Territorial Jurisdiction of the International Commission of the River Oder*; *The Oscar Chinn Case*, PCIJ (1934), Ser. A./B. No. 63; *The Diversion of Water from the Meuse*, PCIJ (1937), Ser. A./B. No. 70; and *Jurisdiction of the European Commission of the Danube between Galatz and Braila*.

[87] Art. 23: 'Subject to and in accordance with the provisions of international conventions existing or hereafter to be agreed upon, the Members of the League: . . . (e) will make provisions to secure and maintain freedom of communications and of transit and equitable treatment for the commerce of all Members of the League'. Art. 338: 'The [navigation] regime set out in Arts. 332–337 above shall be superseded by one laid down in a General Convention drawn up by the Allied and Associated Powers, and approved by the League of Nations, relating to the waterways recognized in such Convention as having international character'.

[88] Vitànyi, *The International Regime of River Navigation*, p. 102.

[89] Art. 3, 1921 *Barcelona Convention*: 'Subject to the provisions contained in Articles 5 and 17, each of the Contracting States shall accord free exercise of navigation to the vessels flying the flag of any one of the other Contracting States on those parts of navigable waterways specified above which may be situated under its sovereignty or authority'.

[90] S. Salman, 'From Marrakech through The Hague to Kyoto: Has the Global Debate on Water Reached a Dead End? Part One' (2003) 28 *Water International* 491–500; and S. Salman, 'From Marrakech through The Hague to Kyoto: Has the Global Debate on Water Reached a Dead End? Part Two' (2004) 29 *Water International* 11–19.

of political decision making; the latter is a partnership of international organisations, governments, non-governmental organisations and private companies that promotes integrated water resources management. The policy focus on water was part of a wider concern about the degradation of the human environment. The 1972 *Stockholm Declaration on the Human Environment* urges States to safeguard water, air, land, flora and fauna 'for the benefit of present and future generations through careful planning or management', and it advocates a cooperative approach to those challenges 'by all countries, big and small, on an equal footing'.[91] Other international conferences were dedicated solely to the issue of water resources, most notably the 1977 UN Water Conference in Mar del Plata[92] and the 1992 International Conference on Water and Environment,[93] which took place in Dublin. The 1992 United Nations Conference on Environment and Development (UNCED) in Rio de Janeiro adopted a comprehensive Program of Action, *Agenda 21*, with the objective of ensuring environmentally sustainable economic development. Chapter 18 of the Program of Action is dedicated entirely to the 'protection of the quality and supply of freshwater resources' and the 'application of integrated approaches to the development, management and use of water resources'.[94]

As successor to the LN, the UN took note of the increasing concern for regulation of non-navigational uses of transboundary waters in international treaty practice and engaged in codification of these norms on its part. Legal issues relating to the utilisation of international rivers had been a concern of the UN since the 1960s. In 1959, the GA in *Resolution 1401 (XIV)* requested the Secretary-General to initiate studies on the laws and legislation in force to determine whether the subject was appropriate for codification. In 1963, the Secretary-General responded with the submission of a report on 'Legal Problems in Relation to the Non-Navigational Uses of International Watercourses'.[95] As a consequence, in

91 Principles 2 and 24, *Report of the United Nations Conference on the Human Environment, Stockholm, 5–16 June 1972* (New York: UN, 1972).

92 *Report of the United Nations Water Conference, Mar del Plata 14–25 March, 1977* (New York: UN, 1977).

93 The 1992 Dublin conference issued a statement comprising four principles, the so-called Dublin Principles, which continue to frame the global water debate: (I) freshwater is a finite and vulnerable resource; (II) water development and management should be based on a participatory approach; (III) women play a central part in the provision, management and safeguarding of water; and (IV) water has an economic value.

94 Chapter 18, UN Doc. A/CONF.151/26 (Vol. II), paras. 18.1–18.9.

95 UN Doc. A/5409. The Secretary-General also compiled a second document, *Legislative Texts and Treaty Provisions Concerning the Utilization of International Rivers for Other Purposes than Navigation*, United Nations, 1963, UN Doc. ST/LEG/SER.B/12.

1970, the GA recommended that the ILC 'should take up the study of the non-navigational uses of international watercourses'.[96] In 1994, after more than twenty years of deliberation, the ILC presented to the GA a set of draft articles on the law of non-navigational uses of international watercourses and a resolution on confined transboundary groundwater.[97] After deliberations in the Sixth Committee convening as a working group of the whole, the *Convention on Non-Navigational Uses of International Watercourses* was adopted by the GA during its fifty-first session on 21 May 1997. The *Convention* was adopted in *Resolution 51/229* by a vote of 103 States in favour, 3 against, and 27 abstentions. The States that voted against the adoption of the *Convention* were Turkey, China and Burundi. Thirty-three states were absent for the vote.[98] The *Convention* has not yet entered into force because, of the thirty-five required instruments of ratification, only thirty had been deposited by March 2013.[99]

The *Convention on Non-Navigational Uses of International Watercourses* aims to ensure sustainable utilisation and protection of international watercourses and to encourage cooperative management. The *Convention*'s reach is wider than the title seems to indicate; Article 1 (2) specifies that navigation is within the scope of the *Convention* as far as 'other uses affect navigation or are affected by navigation'. Hence, the impacts of navigation on water quality, consultation requirements relative to navigation infrastructure, and the impact of this infrastructure on a river's flow regime are also governed by the *Convention*. Geographically, the *Convention* applies to all transboundary watercourses, defined as 'a system of surface waters and groundwaters constituting by virtue of their physical relationship a unitary whole and normally flowing into a common terminus'.[100] It does not apply to confined groundwater.[101]

The gap in universal codification has since been addressed by the ILC; a Special Rapporteur on the topic of shared natural resources, including

[96] GA, *Res. 2669 (XXV)*, 8 December 1970, UN Doc. A/RES/2669(XXV).

[97] ILC, *Draft Articles on the Law of the Non-Navigational Uses of International Watercourses and Commentaries thereto and Resolution on Transboundary Confined Groundwater* (1994) II(2) *YBILC* 89–135.

[98] UN Press Release, 21 May 1997, UN Doc. GA/9248.

[99] The drafting process of the ILC and the deliberations of the Sixth Committee were described by numerous authors and widely commented on in legal literature; therefore, they are not presented herein. See McCaffrey, *The Law of International Watercourses*; Tanzi and Acari, *The United Nations Convention*.

[100] Art. 2 (a), *UN Watercourse Convention*.

[101] See ILC *Resolution on Transboundary Confined Groundwater*, which points out the need for 'continuing efforts to elaborate rules pertaining to confined transboundary groundwater'; (1994)II(2) *YBILC* 135.

natural gas, oil and groundwater, was appointed in 2002.[102] In 2008, the GA took note of the ILC *Draft Articles on the Law of Transboundary Aquifers* (hereinafter, 2008 ILC *Draft Articles*) and encouraged States to take appropriate measures for the bilateral and multilateral management of their transboundary aquifers.[103] The nineteen articles apply to all transboundary aquifer systems – those that are and are not linked to surface water systems.[104] Thus, there is overlap between the *UN Watercourses Convention* and the 2008 ILC *Draft Articles*. This is unfortunate because the rules of the two documents are similar but not identical. There is a risk of legal uncertainty with regard to which rules would apply in case both documents become legally binding instruments.[105] Despite the criticism voiced about the 2008 ILC *Draft Articles*,[106] the process of their development has had considerable positive impact; their elaboration by the ILC and recognition by the GA have moved the management of groundwater resources onto the agenda of the global-water dialogue.[107] This is an important contribution with regard to raising public awareness of the growing problem of groundwater depletion and the significance of aquifers as a source of freshwater supply.

Interaction between universal instruments and treaties at the regional and basin levels

The UN recognised the trend towards increasing multilateral codification and harmonisation of international water law that began in the nineteenth century and, with its almost global membership, elevated these efforts to a truly universal level. However, at the regional and basin levels, bilateral treaties were also complemented with an increasing number of multilateral documents.[108] The two UN documents comprise a reflection

102 *Official Records of the General Assembly, Fifty-Seventh Session, Supplement No. 10 and Corrigendum*, UN Doc. A/57/10 and Corr. 1, para. 518.

103 GA, *Res. 63/124*, 11 December 2008, UN Doc. A/RES/63/124.

104 ILC, *Fifth Report on Shared Natural Resources: TransboundaryAquifers by Chusei Yamada, Special Rapporteur*, 21 February 2008, UN Doc. A/CN.4/591, p. 6.

105 S. McCaffrey, 'International Law Commission Adopts Draft Articles on Transboundary Aquifers' (2009) 103 *AJIL* 283f.

106 In addition to the problem of overlap, the *Draft Articles* have been criticised, for instance, for establishing sovereignty over aquifers as a guiding principle, whereas the *UN Watercourses Convention* managed to avoid such a reference to the principle of sovereignty.

107 K. Mechlem, 'International Law Commission Adopts Draft Articles of a Transboundary Aquifers Convention' (2008) 12 *ASIL Insights* 18.

108 Of the treaties analysed in Annex, 49 treaties of 129 that were concluded after 1945 were multilateral; this contrasts to 5 multilateral treaties of 38 concluded between 1900 and 1945.

of this treaty practice; they are, in large part, codifications of universal customary norms that emerged from State practice at the basin level. At the same time, the two documents informed the negotiation process of a considerable number of treaties at the local and regional levels.[109] This interplay between regional and basin agreements and the documents codified at the universal level has been defined as a 'double process of nurturing'.[110]

At the regional level in particular, a series of instruments has been influenced by and developed alongside of the ILC documents,[111] including the *Convention on the Protection and Use of Transboundary Watercourses and International Lakes*,[112] which was adopted in 1992 within the framework of the UNECE (hereinafter, 1992 *UNECE Water Convention*); the 2000 *Water Framework Directive (WFD) of the EU*;[113] and the 1995 *Protocol on Shared Watercourses* developed by the SADC. Whereas the SADC document establishes general principles for cooperative management and development at the basin level, both the 1992 *UNECE Water Convention* and the *WFD* have a strong focus on water quality management and pay tribute to the ecological interdependence of water and land resources. The 2000 *SADC Revised Protocol*, similar to the *UN Watercourses Convention*, recommends the establishment of joint management mechanisms and river commissions at the basin level.[114] In contrast, the 1992 *UNECE Water Convention* and the *WFD* – with a view to obliging riparian States to engage in a continued and structured process of cooperation on the resources they share – both stipulate obligations to establish such joint mechanisms.[115]

The outcome of the process of mutual nurturing – and the fact that multilateral documents moved from the subject of navigation towards

109 Boisson de Chazournes, '*Eaux internationales et droit international*', p. 24; McCaffrey, *The Law of International Watercourses*, pp. 376f.

110 L. Boisson de Chazournes, *Freshwater and International Law: The Interplay between Universal, Regional and Basin Perspectives* (Paris: United Nations Educational, Scientific and Cultural Organization, 2009), p. 1.

111 At the basin level, instruments such as the 1991 *Protocol on Common Water Resources Concluded between Argentina and Chile*, the 1995 *Agreement on the Cooperation for Sustainable Development of the Mekong River Basin*, and the 2003 *Protocol for Sustainable Development of Lake Victoria Basin* can be cited among those that were informed by the development of the *UN Watercourses Convention*. McCaffrey, *The Law of International Watercourses*, pp. 376f.

112 *Convention on the Protection and Use of Transboundary Watercourses and International Lakes* (1992 *UNECE Water Convention*), 17 March 1992, 31 ILM 1312. The *Convention* was ratified by 38 States and the EU (status as of October 2012).

113 *Directive 2000/60/EC of the European Parliament and the Council*, 2000 OJ L 327/1.

114 Art. 4 (3), 2000 *SADC Revised Protocol*.

115 Art. 9, 1992 *UNECE Water Convention*; Art. 3, *WFD*.

increasingly comprehensive water resources management – reflects another trend in the development of international water law: a widening in the thematic scope of water treaties. This development illustrates that once riparian States increased their economic activities and their uses of international water systems, their mutual interdependence became increasingly more pronounced.

2.2.3 Widening thematic scope

As early as the nineteenth century – when navigation still featured as the most prominent use regulated by international agreements – other water uses had become the object of treaties in some basins, such as fisheries in the Rhine River and Lake Constance Basin, and irrigation, hydropower production and water diversion in the boundary basins of the United States and Canada.[116] The increasing importance of international rivers for exploitation of hydropower and the need for rules that would ensure cooperation of riparian States with respect to large infrastructure developments were reflected in the effort expended in the codification of the 1923 *Convention on Hydraulic Power*. This trend of expanding subject areas of international water law gained increasing momentum starting in the 1950s.[117]

The growing intensity of economic activities other than navigation and the negative impacts on riverine ecosystems urgently demanded their regulation. The impact of large infrastructure works on transboundary water systems, and the location of some on contiguous stretches of rivers, necessitated treaties among riparian States. When the effects of increasing industrial activity and the pollution of rivers became a problem for the fisheries, water quality and ecologic balance of riverine and coastal ecosystems, the result was the adoption of numerous treaty instruments with the goal of environmental and water quality protection.[118] Analysis of treaty development shows that the number of water treaties addressing issues other than navigation (and demarcation) has drastically increased

[116] Lammers, *Pollution of International Watercourses*, pp. 124–126, 137; see also *Treaty between the United States and Great Britain Relating to Boundary Waters, and Questions Arising between the United States and Canada*, 11 January 1909, *British Treaty Series 1910*, No. 23.

[117] E. Brown Weiss, 'The Evolution of International Water Law' (2007) 331 *RCADI* 238; Boisson de Chazournes, '*Eaux internationales et droit international*', p. 17; Caflisch, '*Règles Générales*', pp. 42f.

[118] For detail on the development of international water law by treaty practice in general and according to region, see Brown Weiss, *ibid.*, pp. 238–257.

since the 1950s.[119] In this process, the international law of transboundary watercourses has evolved into a legal system in which cooperation occupies an ever-widening space.[120]

This expansion in thematic scope of international water law can be traced, for example, to the evolution of the legal regime in the Rhine River Basin. The Rhine was one of the first rivers internationalised for navigation, and the first international river commission was established for it. Fisheries soon emerged as an additional water use, raising the need for international regulation – specifically with respect to contiguous river stretches and Lake Constance.[121] When re-industrialisation and economic development gained new momentum after World War II, subsequent conventions between the key riparian States first targeted pollution reduction[122] and later expanded their scope to include sustainable development and protection of the riverine ecosystem and adjacent coastal areas, as well as flood prevention.[123] This occurred in response to increasing water pollution through agricultural runoff and industrial activities.[124] Development of legal regimes in other basins followed comparable patterns, with each characterised by particular regional settings and individual requirements of riparian States.

In a small number of basins, riparian States went beyond the focus on water management and began to view river basins as suitable geographic units for regional economic development and integration. The *Treaty for*

[119] See Annex.

[120] Caflisch, '*Règles Générales*', p. 37.

[121] *Convention Establishing Uniform Regulations Concerning Fishing in the Rhine between Constance and Basle between Baden and Switzerland*, 9 December 1869 (1976), 140 Consol. TS 137–143; *Convention Establishing Uniform Regulations Concerning Fishing in the Rhine and its Tributaries, Including Lake Constance between Baden and Switzerland*, 25 March 1875 (1977), 149 Consol. TS 139–145; *Convention between France and Switzerland for the Regulation of Fishing in Frontier Waters*, 28 December 1880 (1977), 157 Consol. TS 295–301; *Convention Establishing Uniform Regulation for Fishing in Lake Constance between Austria-Hungary, Baden, Bavaria, Württemberg, Switzerland and Liechtenstein*, 5 July 1893 (1979), 179 Consol. TS 37–52.

[122] *Agreement on the International Commission for the Protection of the Rhine against Pollution*, 29 April 1963, 994 *UNTS* 3; *Convention on the Protection of the Rhine against Chemical Pollution*, 3 December 1976, 1124 *UNTS* 423.

[123] *Convention on the Protection of the Rhine* (1999 *Rhine Protection Convention*), 12 April 1999, OJ 2000 L289/31.

[124] In November 1986, a fire at a Sandoz agrochemical warehouse near Basel caused a toxic chemical spill, which killed large proportions of fish in the river. Rhine River pollution was also the subject in an international arbitration case, the *Arbitration in the Application of the Convention of December 3, 1976, on the Protection of the Rhine against Pollution by Chloride and the Additional Protocol of September 25, 1991*, 12 March 2004, www.pca-cpa.org.

Amazonian Cooperation[125] and the *Agreement for the Establishment of the Organization for the Management and Development of the Kagera River Basin*[126] stand out among those agreements. Regional development activities envisaged by these two treaties include road construction, trade and mineral exploration. In other basins, cooperative water resources management, utilisation and planning were also used as vehicles to enhance regional cooperation. The riparian States of the Senegal River, for example, created one of the most elaborate regimes, which include the co-ownership of infrastructure assets for hydropower development and the agreement to distribute the benefits of joint water uses across the basin States independent of the territory in which they are generated.[127] Hydroelectricity production spurs regional integration, extending even beyond the basin level in both the Mekong Basin and the Nile Basin.[128] Joint basin management mechanisms cooperate with energy companies in the establishment of regional interconnection networks and electricity markets that expand beyond the watersheds – for example, from the Nile Basin into the SADC region.[129]

125 *Treaty for Amazonian Cooperation*, Brasília, 3 July 1978, 1202 *UNTS* 71. With regard to Amazonian cooperation, see B. Garcia, *The Amazon from an International Law Perspective* (Cambridge: Cambridge University Press, 2011).

126 *Agreement for the Establishment of the Organization for the Management and Development of the Kagera River Basin*, Rusumo, 24 August 1977, 1089 *UNTS* 171. The Kagera Basin Organization (KBO) became defunct after 1995. Its assets were taken over by the Nile Basin Initiative (NBI) – Equatorial Lakes Subsidiary Action Program. This Program was established in 2001 with the objective of coordinating regional water resources development and management in the Equatorial Lakes sub-basin of the Nile River.

127 *Convention Relating to the Statute of the Senegal River*, 11 March 1972, *Basic Documents of African Regional Organizations* (New York: Oceana, 1972), Vol. III; 1972 *Convention Establishing the Senegal River Basin Organization*; 1978 *Convention Concerning the Legal Statute of Common Works*; *Conventions for the Establishment of the Management Agencies for the Operation of the Diama and Manantali Dams* were concluded on 7 January 1997, OJ of Mali, Décret No. 97–110/P-RM and Décret No. 97–111/P-RM, 11 March 1997. All treaties were established among Mali, Mauritania and Senegal. The adoption of the 2002 *Senegal Water Charter* paved the way for extending water resources cooperation across the entire basin again. The Charter recognises the interests of Guinea, the fourth riparian, in joint basin development; the latter State re-joined the basin management regime in 2005.

128 The concept of 'powershed' was introduced (as a correlative term to 'watershed') to describe adequate management units for regional power development. D. Magee, 'Powershed Politics: Yunnan Hydropower under Great Western Development' (2006) 185 *The China Quarterly* 23–41.

129 International Rivers Network, *Trading Away the Future – The Mekong Power Grid* (Berkeley, CA: International Rivers Network, 2006); African Union (AU), *The AU/NEPAD African Action Plan 2010–2015: Advancing Regional and Continental Integration in Africa* (Addis Ababa: African Union, 2009).

2.3 Conclusion

Today, it is widely recognised that territorial sovereignty over shared resources is limited not only by facts of nature – because shared resources do not permanently form part of a State's territory – but also in legal terms. The sovereignty of one State over transboundary resources is limited by the rights of other States to use the same resources. International law restricts sovereignty. Over time, States have continued to expand this area of law by concluding international agreements concerning the development of transboundary waters; and they have agreed on legal principles and rules of water management not only with direct neighbours but also at the basin, regional and universal levels.

Initially, water use was first and foremost regulated through bilateral agreements. Since the nineteenth century, multilateral treaties increasingly complemented bilateral regimes. At first, the multilateral regimes were established mainly to avoid conflicts; later, they increasingly served the purpose of harmonising norms and codifying regionally and universally applicable standards for water management. With accelerating economic development, the breadth and intensity of water uses increased; along with this development, the negative impacts of resource overuse became visible, resulting in the need for more and new regulation. Water treaties became more comprehensive and began to address management aspects in an integrated manner. Regulation issues expanded from an initial focus on navigation to a multitude of other uses and to comprehensive basin management. In a few cases, States sharing river basins realised that, indeed, the community to which they belong because of hydrologic interconnectedness offers greater opportunities than merely cooperative water management. As a result, the hydrographic basin area in some cases was adopted as the natural management unit for regional economic integration, including sectors that reach beyond water management (e.g., transport and power trade).

The community of interest of riparian States (and, in some cases, of non-riparian States regarding specific interests such as navigation or international stability) was recognised as the factual basis for equality of rights of riparian States and, subsequently, for the principle of equitable and reasonable utilisation of shared water resources. The evolution of treaty practice and the expansion of the scope of international water law highlight that the community that exists among States as a consequence of hydrologic interdependence also served as a basis for an increase in factual- as well as treaty-based cooperation. Over time, this has led to the crystallisation of a general duty to cooperate. This normative development is the subject of analysis in Part II, which describes the legal nature of this general duty and the evolution of its codification.

PART II · A GENERAL DUTY TO COOPERATE: WHAT DOES IT MEAN?

"If you don't know where you are going, you will probably end up somewhere else".

Laurence J. Peter

Part I demonstrated that international cooperation as well as its regulation has intensified over the course of history as States increasingly felt a need to establish rules to manage their interdependence, which exists due to both natural causes and human conduct. In line with this trend, the scope and frequency of cooperation also widened with respect to development and management of transboundary water systems. Over time, continuous and coherent State and treaty practice gave rise to the crystallisation of a general duty to cooperate. Part II analyses the legal scope and the process that led to recognition of the duty as one of the 'general principles' of international water law. Chapter 3 explores the relationship of the duty to cooperate with other principles applicable in this field, demonstrating that these principles are functionally linked and mutually reinforcing. Chapters 4 and 5 analyse the role of specific cooperation obligations and rights – as well as the role of joint measures, which have been adopted by basin States to manage specific water challenges – in achieving sustainable and equitable management of transboundary freshwater resources.

3 The duty to cooperate and concurrence of principles

Clarification of the use of the term 'general principles' in international water law instruments is warranted before beginning the analysis. Not all 'general principles' of international water law have the characteristics of 'legal principles' in the strict sense of the terms. As Fitzmaurice explains, 'by a [legal] principle, as opposed to the rule, even a general rule, of law is meant chiefly something which is not itself a rule, but which underlies a rule, and explains and provides a reason for it'.[1] A distinction is drawn here between 'legal principles' and 'general principles' of international water law. 'Legal principle' as applied in this book follows the definition provided by Riedel and is in line with Fitzmaurice's characterisation of legal principles. According to Riedel, legal principles represent legal standards that form a 'normative umbrella', which shapes State behaviour; they provide direction and give rise to specific obligations and rights.[2] This distinction between principles and specific rights and obligations is not consistently applied in instruments of international water law. Part II of the *UN Watercourses Convention*, which is entitled 'General Principles', includes both: more general norms that are closer to the character of legal principles, and norms of a more specific nature, such as the obligation not to cause significant harm and to regularly exchange data and information.[3] Similarly, the 2000 *SADC Revised Protocol on Shared Watercourses* lists information exchange among the principles outlined in Article 3,[4] as do the 2008 ILC *Draft Articles* in the section

1 Fitzmaurice, 'General Principles of International Law', p. 7.
2 E. Riedel, 'International Environmental Law – A Law to Serve the Public Interest? An Analysis of the Scope of the Binding Effect of Basic Principles (Public Interest Norms)', in J. Delbrück (ed.), *New Trends in International Lawmaking – International 'Legislation' in the Public Interest* (Berlin: Duncker & Humblot, 1997), p. 76.
3 Arts. 7 and 9, 1997 *UN Watercourses Convention*. 4 Art. 3 (6), 2000 *SADC Revised Protocol*.

on 'General Principles'.[5] Yet, 'legal doctrine' generally refers to this norm as an obligation.[6] The titles of sections or provisions in water law instruments, which refer to 'principles', are therefore not always representative of the true legal character of the norms contained therein. The norms included in Part II ('General Principles') of the *UN Watercourses Convention* and of the 2008 ILC *Draft Articles* are, without doubt, fundamental to and represent key guiding norms of transboundary water law. However, at the same time, they do not all correspond to the strict definition of 'legal principles' outlined herein.

3.1 The general duty to cooperate

The general duty to cooperate had initially not been identified as a general principle of international water law and, given its unique characteristics, it still evades a clear categorisation as principle.

3.1.1 Crystallisation of an autonomous duty

As highlighted in Part I, it seems to follow naturally that the general duty to cooperate would constitute a fundamental rule of any body of norms that applies to the development and management of shared resources.[7] It is therefore interesting that from the outset, scientific and expert bodies did not reflect this fundamental importance of cooperation in the codification documents they developed. They only gradually accepted that the role of cooperation as a precondition for the effective management, protection and utilisation of transboundary water resources must be translated also into adequate legal terms.[8]

Codification by scientific commissions

The incorporation of provisions stipulating a general duty to cooperate into instruments that codify universal norms of international water law developed by the IIL and the ILA[9] took a relatively long time – despite

[5] Art. 8, 2008 ILC *Draft Articles*.

[6] McCaffrey, *The Law of International Watercourses*, p. 478; McIntyre, *Environmental Protection*, p. 235; Tanzi and Arcari, *The United Nations Convention*, pp. 196f.

[7] H. Mosler, 'The International Society as a Legal Community' (1974) 140 *RCADI* 20f.

[8] ILA, *Report of the Fifty-Sixth Conference Held at New Delhi, December 29th, 1974, to January 4th, 1975* (London: ILA, 1976), p. 116; Finnish Branch of the ILA, *The Work of the International Law Association on the Law of International Water Resources* (Helsinki: Finnish Branch of the ILA, 1988), p. 249; and commentary to Art. 11, 2004 *Berlin Rules on Water Resources*, in ILA, *Report of the Seventy-First Conference (Berlin)* (London: ILA, 2004), p. 361.

[9] Both the IIL and the ILA were established in 1873 as independent legal expert societies.

the fact that a considerable number of water treaties already stipulated numerous and specific cooperation obligations.

The IIL contributed to the formation of the body of international water law through various documents. Of interest here, relative to the evolution of cooperation, are the early documents on non-navigational uses of non-maritime international waters (adopted in 1911 and 1961) and on the pollution of rivers and lakes (adopted in Athens in 1979).[10] The first set of regulations concerning non-navigational uses of transboundary watercourses proposed by the IIL was conceived of at a time when economic uses other than navigation had begun to gain importance as an object of international water treaties. *The Declaration on the International Regulation Regarding Uses of International Watercourses for Purposes Other than Navigation*, adopted in Madrid on 20 April 1911,[11] therefore still seems rudimentary in content when compared to the *Articles on Utilization of Non-Maritime International Waters* (except navigation), adopted in 1961 in Salzburg.[12] The 1911 *Madrid Declaration* is divided into two Articles that recommend the establishment of permanent joint commissions and prohibit utilisation of transboundary watercourses in a way that would harmfully affect co-riparian States. The first Article, which applies to rivers that form the border between two States and to lakes that extend over the territory of two or more States, forbids States and their subjects to carry out activities that might harmfully alter the shore of the co-riparian States without their consent. It further prohibits uses that might impede utilisation of the same waters by co-riparian States and their subjects (e.g., individuals and private enterprises). The second Article applies to successive watercourses and proscribes uses that damage water quality and cause substantive quantity abstractions, as well as installations that lead to the inundation of other States' territory. Rather than prohibiting certain activities, the nine Articles of the 1961 *Salzburg Resolution on the Utilization of Non-Maritime International Waters* display a preponderance of norms prescribing cooperation among basin States. Following a definition of the scope of application in Article 1 and the outline of initial formulations of the principles of limited territorial sovereignty, equitable utilisation,

[10] Others include the *Projet de règlement international de navigation fluvial*, 9 September 1887; and *Règlement pour la navigation des fleuves internationaux*, 18 October 1934, reprinted in H. Wehberg, *Tableau Général des Résolutions (1873–1956)* (Basle: Verlag für Recht und Gesellschaft AG, 1957), pp. 70–80.

[11] *Ibid.*, pp. 81f.

[12] IIL, *Annuaire de l'Institut de Droit International – Session de Salzbourg, September 1961* (Basle: Verlag für Recht und Gesellschaft AG, 1961), pp. 381–384.

and the no significant harm obligation in Articles 2 through 4, Articles 5 through 8 provide rules for notification, consultation and negotiation for States that want to engage in works or utilise waters that seriously affect the possibility of use by other States of the same waters.

The *Salzburg Resolution* does not contain a specific Article stipulating a general duty to cooperate, yet cooperation is the underlying tenor of the document. The preamble recognises the interdependence of basin States and their common interest in the maximum utilisation and rational exploitation of available natural resources. The Articles in the *Salzburg Resolution* focus on the limits of the rights of one State as determined by the rights of other States. Overall, the *Resolution* regulates how States are to coordinate their respective uses with one another. The IIL first introduced a general duty to cooperate with respect to pollution of rivers and lakes in the 1979 *Athens Resolution*.[13] Specific measures on how to implement this general duty are identified, including regular exchange of data, coordination of research and monitoring programs and provision of technical and financial aid to developing countries – in addition to the previously discussed obligations of notification and consultation.[14]

Another set of contributions to the codification of international water law came from the ILA. The 1966 *Helsinki Rules on the Uses of the Water of International Rivers* is the most cited document in this context.[15] It had a significant impact on subsequent codifications of international water law and treaties. The *Helsinki Rules*, which apply to both navigational and non-navigational uses of the waters of international drainage basins, represented the most comprehensive codification of existing international water law until 1997.[16] The primary focus of the *Helsinki Rules* is the principle of equitable and reasonable utilisation. Yes, the *Rules* do not specifically mention cooperation obligations; rather, they focus on defining the rights and obligations of individual States with respect to equitable and reasonable utilisation, and they remain rather mute on basin States'

[13] Art. IV IIL *Athens Resolution*: '*Afin de se conformer aux obligations énoncées . . . , les Etats utiliseront notamment les moyens suivants: . . . b) sur le plan international, l'exercice d'une coopération de bonne foi avec les autres Etats intéressés*'. IIL, *Annuaire de l'Institut de droit international, Session d'Athènes 1979* (Paris: S. Karger, 1979), pp. 265–267.

[14] Art. VI, IIL *Athens Resolution*.

[15] ILA, *Report of the Fifty-Second Conference*, pp. 477–531.

[16] Finland, which introduced the 1970 resolution on international watercourse law in the Sixth Committee of the GA, suggested that the *Helsinki Rules* be considered as a model. This suggestion was not taken up by the Committee, and the resolution to refer the topic to the ILC was adopted without explicit reference to the 1966 *Rules*. J. Wescoat, 'Beyond the River Basin: The Changing Geography of International Water Problems in International Watercourse Law' (1992) 3 *CJIELP* 306f.

duties relative to one another. It is only with regard to dispute prevention and timber floating that the *Helsinki Rules* recommend certain cooperative acts, such as information exchange, notification, consultation and negotiation.[17]

The ILA included a general obligation to cooperate for the first time in the 1972 *Supplementary Rules Applicable to Flood Control*.[18] Article 2 stipulates that '[b]asin States shall cooperate in measures of flood control in a spirit of good neighborliness, having due regard to their interests and well-being as co-basin States'. Yet, the document that is more frequently cited with respect to establishing and confirming the existence of a general duty to cooperate is the 1982 *Montreal Rules on Water Pollution in an International Drainage Basin*.[19] Article 4 states: 'In order to give full effect to the provisions of these Articles, States shall cooperate with the other States concerned'. In the commentary to this Article, the ILA justified the inclusion of a general duty to cooperate by arguing that it is now considered generally accepted as a fundamental principle.[20]

The IIL and ILA both specify a general duty to cooperate in their resolutions concerning water pollution, evidencing that due to increasing deterioration of the quality of shared waters, the inevitability of interdependence among basin States and the need for cooperation were recognised as matters of urgency.

Codification by the International Law Commission

The ILC drew from the work of the IIL and ILA in its process of developing the *Draft Articles Concerning the Law of Non-Navigational Uses of International Watercourses*, which later formed the basis for the *UN Watercourses Convention*. When the GA accorded the ILC the mandate to codify the law of non-navigational uses of transboundary water resources, it requested that the Secretary-General supplement his study on legal problems relating to utilisation and use of international rivers – which was initiated based on GA *Resolution 1401 (XIV)* – by considering *inter alia* non-governmental studies and to then forward all relevant records on the topic to the ILC.[21] The evolution of the ILC work on the *Draft Articles* illustrates the progressive recognition of cooperation as a general principle; as with the work of the IIL and ILA, the general duty to cooperate made its way only gradually into

17 Arts. XXIV, XXV, XXIX–XXXI, 1966 *Helsinki Rules*.

18 ILA, *Report of the Fifty-Fifth Conference, New York 1972* (London: ILA, 1974), p. 47.

19 ILA, *Report of the Sixtieth Conference, Montreal 1982* (London: ILA, 1982), pp. 535–548.

20 *Ibid.*, pp. 539f. 21 GA, *Res 2669 (XXV)*, 8 December 1970, UN Doc. A/RES/2699(XXV).

the ILC *Draft Articles*. A first step was taken in 1981 by the second Special Rapporteur working on this topic. Schwebel proposed introducing the concept of 'equitable participation' to reflect that 'conditions and expectations have tended to move the international community to a position of affirmative promotion of cooperation and collaboration with respect to shared water resources'.[22] According to his views, States not only have a right to reasonable and equitable sharing but also a duty to participate in the protection and management of watercourses.[23] Furthermore, as a corollary to the duty to participate, basin States have a right to the cooperation of other States sharing a transboundary water system.[24] In contrast to the 1966 *Helsinki Rules*, which include no particular procedural provisions on how the principle of reasonable and equitable utilisation would be implemented, Schwebel introduced procedural components of cooperation by stipulating a duty to participate.

Evensen, the following Special Rapporteur, was the first to include an Article in the ILC project, which explicitly defined the general principle of cooperation.[25] He dedicated an entire section of his proposed expanded set of Articles to cooperation by introducing a new Chapter III on 'Cooperation and Management in Regard to International Watercourse Systems'. Evensen emphasised two reasons for this consideration: (1) he argued that it followed from the nature of watercourses as 'indivisible units' that cooperation among States is essential for effective management and optimal utilisation, as well as for reasonable and equitable sharing in this utilisation; and (2) this inclusion would echo the conclusions of the 1977 *United Nations Mar del Plata Conference on Water* and the 1981 *Interregional Meeting of International River Organizations of Dakar*, both of which stressed the importance of State cooperation.[26] The political commitment to cooperation on transboundary waters was reaffirmed in 1992 with the adoption of *Agenda 21* at the UNCED. In Chapter 18 of this Program of Action, States commit to the implementation of integrated approaches and protection of the quality and supply of the world's freshwater resources, through both national

[22] ILC, 'Third Report on the Law of the Non-navigational Uses of International Watercourses, by Mr. Stephen M. Schwebel, Special Rapporteur' (1982) II(1) *YBILC* 85, para. 85.

[23] *Ibid.*; Tanzi and Arcari, *The United Nations Convention*, pp. 182f. [24] *Ibid.*

[25] Art. 10, *General Principles of Cooperation and Management*, ILC, 'Second Report on the Law of Non-navigational Uses of International Watercourses, by Mr. Jens Evensen, Special Rapporteur' (1982) II(1) *YBILC* 113, para. 64.

[26] ILC, 'First Report on the Law of the Non-navigational Uses of International Watercourses, by Mr. J. Evensen, Special Rapporteur' (1983) II(1) *YBILC* 173f., paras. 103–106.

and international cooperation.[27] The remaining Articles of the new Chapter III introduced by Evensen stipulated specific cooperation obligations and rights, including consultation, negotiation and prior notification of planned measures.

The proposal to include Chapter III in the *Draft Articles* reflected the increasing acceptance of cooperation as not only a necessary political paradigm but also a principle of international water law at that time.[28] In a later version of the *Draft Articles*, the provision concerning the general duty to cooperate was moved to Part II on 'General Principles' (previously Chapter II), in full recognition of the nature of this rule as a principle and fundamental rule of international water law.[29] In accordance with that decision, the general duty to cooperate was also later included as a 'general principle' of the law of transboundary aquifers in the 2008 ILC *Draft Articles* on this topic.[30]

The legal nature of the general duty to cooperate

The inclusion of the general duty to cooperate among the 'general principles' outlined in Part II of the *UN Watercourses Convention* and in the 2008 *Draft Articles* acknowledges that it represents a guiding norm of international water law. Yet, as previously pointed out, it does not clarify per se the legal status of the norm because the provisions included in this section have a differing legal nature. Therefore, the question of whether the general duty to cooperate is a principle of international law that gives rise to more specific obligations but is not in itself an independent obligation or whether it represents an autonomous legal obligation and, if so, of what nature requires further analysis.[31]

This issue was thoroughly discussed by the ILC: several members expressed doubt that cooperation represented anything more than a

27 UN Doc. A/CONF.151/26/REV.1 (Vol. II).

28 ILC, 'Second Report, Evensen', p. 112, para. 59.

29 ILC, 'Draft Articles on the Law of the Non-navigational Uses of International Watercourses. Titles and Texts Adopted by the Drafting Committee: Titles of Parts II and III of the Draft Articles; Articles 8 to 21 – Reproduced in A/CN.4/SR.2070 to SR.2073' (1988) I *YBILC* 182, para. 75.

30 Art. 7, 2008 ILC *Draft Articles*.

31 See, for example, the decision in the *Case Concerning Land and Maritime Boundary between Cameroon and Nigeria*, in which the ICJ observed regarding to the principle of good faith that although the principle is 'one of the basic principles governing the creation and performance of legal obligations . . . it is not in itself a source of obligation where one would otherwise exist'. *Case Concerning Land and Maritime Boundary between Cameroon and Nigeria (Cameroon v. Nigeria) (Preliminary Objections)* (1998) ICJ Reports, pp. 290ff., paras. 21–47.

principle of international law; others argued that it did indeed represent a legal duty for States, particularly with respect to the management of shared resources.[32] For one member, the debate itself highlighted that the existence of a general legal obligation was not established as of yet.[33] Another colleague pointed out that 'neither doctrine nor practice had succeeded in properly defining a dividing line between a legal rule and a legal principle', which may have contributed to the difficulties in identifying the true legal character of the norm.[34] For this colleague, the general duty to cooperate fits somewhere in the 'grey zone' that constitutes the unidentifiable dividing line between the two. The majority of ILC members who intervened in the discussion agreed that the norm was of a fundamental character and therefore belonged under the 'general principles' of international water law.

Commentators of the *UN Watercourses Convention* support the point of view that international cooperation on transboundary waters 'is not only necessary; but is probably now required by general international law'[35] and that 'it stands as a legal obligation of its own'.[36] Tanzi and Arcari cited Article 30 of the *UN Watercourses Convention* as additional evidence that States accept the existence of a general obligation to cooperate.[37] Article 30 provides for States to comply with the cooperation obligations of the *Convention*, even when direct contact between the States is not possible – for example, due to a breakdown of diplomatic relations. In such a situation, States shall adopt indirect procedures for cooperation.

At the same time, these commentators agree that the general duty to cooperate is an abstract rule; it provides a general framework that requires specification through more exact obligations.[38] McCaffrey argued that 'it is of little use to speak of an "obligation to cooperate" in the abstract';[39] norms that speak of a 'general duty to cooperate' should be understood as 'umbrella terms' embracing specific individual rights and duties to cooperate (e.g., obligations to consult and negotiate) and obligations to notify others of emergencies and planned measures.[40] These obligations already form an established customary content for implementation of the general duty to cooperate. At the same time, the general duty also provides

[32] ILC, 'Summary Records of the Meetings of the Thirty-Ninth Session, 4 May–17 July 1987' (1987) I *YBILC* 70–95.

[33] Commentary of Njenga, *ibid.*, 89. [34] Commentary of Sepúlveda Gutiérrez, *ibid.*, 87.

[35] McCaffrey, *The Law of International Watercourses*, p. 471.

[36] Tanzi and Arcari, *The United Nations Convention*, p. 183.

[37] *Ibid.*, p. 185. [38] *Ibid.*, p. 183.

[39] McCaffrey, *The Law of International Watercourses*, pp. 470f. [40] *Ibid.*

a normative 'umbrella' for further cooperative action according to the specific circumstances of transboundary water bodies and according to the interests of their riparian States. Hence, States comply with the general duty to cooperate in many different ways: through compliance with customary obligations that arise from the general duty and through other unilateral acts of cooperative nature and mutual engagement in negotiations, consultations, planning and joint management.[41] In addition to these forms of horizontal cooperation directly among States, cooperation can be carried out by establishing and working together through institutions.[42]

The general duty to cooperate, while being an abstract obligation itself, must be made operational through more specific obligations and rights in the same way as principles of international law. The legal nature of the general duty to cooperate therefore resides somewhere in the grey zone between definitions of the concepts of 'specific obligation' and 'legal principle'; it is neither one nor the other but rather includes elements of both. The general duty to cooperate is a general obligation with a legal nature of its own: it has all of the attributes of a legal principle and yet is an obligation of general nature.

3.1.2 Terms and objectives of the general duty to cooperate

The general framework for the more specific rights and obligations of cooperation is determined by the terms and objectives of the general duty to cooperate. The provisions of the *UN Watercourses Convention* and the 2008 ILC *Draft Articles* outline the principles and elements that form the basis for this rule. A close reading of the respective Articles helps to identify its content and scope. Article 8 (1) of the *UN Watercourses Convention* provides that '[w]atercourse States shall cooperate on the basis of sovereign equality, territorial integrity, mutual benefit and good faith in order to attain optimal utilization and adequate protection of an international watercourse'. The counterpart Article in the 2008 ILC *Draft Articles* adds 'sustainable development' to the concepts that should guide cooperation, and it reformulates the objective, replacing the terms 'optimal utilization' and 'adequate protection' with the objective 'to attain equitable and reasonable utilization and appropriate protection of [the] transboundary aquifer or aquifer system'.[43] A provision on the general duty to cooperate was also included in the ILA's most recent codification project of universal

[41] See Chapters 4 and 5 for more details.
[42] ILC, 'Summary Records, Thirty-Ninth Session', p. 78.
[43] Art. 7 (1), 2008 ILC *Draft Articles*.

norms applicable to freshwater resources management, the 2004 ILA *Berlin Rules*.[44] This document stipulates the general duty in abstract terms without further defining the objective of cooperation relative to the context of transboundary waters. Article 11 states: 'Basin States shall cooperate in good faith in the management of waters of an international drainage basin for the mutual benefit of the participating States'. Thus, the objective is solely defined with respect to the interests of States – that is, their mutual benefit. Yet, this Article must be interpreted in the context of the principles of water management otherwise defined in the *Berlin Rules* – for example, sustainable water use and prevention of environmental harm in Articles 7 and 8, respectively.

Basic principles and elements of cooperation

The terms under which States are to cooperate according to the provisions cited herein reflect fundamental principles on which the international legal order is built: sovereign equality, territorial integrity and good faith. It can be argued that these principles do not add any normative value because they apply to the exercise of any international legal act.[45] Yet, their explicit mention in the texts of the provisions that define the general duty to cooperate can – or must – be regarded as reemphasising that cooperation will be meaningful only if these principles are respected and applied.

The mention of 'sovereign equality' is a reminder that cooperation is most sustainable when those who engage in it regard one another and act as equals. Although this may not reflect the political reality, the underlying assumptions of the international legal order are that States interact on an equal footing, that they are vested with equality in rights, and that no one State can impose its will on another. To accept an offer after negotiations and conclude a legal agreement are independent sovereign acts by the agreeing State. Regardless of whether there are differences in the political or economic power among States, in their sovereign decisions about legal engagement, they act as equals. Furthermore, because transboundary water resources traverse and interact with the territory of basin States, cooperation must be based on the premise of mutual respect for one another's territorial integrity, within the limits of international law. As discussed previously, the meaning of 'territorial integrity' is no longer understood in absolute but rather relative terms; analogous to

44 ILA, *Report of the Seventy-First Conference*, pp. 340–411.
45 Tanzi and Arcari, *The United Nations Convention*, p. 184.

sovereignty rights, the territorial integrity of one State ends where the territorial integrity of the other State begins.[46]

It is interesting that sustainable development is mentioned as an additional concept correlated with implementation of cooperation in Article 7 of the 2008 ILC *Draft Articles*. Sustainable development was recognised as an objective in Principles 3 and 4 of the 1992 *Rio Declaration on Environment and Development*; the underlying rationale is to achieve a balance between environmental and economic needs in the development process.[47] Opinions and assessments concerning the legal character of this concept differ: some authors favour recognition as a legal principle, whereas others do not attribute legal value to the concept but instead consider it a political paradigm.[48] The ICJ referred to sustainable development as a concept in its decision in the *Gabčíkovo–Nagymaros Case*.[49] Yet, the ILC goes further in referring to it as a general principle in the commentary to Article 7 of the 2008 ILC *Draft Articles*: 'The term "sustainable development" denotes the general principle of sustainable development and should be distinguished from the concept of "sustainable utilization"'.[50]

Implementation of cooperative acts in the absence of good faith would reduce those acts to the mere exercise of procedural requirements or empty transactions. As stated in the *Lac Lanoux* arbitration award, entering into negotiations without the intention to attempt a mutual agreement is meaningless and against international practice.[51] It is interesting that

46 See Part I, Chapter 2, in this volume.

47 *Rio Declaration on Environment and Development*, 12 August 1992, UN Doc. A/CONF.151/26 (Vol. I).

48 In favour of a legal principle: C. Voigt, *Sustainable Development as a Principle of International Law: Resolving Conflicts between Climate Measure and WTO Law* (Leiden: Martinus Nijhoff Publishers, 2009); G. Hafner, 'The Optimum Utilization Principle and the Non-Navigational Uses of Drainage Basins' (1993) 45 *Austrian Journal of Public and International Law* 130f. Of different opinion: P. Birnie and A. Boyle, *International Law and the Environment*, 3rd edition (New York: Oxford University Press, 2009), pp. 125–127; and J. Ellis, 'Sustainable Development as a Legal Principle: A Rhetorical Analysis', Working Paper Series, McGill University (2008).

49 *Gabčíkovo–Nagymaros Project (Hungary/Slovakia)* (1997) ICJ Reports, para. 140.

50 ILC, *Report on the Work of its Sixtieth Session (5 May to 6 June and 7 July to 8 August 2008)*, UN Doc. A/63/10, p. 49.

51 *Sentence du tribunal franco-espagnol du 16 Novembre 1957 dans l'affaire de l'utilisation des eaux du lac Lanoux* (1958) 29 *RGDIP* 108. The notion of good faith is also frequently referred to in legal provisions in the same breath as cooperation, consultation and negotiation. In the context of the 1997 *UN Watercourses Convention*, see, for example, Art. 3 (5): '... watercourse States shall consult with a view to negotiating in good faith for the purpose of concluding ... agreements'; Art. 17 (2): '... consultations and negotiations shall be conducted on the basis that each State must in good faith pay reasonable

the ILC removed the term 'good faith' from the original version [52] to avoid redundancy and did not include it in Article 8 of the 1994 ILC *Draft Articles*.[53] Nevertheless, the ILC listed the principle of good faith in the commentary to Article 8 as one among other principles considered relevant to cooperation.[54] The principle was subsequently reinserted by the States that were negotiating the text of the *UN Watercourses Convention* in the Drafting Committee of the Sixth Committee of the GA.[55] Several States proposed to include the principle of good faith because it was considered that its mention was important in its own right,[56] and by explicitly mentioning the majority of principles that intertwined with the principle of cooperation, its importance would be further highlighted.[57] Therefore, the inclusion of 'good faith' by the Drafting Committee in the text that became the 1997 *UN Watercourses Convention* is recognition of the level of significance that States attribute to this principle regarding cooperation and the fundamental value for the international legal order. This also emphasises the importance of the role of the general duty to cooperate in the context of transboundary water resources.

Objective of the general duty according to codifications of universal rules

The duty to cooperate follows from the indivisibility of the waters shared by riparian States.[58] Consequently, the general duty to cooperate applies to all riparian States. However, this does not mean that all States of a multi-riparian hydrologic system are obligated to cooperate at all times, which would not be practical. Yet, in situations when only two States of a multi-riparian system engage in cooperation, they must consider the needs of the entire system as well as other concerned riparian States. Thus, the general duty to cooperate becomes a way to prevent exploitation of shared resources by individual States unilaterally.[59]

regard to the rights and legitimate interests of the other State'; and Art. 31: '... that State shall cooperate in good faith with the other ... with a view to providing as much information as possible under the circumstances'.

52 ILC, 'Summary Records, Thirty-Ninth Session', p. 63.

53 ILC, 'Draft Articles ... and Commentaries', p. 105. 54 *Ibid*., p. 106.

55 *Report of the Drafting Committee*, 22 October 1996, UN Doc. A/C.6/NUW/WG/L.1, p. 3.

56 *Convention on the Law of Non-navigational Uses of International Watercourse, Report of the Secretary-General*, 6 August 1996, UN Doc. A/51/275, p. 48.

57 *Ibid*., p. 49.

58 ILC, 'First Report, Evensen', pp. 173f., paras. 103–108; H. Smith, *The Economic Uses of International Rivers* (London: P. S. King, 1931), pp. 150f.

59 B. Hettne and F. Söderbaum, *Regional Cooperation* (Gothenburg: International Task Force on Public Goods, Göteborg, 2004), p. 33.

In line with the generic definition of the term 'cooperation', States are under the general duty to aim for achieving mutual benefits.[60] Each participant should be able to 'walk away' with absolute gains when cooperation is successful. Article 11 of the ILA *Berlin Rules* stipulates this abstract objective, however, without repeating *expressis verbis* what these mutual benefits should be. Reflecting State practice, the provisions elaborated by the ILC are more specific about what comprises the mutual benefits and the objective of cooperation. At the same time, and in consideration of the diversity of transboundary water systems and the wide variety of uses thereof, the objectives of cooperation remain stated in general terms.[61] Article 8 of the *UN Watercourse Convention* addresses 'optimal utilization', and Article 7 of the 2008 *Draft Articles on the Law of Transboundary Aquifers* refers to 'equitable and reasonable utilization'. A contextual analysis of these documents reveals that the notion of optimal utilisation is applied such that it is almost synonymous with the concept of equitable and reasonable utilisation.[62] As stated by the ILC, 'attaining optimal utilization and benefits does not mean achieving the "maximum" use, the most technologically efficient use, or the most monetarily valuable use much less short-term gain at the cost of long-term loss.... Rather, it implies attaining maximum possible benefits for all watercourse States and achieving the greatest possible satisfaction of all their needs, while minimizing the detriment, or unmet needs for each'.[63] In addition to the objective of equitable and reasonable utilisation, the ILC documents are in agreement that 'adequate' and 'appropriate' protection of shared freshwater resources constitutes a benefit to all system States. Combining these objectives, the mutual benefits to be achieved through cooperation can be summarised as sustainable and equitable development and management of the shared resources, in which the concept of sustainable management includes in its objectives the protection of those resources.

These objectives of the general duty to cooperate highlight the interrelationship between the 'general principles' of international water law. The objectives (i.e., equitable and reasonable utilisation and adequate protection of shared water resources) derive directly from the content of

60 For a definition of international cooperation, see Zartman, 'Concept: Cooperation', pp. 5–7; and Keohane, *After Hegemony*, pp. 51f.

61 ILC, 'Draft Articles ... and Commentaries', p. 106.

62 Article 5 of the *UN Watercourses Convention* on the principle of equitable utilisation stipulates that '[i]n particular, an international watercourse shall be used and developed by watercourse States with a view to attaining optimal and sustainable utilization thereof and benefits therefrom ...'.

63 ILC, 'Draft Articles ... and Commentaries', p. 97.

the principle of equitable and reasonable utilisation, and they are best achieved by applying the general principles as an ensemble.

3.2 Considering legitimate interests of others

Cooperation promotes the consideration and integration of the interests of the participating riparian States and, as such, is a means to collaboratively work towards achievement of the objectives of other principles.

3.2.1 *The principle of equitable and reasonable utilisation*

The principle of equitable and reasonable utilisation is characterised by many authors as the governing principle of international water law.[64] This characterisation as the governing principle, or the fundamental rule, has been interpreted by some to mean that it should therefore be considered 'the prevailing rule of international law relating to the non-navigational uses of international [waters]';[65] however, the principle of equitable and reasonable utilisation is, in fact, one of several that intervene simultaneously and in conjunction with one another. This was indicated by the ILC in the commentary to the 1994 *Draft Articles* stating that '[a] survey of all available evidence of the general practice of States ... reveals that there is overwhelming support for the doctrine of equitable utilization as *a* general rule of law for the determination of the rights and obligations of States in this field' [emphasis added].[66] Furthermore, the same authors who referred to the principle as 'the fundamental rule' also elaborated on the fact that different principles interact and complement one another.[67]

The instrument accredited with firmly establishing the principle of equitable and reasonable utilisation as a core principle applicable to international watercourses is the 1966 ILA *Helsinki Rules*.[68] The *Rules* state that

[64] See, for example, Y. He, 'On the Performance of the Principle of Equitable and Reasonable Utilization in the Practice of Utilization of Transboundary Water Resources' (2009) 1 *iBusiness* 40; McCaffrey, *The Law of International Watercourses*, p. 384; S. Salman and K. Upreti, *Conflict and Cooperation on South Asia's International Rivers – A Legal Perspective* (The Hague: Kluwer Law International, 2002), p. 22.

[65] McIntyre, *Environmental Protection*, p. 85; I. Kaya, *Equitable Utilization: The Law of Non-Navigational Uses of International Watercourses* (Aldershot: Ashgate, 2003), p. 81.

[66] ILC, 'Draft Articles ... and Commentaries', p. 98.

[67] See, for example, McCaffrey, *The Law of International Watercourses*, pp. 407f., 436 and 445.

[68] Arts. IV and V, 1966 *Helsinki Rules* in ILA, *Report of the Fifty-Second Conference*, pp. 486–491. The roots of the principle are traced to US Supreme Court decisions in the early twentieth century. See, in more detail, McCaffrey, *The Law of International Watercourses*, pp. 384–401.

watercourse States have the duty to utilise transboundary waters in an equitable and reasonable way, considering the interests of other concerned watercourse States.[69] The principle applies to all uses, including allocation and non-consumptive uses and protection of water resources. Its application to ecosystem protection was confirmed by the ICJ in its decision on the *Gabčíkovo–Nagymaros Project*.[70] The identification of what equitable and reasonable use consists of in individual water systems depends on specific conditions that are prevalent in each basin. Non-exhaustive lists of criteria that have to be considered for this identification were provided in a number of legal instruments, such as Article V of the 1966 *Helsinki Rules*, Article 6 of the 1997 *UN Watercourses Convention*, Article 3 (8) of the 2000 *SADC Revised Protocol*, and Article 5 of the 2008 ILC *Draft Articles*. Criteria include geographic, hydrographic, hydrological, climatic and ecological factors; social and economic needs of riparian States; populations dependent on shared water resources; effects of uses on different riparian States; existing and potential uses; conservation, protection and economic efficiency of use of the shared resources; and potential alternative uses that are comparable to a planned or existing use.

Multiple factors and subjective objectives

These criteria are to be considered together and must be evaluated according to specific circumstances of an individual water system in question.[71] Customary law does not provide any further guidance to States on how much weight to attribute to individual factors; it leaves this decision to the States. Ultimately, weighing the factors and identifying what constitutes equitable and reasonable use depend in large part on subjective value judgements: What does equity mean for individual States and how does each State rank the importance of individual criteria? Individual factors can be valued differently by each State concerned, and the notion of equity itself also can be understood in various ways. For example, if States in a shared basin are at different levels of socioeconomic development and a question arises concerning water allocation to meet economic needs, one State may consider that equitable allocation is achieved if each

[69] Arts. 5 and 6, 1997 *UN Watercourses Convention*; Arts. 4 and 5, 2008 ILC *Draft Articles*. Note that the ILC reports point to the fact that applicability of the principle to aquifers is disputed by some States. *Third Report by the Special Rapporteur Chusai Yamada on Shared Natural resources; Transboundary Groundwaters*, 11 February 2005, UN Doc. A/CN.4/551, p. 9.

[70] *Gabčíkovo–Nagymaros Project (Hungary/Slovakia)* (1997) ICJ Reports, p. 7.

[71] Art. 6 (3), 1997 *UN Watercourses Convention*; Art. 5 (2), 2008 ILC *Draft Articles*; Art. V (3), 1966 *Helsinki Rules*.

riparian receives the water required to maintain its current level of economic development. Another State may consider that equity requires those States that are at a relatively lower level of socioeconomic development are entitled to a water allocation that allows them to 'catch up' with the relatively stronger States. The former State understands equity as a distributive concept, whereas the latter State focuses on its corrective aspects.[72] The issue in question here is not whether each riparian should be allocated the same amount of water according to the former State, or whether weaker States should receive relatively more water according to the latter State. Rather, the issue is the respective understanding of the concept of equity: depending on underlying value assumptions, the responses to questions about allocation may be very different.

Identification of the exact contours and content of the principle is further complicated by the fact that additional terms used in legal instruments to clarify the objective of the principle are vague (e.g., 'beneficial' and 'reasonable' use and 'optimal' and 'sustainable' utilization). These terms, although containing quantifiable components, escape precise definition. In an economic sense, the question of what constitutes beneficial use can have different responses depending on which time horizon is applied. For example, with respect to hydropower infrastructure, in the short to medium term, an increase in electricity supply may be more beneficial even if it causes environmental damage or increases evaporation rates; in the long term, uses that emphasise preservation of the natural environment that humans depend on may be more beneficial. The ILA notes that a 'beneficial use', according to the principle of equitable utilisation, 'need not be *the* most productive use . . . nor need it utilise the most efficient methods known in order to . . . insure maximum utilization' [emphasis in the original].[73] As mentioned previously, the ILC takes a similar stance on the meaning of optimal use in the commentary to Article 5 of the 1994 *Draft Articles*; it considers the objective to be to attain 'maximum possible benefits for all watercourse States and achieving the greatest possible satisfaction of all their needs, while minimizing the detriment to, or unmet needs, of each'.[74] This comment is valuable because it confirms that the objectives of the principle

[72] On distributive and corrective equity, see E. Louka, *International Environmental Law: Fairness, Effectiveness and World Order* (New York: Cambridge University Press, 2006), pp. 68 and 100.

[73] ILA, *Report of the Fifty-Second Conference*, p. 487.

[74] ILC, 'Draft Articles . . . and Commentaries', p. 97.

are not to be understood in only financial- and efficiency-related terms but rather that a balance must be achieved among multiple needs. The Article was later expanded to also include 'sustainable use' in the text of Article 5 (1) of the *UN Watercourses Convention*. This insertion and the specific reference to the protection of watercourses further establishes that in weighing factors and in assessing equity, needs of the environment also must be considered.[75] A response to the question of how different needs should be evaluated, however, is again left to those who implement the principle.

3.2.2 *Implementation of the principle*

As a consequence, questions arise about who has the authority to decide which concept of equity applies and what implementation of the principle means in a given case. As a principle, the rule of equitable and reasonable utilisation is not sufficiently specific and does not provide enough information to a State planning or carrying out a utilisation in order to assess whether it is also equitable and reasonable as perceived by other concerned States.[76] Article 5 of the *UN Watercourses Convention* stipulates that in using a watercourse in an equitable and reasonable manner, the interests of other concerned watercourse States must be taken into account. It follows that not only the State planning a new use or carrying out a use but also all States sharing a transboundary water system and affected by individual uses are authorised to decide and must be involved in the assessment of equity and reasonableness of the specific use. Equally, the assessment of different factors requires information exchange and communication among concerned States; equity and reasonableness can be assessed based only on information about conditions of the entire system. Often, this information is available only within individual States.

For these reasons, Article 6 (2) of the *UN Watercourses Convention* stipulates that concerned watercourse States 'shall, when the need arises, enter into consultations in a spirit of cooperation'. Implementation of the principle requires cooperation among concerned States due to the multiplicity of factors that intervene. The determination of what constitutes equitable and reasonable utilisation in a specific water system requires a cooperative process of reciprocal data exchange among the concerned States and information about utilisation preferences, interests and needs

[75] See also Art. 3 (8), 2000 *SADC Revised Protocol*.
[76] Tanzi and Arcari, *The United Nations Convention*, pp. 96f.

of co-riparian States.[77] It should be a continuous process because adjustments may be required due to underlying circumstances that are subject to change.

The ICJ also follows the maxim that equitable solutions are best found through cooperation among concerned States. In the *North Sea Continental Shelf Case*, which concerned the maritime boundary delimitation along a concavely shaped coastal line shared by three States, the ICJ ordered the Parties to settle their dispute by negotiation and provided guidance on the criteria to be considered:

> ... the principles and rules of international law applicable to the delimitation between the Parties ... are as follows:
>
> (1) Delimitation is to be effected by agreement in accordance with equitable principles, and taking account of all relevant circumstances ...;
> (2) If, in application of the preceding sub-paragraph, the delimitation leaves to the Parties areas that overlap, these are to be divided between them in agreed proportions, or, failing agreement, equally, unless they decide on a régime of joint jurisdiction ...;
>
> (D) In the course of the negotiations, the factors to be taken into account are to include:
>
> (1) the general configuration of the coasts of the Parties ...;
> (2) so far as known or readily ascertainable, the physical and geological structure, and natural resources, of the continental shelf areas involved;
> (3) the element of a reasonable degree of proportionality ...[78]

The Court ruled similarly in two other cases that concerned equitable rights to shared resources.[79] In 1972, Iceland adopted a law that unilaterally extended the zone where it would have exclusive fishing rights in the Atlantic Ocean. Both Germany and the United Kingdom challenged the opposability of this regulation to fishing vessels flying under their flags. In the decisions on these cases, the ICJ held that the parties were 'under mutual obligations to undertake negotiations in good faith for the equitable solution of their differences concerning their respective fishery

[77] ILC, 'Draft Articles ... and Commentaries', p. 97; McCaffrey, *The Law of International Watercourses*, pp. 401–404.

[78] *North Sea Continental Shelf Case (Federal Republic of Germany/Denmark, Federal Republic of Germany/Netherlands)*, (1969) ICJ Reports 3, para. 101.

[79] *Fisheries Jurisdiction (Federal Republic of Germany v. Iceland)* (1974) ICJ Reports, p. 175; and *Fisheries Jurisdiction (United Kingdom v. Iceland)* (1974) ICJ Reports, p. 3.

rights', and it again established criteria to be taken account of in this process.[80]

The Senegal River Basin is an example of a basin in which riparian States applied the maxim that equitable solutions are best found by way of cooperation and thus negotiated a complex legal framework based on the idea of equitable use and sharing of benefits. The legal framework for joint management initially established by Mali, Mauretania and Senegal – with the recent addition of the fourth riparian, Guinea[81] – is one of the most prominent examples of regimes that established equitable utilisation at the basin level.[82] The initial motivation was to increase flood protection as well as hydropower exploitation, irrigation and navigation through joint flow management.[83] A joint management mechanism was set up, the *Organisation pour la Mise en Valeur du Fleuve Sénégal* (OMVS),[84] and two jointly owned[85] multi-purpose reservoirs, at Diama and at Manantali, were constructed for this purpose. The reservoirs are operated according to jointly agreed-to principles, the so-called keys, which determine the amount of irrigation, hydropower, navigation and water benefits attributed to each country. This allocation is regularly revised by the benefitting countries.[86]

Equitable participation

In reflecting on the role that cooperation plays in implementing the principle of equitable and reasonable utilisation, the ILC introduced the concept of equitable and reasonable participation 'in the use, development and protection of an international watercourse'.[87] The provision of

80 *Fisheries Jurisdiction (Federal Republic of Germany v. Iceland)*, para. 77; and *Fisheries Jurisdiction (United Kingdom v. Iceland)*, para. 79.

81 Guinea joined the OMVS and the basin-wide legal framework, including the 2002 *Senegal Water Charter*, by virtue of the *Treaty of Adhesion*, 17 March 2006, LOI No. 2006–30, OJ of the Republic of Senegal, No. 6307 (2006).

82 W. Yu, *Benefit Sharing in International Rivers: Findings from the Senegal River Basin, the Columbia River Basin, and the Lesotho Highlands Water Project* (Washington, DC: World Bank, 2008); R. Paisley, 'Adversaries into Partners: International Water Law and the Equitable Sharing of Downstream Benefits' (2002) 3 *Melbourne Journal of Law* 280–300.

83 1972 *Convention Concerning the Statute of the Senegal River.*

84 1972 *Convention Establishing the Senegal River Basin Organization.*

85 1978 *Convention on the Legal Statute of Common Works.*

86 For a summary presentation of the 'key', see Yu, *Benefit Sharing in International Rivers*, pp. 17–20.

87 Art. 5 (2), 1997 *UN Watercourses Convention*; C. Behrmann, *Das Prinzip der angemessenen und vernünftigen Nutzung und Teilhabe nach der VN-Wasserlaufkonvention* (Berlin: Duncker & Humblot, 2008), p. 28; S. McCaffrey, 'The UN Convention on the Law of the Non-Navigational Uses of International Watercourses: Prospects and Pitfalls', in Salman and Boisson de Chazournes, *International Watercourses – Enhancing Cooperation and Managing Conflict* (Washington, DC: World Bank, 1998), p. 20.

Article 5 (2) of the *UN Watercourses Convention* outlines the right to utilise as well as the duty to cooperate in the protection and development of the watercourse. In the commentary to the *Draft Articles*, the ILC pointed out that cooperation forms the core duty with respect to equitable participation and added that States have a 'right to cooperation of other watercourse States'.[88] The explicit reference to the duty to cooperate in this context emphasises that 'cooperative action by watercourse States is necessary [in order] to produce maximum benefits'.[89] This adds authority to the claim that States cannot achieve equitable and reasonable utilisation in isolation; participation of the other State(s) sharing a hydrologic system is required.

Furthermore, this provision of the *Convention* highlights the importance of considering protection of hydrologic systems when States plan or carry out uses of the same and the need for cooperation in the protection of the watercourse. The ICJ considered the delicate balance between economic utilisation and environmental protection in the *Gabčíkovo–Nagymaros Case*. In this case, the ICJ recommended adapting the operation of the system of locks in such a way that 'both the economic operation of the system of electricity generation and the satisfaction of essential environmental concerns' can be accommodated.[90] Mandating the States to find a negotiated solution, the Court went on to state that the '[r]e-establishment of the joint régime will also reflect in an optimal way the common utilization of shared water resources . . . in concordance with Article 5 (2) of the Convention of the Law of Non-navigational Uses of International Watercourses'.[91]

The concept of participation in equitable utilisation was reflected in a limited number of subsequent treaties. It is reproduced almost verbatim in Article 3 (7) (b) of the 2000 *SADC Revised Protocol*; in accordance with this regional framework agreement, the rule is also reflected in Article 3 (b) of the *Tripartite Interim Agreement for Cooperation on the Protection and Sustainable Utilization of the Water Resources of the Incomati and Maputo Watercourses*.[92] It is also reflected in the 2008 *Niger Basin Charter*.[93] The 2004 *Berlin Rules* stipulate the principle of equitable utilisation and the rule of equitable participation in separate articles: Article 10 grants a right to

[88] Commentary to Art. 5, in ILC, 'Draft Articles . . . and Commentaries', p. 97. [89] *Ibid.*
[90] *Gabčíkovo–Nagymaros Project (Hungary/Slovakia)* (1997) ICJ Reports, para. 146.
[91] *Ibid.*, paras. 147 and 155.
[92] Signed by Mozambique, South Africa and Swaziland, 29 August 2002, LEX-FAOC034943.
[93] Art. 4, *Niger Basin Water Charter*, 30 April 2008, www.abn.ne.

participate in an equitable, reasonable and sustainable manner in the management of waters of an international drainage basin; and Article 12 establishes the principle of equitable utilisation. Yet, not all treaties adopted after 1997 reflect the concept of equitable participation *expressis verbis*, and they are limited to the principle of equitable utilisation, such as the 2002 *Framework Agreement on the Sava River Basin*[94] and the 2002 *Senegal Water Charter*. A possible reason is that it can be considered that the rights and duties embodied in the concept of equitable participation are covered by the general duty to cooperate and the obligations related to this principle.

3.2.3 Inter-generational equity

Closely related to the principle of equitable and reasonable use is the concept of 'inter-generational equity', which has recently found more prominence in the context of transboundary aquifers. The objective of the principle of equitable utilisation as one seeking intra-generational equity (i.e., to achieve equity among the present generation) emerges as a clear goal by the way it is formulated in legal instruments. Yet, the question arises of whether the principle also applies in an intertemporal context across different generations.

The rationale for the theory of inter-generational equity is that present generations hold natural resources in trust for future generations. This concept is included in Principle 2 of the 1972 *Stockholm Declaration on the Human Environment*: 'The natural resources of the earth, including air, water, land, flora, and fauna . . . must be safeguarded for the benefit of present and future generations through careful planning and management, as appropriate'.[95] Each generation has the duty to maintain or improve environmental conditions of the planet during its time of use – hence, to protect the life support system for future generations.[96] Each generation is entitled to equitable use in its own time and has the duty to ensure equitable use across generations.[97]

[94] 3 December 2002, 2366 *UNTS* 479.

[95] *Report of the United Nations Conference on the Human Environment* (New York: United Nations, 1972). See also Principle 3 of the 1992 *Rio Declaration on Environment and Development*: 'The right to development must be fulfilled so as to equitably meet developmental land environmental needs of present and future generations,' UN Doc. A/CONF.151/26 (Vol. I). See also Sands, *Principles of International Environmental Law*, pp. 256f.

[96] E. Brown Weiss, 'Our Rights and Obligations to Future Generations for the Environment' (1990) 84 *AJIL* 200.

[97] *Ibid.*

This premise is recognised in universal declarations and preambles of international agreements; it has been recognised as a principle *inter alia* in the *UNFCCC*[98] and the 1992 *UNECE Convention on the Protection and Use of Transboundary Watercourses and International Lakes*.[99] Consideration of the legal interests of future generations is implied in the provisions concerning the principle of equitable and reasonable utilisation of the *UN Watercourses Convention* and the 2008 ILC *Draft Articles*; both refer to existing as well as potential uses of the water systems as factors to be considered in the assessment.[100] The term 'potential' applies to both the intra-generational context and uses by future generations, thus acknowledging the likelihood of future water use interests. The 2008 ILC *Draft Articles* place additional emphasis on inter-generational considerations; Article 4 (c) stipulates that States 'shall establish individually or jointly a comprehensive utilization plan, taking into account *present and future needs* of, and alternative water sources for, the aquifer States' [emphasis added]. In addition, Article 5 emphasises that with respect to social, economic and other needs of aquifer States, both the present and future must be considered.

The interests of future generations are thus considered to a certain extent by the principle of equitable and reasonable utilisation. The environmental components of the theory of inter-generational equity are additionally supported by the rules concerning ecosystem protection and prevention of harm – obligations that are firmly anchored in international water law. Environmental protection is a way to achieve one of the core aspects of the theory – namely, that present generations should pass on the planet to the next generation in a condition that is no worse than it was when they received it.[101]

Closely related to this is another principle that intervenes for the benefit of future generations – that is, the precautionary principle. According to this principle, lack of full scientific certainty should not be a reason to postpone measures to prevent environmental degradation where there is a threat of serious or irreversible damage. The Ministers of the UNECE Region recognised the principle in the 1990 *Bergen Ministerial Declaration on Sustainable Development in the ECE Region*.[102] The principle

[98] Art. 3, 1992 *UNFCCC*. [99] Art. 2 (5) (c), 1992 *UNECE Water Convention*.

[100] Art. 6 (2), 1997 *UN Watercourses Convention*, and Arts. 4 (c) and 5 (2), 2008 ILC *Draft Articles*. Note that Art. V, 1966 *Helsinki Rules*, does not make reference to potential future uses, only to past and existing utilisation.

[101] Brown Weiss, 'Our Rights and Obligations', p. 200.

[102] UN Doc. A/CONF.151/PC/10 (1990).

is also reflected in the *Rio Declaration* as Principle 15 in a more modest version that limits the obligation of States to apply the precautionary approach 'according to their capabilities'.[103] The principle is sensitive to future consequences of human activity, thereby considering the interests of future generations.[104]

Although firmly established in Europe, the precautionary principle is considered to be only an emerging principle elsewhere, and its universal applicability to international water law is debated.[105] The ICJ missed the opportunity to pronounce on the applicability of the principle to transboundary water law when the Hungarian pleadings raised the point in the *Gabcíkovo–Nagymoros Case*.[106] Regarding the *Pulp Mills on the River Uruguay Case*, four judges again noted the missed opportunity to recognise the existence of the precautionary principle. In particular, Judge Cançado Trinidade pointed to the fact that both parties to the dispute accepted the existence of a precautionary principle in their arguments.[107] The 1997 *UN Watercourses Convention* does not explicitly reference the principle. However, according to the ILC commentary on the 1994 *Draft Articles*, 'the principle of precautionary action' is inherent in Articles 20, 21 and 22. The ILC argued that wording such as 'pollution that may cause significant harm' and 'the introduction of species which may have detrimental effects' establishes an implicit link to the principle of precautionary action.[108] The 2008 ILC *Draft Articles* are more specific; although not recognising the principle per se, they stipulate that 'Aquifer States shall take a

[103] Principle 15, 1992 *Rio Declaration*: 'In order to protect the environment, the precautionary approach shall be widely applied by States according to their capabilities. Where there are threats of serious or irreversible damage, lack of full scientific certainty shall not be used as a reason for postponing cost-effective measures to prevent environmental degradation'.

[104] L. Boisson de Chazournes, 'Precaution in International Law: Reflection on its Composite Nature', in T. M. Ndiaye and R. Wolfrum (eds.), *Law of the Sea, Environmental Law and Settlement of Disputes – Liber Amicorum Judge Thomas A. Mensah* (Leiden/Boston: Martinus Nijhoff Publishers, 2007), pp. 32f.

[105] Sands, *Principles of International Environmental Law*, p. 279.

[106] *Gabcíkovo–Nagymoros Project (Hungary/Slovakia)* (1997) ICJ Reports, paras. 54, 97 and 113. Judge Koroma recognised the inclusion of the precautionary principle in the 1997 *UN Watercourses Convention* in his *Separate Opinion*, *ibid.*, p. 152.

[107] *Case Concerning Pulp Mills on the River Uruguay (Argentina v. Uruguay), Separate Opinion of Judge Cançado Trinidade* (2010) ICJ Reports, paras. 51 and 53; *Joint Dissenting Opinion Judges Al-Khasawneh and Simma*, para. 23; and *Dissenting Opinion Judge ad hoc Vinuesa*, para. 100.

[108] ILC, 'Draft Articles . . . and Commentaries', pp. 120, 122 and 124. J. Brunée and S. Toope, 'Environmental Security and Freshwater Resources: Ecosystem Regime Building' (1997) 91 *AJIL* 49.

precautionary approach in view of uncertainty about the nature and extent of transboundary aquifer or aquifer systems and of its vulnerability to pollution'.[109] The ILC explained that the term 'precautionary approach' rather than 'precautionary principle' was used because it is 'the less disputed formulation' and because, in practice, the two concepts lead to similar results 'when applied in good faith'.[110] This commentary implies that the ILC considered that the approach may have not yet emerged as universally accepted legal principle. Precautionary action is of importance when the dynamics within and the fragility of ecosystems are not yet fully understood by science. This is the case today particularly regarding transboundary aquifer systems; it is therefore consistent that the precautionary approach was included in the 2008 ILC *Draft Articles*. It should also be applied relative to transboundary atmospheric water streams.

Inter-generational equity is promoted through the concurrence of the principles of equitable and reasonable utilisation and environmental protection, as well as through the precautionary approach; furthermore, it necessitates international cooperation to achieve its objectives. The life support system for future generations can best be maintained when States, individually or jointly, protect nature from irreversible damage and share knowledge about the most efficient technologies and methods to do so. The concurrence of existing principles in support of sustainable development for the benefit of future generations imposes a duty on the present generation to show solidarity with those future generations and to cooperate in this regard. Present generations assume the hypothetical interests of the generations to come. In implementing these principles, the present generation therefore is cooperating not only for the benefit of intra-generational equity but also for future generations.

3.3 Minimising negative impact on others

Avoiding or minimising negative impact on others is another way to consider the interests of others; that is, their interests in not being harmed. The role of cooperation, however, is generally much smaller than its contribution to achieving equity, mutual benefits and reasonable use of shared waters.

[109] Art. 12, 2008 ILC *Draft Articles*.

[110] ILC, 'Draft Articles on the Law of Transboundary Aquifers, with Commentaries' (2008) II(2) *YBILC* 59.

3.3.1 Good neighbourliness

Water resources management and utilisation are tightly linked to ways in which States make use of their territory and to the activities they carry out within their territory. Legal principles that give guidance to States on how territorial space should be used (e.g., the 'principle of good neighbourliness') therefore also have an important role in international water law.[111] In reflection of the theory of limited territorial sovereignty, this principle is an expression of the idea that sovereignty over a territory comes not only with rights but also with duties.

The principle of good neighbourliness is linked to the abuse-of-rights doctrine[112] and reflects the maxim *sic utere tuo ut alienum no laedas* (translated in the next paragraph) in more general terms. The relationship between these two concepts is not clearly established in literature. Hinds argues that the two concepts are identical.[113] In contrast, Verdross and Simma argue that the two principles have different origins: whereas the neighbourliness law is rooted in sovereignty, the maxim has its source in the principle of good faith.[114]

The abuse-of-rights doctrine bars States from exercising their rights 'in a way which impedes the enjoyment of other States of their own rights'.[115] In its literal translation, the maxim *sic utere tuo ut alienum non laedas* means that 'one shall use his or her own in a way so as not to cause injury to another'. It also has been translated as 'use your property so as not to damage another's'.[116] Whereas some argue that this maxim can be seen as having given rise and is linked to the principle of equitable utilisation,[117] others argue that 'for the sake of clarity of legal argument [it is better] to avoid the maxim *sic utere tuo ut alienum no laedas* in discussion about the

[111] Finnish Branch of the ILA, *The Work of the International Law Association*, pp. 37–40.

[112] McIntyre, *Environmental Protection*, p. 90; L. Caflisch, 'Regulation of the Uses of International Watercourses', in Salman and Boisson de Chazournes (eds.), *International Watercourses: Enhancing Cooperation and Managing Conflict: Proceedings of a World Bank Seminar* (Washington, DC: World Bank, 1998), p. 12.

[113] C. Hinds, *Umweltrechtliche Einschränkungen der Souveränität – Völkerrechtliche Präventionspflichten zur Verhinderung von Umweltschäden* (Frankfurt am Main: Peter Lang, 1997), p. 131.

[114] Verdross and Simma, *Universelles Völkerrecht*, p. 281.

[115] McIntyre, *Environmental Protection*, p. 90, citing A. Kiss, 'Abuses of Rights', in R. Bernhardt (ed.), *Encyclopedia of Public International Law, Installment 7* (Amsterdam, North Holland Publishing, 1984).

[116] B. Garner (ed.), *Black's Law Dictionary*, 8th edition (St. Paul, MN: Thomson West, 2004).

[117] H. Dickstein, 'International Lake and River Pollution Control: Questions of Method' (1973) 12 *Columbia Journal of Transnational Law*, p. 492 n15; ILA, *Report of the Fifty-Second Conference*, p. 499.

rights and duties of States regarding the use of [transboundary waters]'.[118] On the one hand, a translation that links the maxim to property is inept, particularly in the context of transboundary water law, because water resources that traverse a State's territory do not belong to that State but rather are shared with other riparian States of the respective international hydrologic system. On the other hand, the maxim also can be misleading and vague in the sense that the obligation not to cause any harm at all to another would be too restrictive and almost impossible to observe. It is for these reasons that those authors who accept the existence of the maxim as legal principle do not refer to it in its literal translation but rather in broader terms prohibiting substantial injury to others.[119]

The principle of good neighbourliness captures the latter and wider interpretation of the *sic utere tuo* maxim. According to this principle, which was included in the *Charter of the United Nations* because of its role in facilitating peaceful relations among States,[120] States should exercise their rights in a way that does not prejudice the rights of others.[121] At the same time, each State must tolerate a certain level of (inevitable) interference in its territorial space that is due to the geographic proximity of other States.[122] It is important to recognise that the principle applies not only to immediately bordering States but also to those at a greater distance, depending on the nature of transboundary harm and its vectors.[123] This case was made for acid rain, which can travel even across continents.[124] The focus of the principle of good neighbourliness is on activities that take place within a State's territory that cause harm in another State's territory. Compliance with this principle does not require that States engage in cooperative acts with others; therefore, it does not per se give rise directly to positive cooperation duties. It is also a principle that is not necessarily linked or limited to the utilisation of shared resources; rather, it applies more generally, such as in the case of acid rain or water pollution that causes extinction of vulnerable species in other States' territories. In its implementation, however, the principle relates to equitable and

118 Lammers, *Pollution of International Watercourses*, pp. 571f. 119 *Ibid.*, p. 571.

120 Preamble and Art. 74, 1945 Charter of the United Nations.

121 McIntyre, *Environmental Protection*, p. 90, citing a passage from the decision of the Swiss Federal Tribunal in a dispute between the cantons of Arvovia and Solothurn in 1900.

122 Tanzi and Arcari, *The United Nations Convention*, p. 16; Lammers, *Pollution of International Watercourses*, pp. 563–569.

123 I. van Lier, *Acid Rain and International Law* (Alphen aan den Rijn: Sijthoff & Noordhoff, 1980), p. 104.

124 *Ibid.*

reasonable use. According to Andrassy, the determination of whether a use that causes harm is permissible is carried out through the application of principles of equity.[125] Consequently, effects of a certain utilisation on other watercourse States are listed as one factor to be considered for equitable and reasonable utilisation in Article 6 (1) of the *UN Watercourses Convention*.

Protection of the environment

The principle of good neighbourliness was also linked to the 'principle of environmental protection', which has found expression in numerous international declarations. International courts and tribunals confirmed that it forms a part of customary international law. Principle 21 of the *Stockholm Declaration of the United Nations Conference on the Human Environment* described the application of the principle relative to the use of resources and territory of a State in the following terms:

> States have, in accordance with the Charter of the United Nations and the principles of international law, the sovereign right to exploit their own resources pursuant to their own environmental policies, and the responsibility to ensure that activities within their jurisdiction or control do not cause damage to the environment of other States or of areas beyond the limits of national jurisdiction.[126]

Principle 2 of the 1992 *Rio Declaration on Environment and Development* restates the principle in similar terms, with additional reference to national developmental policies as one of the elements to be respected:

> States have, in accordance with the Charter of the United Nations and the principles of international law, the sovereign right to exploit their own resources pursuant to their own environmental and developmental policies, and the responsibility to ensure that activities within their jurisdiction or control do not cause damage to the environment of other States or of areas beyond the limits of national jurisdiction.

In its *Advisory Opinion on the Legality of the Threat or Use of Nuclear Weapons*, the ICJ confirmed that the principle of environmental protection forms part of customary international environmental law:

[125] '*L'importance du dommage doit être appréciée par rapport aux deux parties en cause. Il faut prendre en considération la proportion ou la disproportion entre le bénéfice tiré par l'une des parties et le désavantage subi par l'autre. La variété des ca n'admet pars de règles fixes, et fait une large place à l'équité.*' J. Andrassy, '*Les relations internationales de voisinage*' (1951) 79 *RCADI* 112.

[126] *Report of the United Nations Conference on the Human Environment*, chap. I.

The Court also recognizes that the environment is not an abstraction but represents the living space, the quality of life and the very health of human beings, including generations unborn. The existence of the general obligation of States to ensure that activities within their jurisdiction and control respect the environment of other States or of areas beyond national control is now part of the corpus of international law relating to the environment.[127]

The ICJ formulated the general rule in a slightly different way than Principle 21 and Principle 2 of the *Stockholm* and *Rio Declarations*, respectively: it limited the application to activities within the jurisdiction *and* control of State, whereas the *Stockholm* and *Rio Principles* refer to 'activities within jurisdiction *or* control' [emphasis added].[128] It has been argued that the Court also broadened the scope; rather than focusing solely on activities that 'cause damage to the environment of other States', it demands 'respect' for the environment 'of other States or of areas beyond national control'.[129] The ICJ referred to this decision and (re)confirmed the rule's applicability for the use of transboundary rivers in its decisions on the *Gabčíkovo–Nagymaros Project Case* and the *Pulp Mills on the River Uruguay Case*, among others.[130] International arbitration tribunals also confirmed the principle in the *Trail Smelter Case*[131] (i.e., transboundary air and land pollution) and in the *Lac Lanoux Case*[132] (i.e., use of shared water resources).

3.3.2 Implementation of the principle: Obligation not to cause significant harm

The principle of good neighbourliness was recognised in several international water treaties as a theme that guides its implementation; 15 per cent of the treaties analysed (see Annex) make explicit

127 *The Legality of the Threat or Use of Nuclear Weapons* (1996) ICJ Reports, 241f., para. 29. In this dictum, the Court also points to the importance of considering impacts of human activity over time, alluding to the concept of inter-generational equity.

128 E. Brown Weiss, 'Opening the Door to the Environment and to Future Generations', in L. Boisson de Chazournes and P. Sands, *International Law, the International Court of Justice and Nuclear Weapons* (Cambridge: Cambridge University Press, 1999), p. 340.

129 *Ibid.*

130 *Gabčíkovo–Nagymaros Project (Hungary/Slovakia)* (1997) ICJ Reports 41, para. 53; *Case Concerning Pulp Mills on the River Uruguay (Argentina v. Uruguay)* (2010) ICJ Reports, p. 38, para. 101.

131 *Trail Smelter Case (United States, Canada)*, 18 April 1938, 11 March 1941 (1941) 3 *RIAA* 1905–1982.

132 *Lac Lanoux Arbitration (France v. Spain)* (1958) 29 *RGDIP* 117.

reference to the principle.[133] The principle also serves as the legal basis for the obligation not to cause significant harm. This obligation was included in Part II of the *UN Watercourses Convention*, in recognition of its character as a fundamental rule of and vital importance within the body of norms of international water law. Article 7 (1) of the *Convention* provides that States 'shall, in utilizing an international watercourse in their territories, take all appropriate measures to prevent the causing of significant harm to other watercourse States'. This provision is of particular importance because any perceptible alteration of the flow regime or of the water quality is likely to impact other riparian States. The 2008 ILC *Draft Articles* on the law of transboundary aquifers go even further by imposing an obligation to prevent not only significant harm in utilising aquifers but also to prevent the implementation of activities that may impact the aquifers and then harm other States through that aquifer or aquifer system.[134]

The requirement and qualification of harm as 'significant' are important; they attest to the fact that the obligation is based firmly on the principle of good neighbourliness and the concept of limited territorial sovereignty, both of which comprise the duty of affected States to tolerate some interference or occurrence of insignificant harm in their territories.[135] According to the ILC, 'significant' means that it is detectable and 'can be established by objective evidence'.[136] The same standard applies as if the notion of 'appreciable' were used; however, it implies a standard that is lower than the term 'serious' or 'substantial' would indicate.[137] At the same time, the ILC highlights the fact that 'the threshold of "significant harm" is a flexible concept. Factual considerations, rather than legal determination, have to be taken into account in each specific case'.[138] That the obligation is one of conduct and not result was confirmed by the ICJ in the *Pulp Mills on the River Uruguay Case*.[139] The exercise of due diligence in this context means that the respective State that has

133 See, for example, 2002 *Framework Agreement on the River Sava*, 2002 *Senegal Water Charter*, 1983 *Agreement on Ad-Hoc Sharing of the Teesta Waters*, and 1954 *Convention Concerning Water Economy Questions Relating to the Drava*.

134 Art. 6, 2008 ILC *Draft Articles*.

135 Tanzi and Arcari, *The United Nations Convention*, p. 148.

136 ILC, 'Draft Articles . . . and Commentaries', pp. 94, 96; ILC, 'Draft Articles on the Law of Transboundary Aquifers', p. 47.

137 ILC, 'Draft Articles . . . and Commentaries', pp. 94, 96.

138 ILC, 'Draft Articles on the Law of Transboundary Aquifers', p. 47.

139 *Case Concerning Pulp Mills on the River Uruguay (Argentina v. Uruguay)* (2010) ICJ Reports, p. 38, para. 101.

caused significant harm has not done so intentionally or negligently but rather has employed the same concern it would usually use in its own domestic affairs and has applied all appropriate measures that are within its means.[140]

Relationship with the principle of equitable and reasonable utilisation

Questions about the relationship between the obligation not to cause significant harm and the principle of equitable and reasonable utilisation have arisen relative to implementation of the two rules, sparking debate during the evolution of the *UN Watercourses Convention* as well as in the legal literature. During the more than twenty years of elaboration of the Articles that form the basis of the *Convention*, drafts discussed by the ILC changed from favouring one principle over another and vice versa to the 1994 *Draft Articles* that attempted to find wording that would balance the two rules.[141] Discussions continued among States in the Working Group of the Sixth Committee that elaborated the *Convention*. The final wording of Article 7 was one of the most contested topics.[142] Finally, balance was sought *inter alia* in the current wording of Article 7 (2) of the *UN Watercourses Convention*:

> 2. Where significant harm nevertheless is caused to another watercourse State, the State whose use causes such harm shall, in the absence of agreement to such use, take all appropriate measures, having due regard for the provisions of articles 5 and 6, in consultation with the affected States, to eliminate or mitigate such harm and, where appropriate, to discuss the question of compensation.

In this context, it is interesting that this provision was apparently ultimately acceptable to States because, on the one hand, those States that are lower riparians on transboundary watercourses considered it 'to be sufficiently neutral not to suggest a subordination of the no-harm rule to the principle of equitable and reasonable utilisation [...while, on the other hand...] upper riparians thought just the contrary, namely that that formula was strong enough to support the idea of such a subordination'.[143]

140 ILC, 'Draft Articles...and Commentaries', p. 103.
141 For more detail, see McCaffrey, *The Law of International Watercourses*, pp. 364–367; P. Wouters, 'The Legal Response to International Water Conflicts: The UN Watercourses Convention and Beyond' (1999) 42 *German Yearbook of International Law* 307–312; and Caflisch, 'Regulation of the Uses', pp. 13f.
142 McCaffrey, *The Law of International Watercourses*, p. 364.
143 Caflisch, 'Regulation of the Uses', p. 15.

It seems that a significant number of States were of the opinion that the wording achieved the necessary balance. Yet, these contradictory positions also highlight the absence of uniform practice and of a common understanding among States concerning the relationship between the two rules. This lack of clarity about the relationship was cited as a core reason for the sluggish ratification of the *UN Watercourses Convention*.[144]

Nevertheless, the debate continued in the literature and various authors proposed different nuances on how the relationship between these rules should be interpreted. Tanzi and Arcari argued that the *Convention* deliberately strikes a delicate balance between the two principles.[145] Differing with this interpretation, Salman argued that a close reading of Articles 5 through 7 of the *Convention* led to the 'prevailing view . . . that the Convention has subordinated the obligation not to cause significant harm to the principle of equitable and reasonable utilisation' and that this was confirmed by the ICJ in its decision in the *Gabčíkovo-Nagymaros Case* because the Court 'emphasized the concept of equitable utilization' and did not mention the obligation not to cause significant harm.[146] Wouters clarified that, in fact, the Court did not pronounce on the relationship between the two rules in the *Gabčikovo-Nagymaros* decision but rather did not mention the obligation not to cause significant harm because it was 'unable to ascertain any significant environmental harm inflicted on Hungary'.[147] She concluded that the 'primary rule of equitable and reasonable utilization [is] supplemented by the requirement of preventive behavior in Article 7'.[148] McCaffrey favoured an interpretation that accords a complementary role to the obligation of no significant harm, stating that the rule 'neither embodies an absolute standard nor supersedes the principle of equitable utilisation where the two appear in conflict with each other [but instead it triggers] discussions between the States concerned'.[149] Citing the *Gabčíkovo-Nagymaros* decision, in which the ICJ ordered the parties to negotiate in good faith with a view to achieving agreement on joint management of the project, McCaffrey further argued in support that the 'facts and circumstances of each case, rather than any *a priori* rule, will

[144] S. Salman, 'The United Nations Watercourses Convention Ten Years Later: Why Has Entry into Force Proven Difficult?' (2007) 32 *Water International* 8.

[145] Tanzi and Arcari, *The United Nations Convention*, pp. 177f.

[146] *Ibid.*, p. 6. [147] Wouters, 'The Legal Response', p. 329. [148] *Ibid.*, p. 335.

[149] Art. 10 (2), 1997 *UN Watercourses Convention*, which provides that conflicts between uses are to be resolved with reference to Arts. 5 through 7, is mentioned as further confirmation that the no significant harm rule does not prevail. McCaffrey, *The Law of International Watercourses*, pp. 366, 408.

ultimately be the key determinants' and 'cooperation and compromise, not... rigid insistence on the rules of law' will ultimately solve disputes that may emerge in the application of these rules.

Given that a uniform understanding is lacking among States and that the literature offers equally diverging interpretations of the relationship, solutions to potential conflicts in implementing these two rules must be and have been sought elsewhere. The *UN Watercourses Convention* indicates how this might be achieved. Article 7 (2) stipulates that where significant harm occurred, States shall try to reverse or mitigate the harm they caused 'in consultation with the affected State' and 'discuss the question of compensation', where appropriate.[150] Article 5 (2) stipulates a 'duty to cooperate in the protection and development' of international watercourses relative to participation in the equitable and reasonable use thereof. The role of cooperation is also emphasised regarding significant harm that might occur due to pollution; Article 21 of the *UN Watercourses Convention* stipulates that States shall, where appropriate, act jointly to prevent, reduce and control pollution, and shall enter into consultations at the request of any of the riparian States 'with a view to arriving at mutually agreeable measures and methods'. Part III of the *Convention* and Article 28 establish notification as well as other cooperation obligations regarding harm that might occur due to planned measures or emergency situations.[151]

Treaty analysis indicates that, in general, States also primarily seek achievement of mutual benefits and equitable use by agreeing to specific cooperation obligations rather than referring *expressis verbis* to the principle of equitable and reasonable utilisation and the obligation not to cause significant harm as guiding rules in their legal agreements. These rules of implementation illustrate that cooperation is often regarded as a key to resolving potential conflicts in the implementation of the principle of equitable utilisation and the objective to avoid significant harm.

3.4 Conclusion: A triangle of principles?

The previous discussion demonstrates that the legal principles applicable to the management of transboundary water resources are not only

[150] Art. 6 (3), 2008 ILC *Draft Articles*, is the equivalent provision.
[151] Although in less detail, see also Arts. 12 and 15 of the 2008 ILC *Draft Articles*.

functionally linked but also are, to a certain extent, inextricably intertwined. The principles of good neighbourliness and equitable and reasonable utilisation complement one another. The former aims to ensure that transboundary harm to third countries is minimised, whereas the latter intervenes to ensure equitable utilisation and distribution of benefits derived from the use of shared resources.[152] Both principles are concerned with the behaviour of States within their territorial space and the transboundary impact, whereas the principle of equitable utilisation provides a normative 'umbrella' for State behaviour with respect to resources shared among multiple States. This latter principle contains the essence of the former and aims to define which level of harm would be considered outside the range of acceptable harm. Given that the weight of factors considered by this principle is to be determined by those who apply it considering the basin as a whole, and due to the lack of a common understanding with regard to hierarchy between these two principles, conflicting claims might emerge in their implementation as to uses that may be considered equitable by one riparian State but as causing significant harm by another.[153] At the same time, the substance of the principle of equitable and reasonable utilisation represents the objective of the general duty to cooperate. This general obligation intervenes as the guiding norm that aims for joint agreement among States on what can be considered equitable by all those concerned and on the level of interference in their territory that therefore would be acceptable. The principle of cooperation also intervenes more particularly to resolve potential conflicts in the implementation of the other principles.

Rather than focusing on the principle of equitable and reasonable utilisation as the cornerstone of international water law, as suggested by some authors,[154] we should speak of (at least) a triangle of cornerstone principles. The general duty to cooperate, the principle of equitable and reasonable utilisation, and the principle of good neighbourliness – expressed in the obligation not to cause significant harm – interact, overlap and mutually support one another as general principles of international water law. This triangle must be understood not as the closed triangle of geometry but rather as the open triangle of the musical instrument. The three principles are functionally related to other principles and maxims (e.g., environmental protection and the precautionary approach); likewise, of

[152] Caflisch, 'Regulation of the Uses', p. 13.
[153] Boisson de Chazournes, '*Eaux internationales et droit international*', p. 26.
[154] McIntyre, *Environmental Protection*, p. 1; Wouters, 'The Legal Response', p. 326.

course, we could conceive of a variety of multi-angular shapes to conceptualise a metaphor for the interplay of legal principles. They interact with one another and with the rights and obligations to which they give rise, similar to the way in which the vibrations of the musical triangle interact with its surrounding environment to unfold the full effect of its chime.[155]

These interactions are described in more detail in Chapters 4 and 5. They outline the ways in which specific rights and obligations, as well as measures adopted by States to jointly manage and develop shared waters, are bringing the general principles of international water law to fruition.

[155] International water treaties often include more extensive lists of principles to be respected in their implementation. The principle of environmental protection has also been considered as one of the cornerstone principles of international water law; even if not prominently reflected as such in Part II of the *UN Watercourses Convention* and the 2008 ILC *Draft Articles*, it has been confirmed in recent decisions of the ICJ in the *Gabčíkovo–Nagymaros Case* and *Pulp Mills of the River Uruguay Case*. Other water treaties, specifically European instruments, refer to the precautionary principle and the polluter-pays principle. See Boisson de Chazournes, '*Eaux internationales et droit international*', p. 25.

4 Informing cooperation

The process of cooperation among riparian States in the management and development of transboundary freshwater systems is framed by specific rights and obligations. Before analysing the content of these rules in more detail, this chapter first explores their characteristics. It pursues the question of whether cooperation obligations merely outline the procedure States must follow relative to the management of shared resources or whether those rules, beyond that, also contain substantive elements.

4.1 Procedure void of substance?

The separation of procedure from substance in the categorisation of rules facilitates their description; however, it can be difficult to draw a clear line between the two, particularly in the context of international law.

4.1.1 Procedural and substantive law

Many authors draw a distinction between procedural and substantive obligations of international water law, and the rights and duties of cooperation are often described as procedural in nature.[1] Judge Higgins labelled the cooperation obligations of international water law more specifically as the 'procedural law of cooperation' in her 'General Course on Public International Law'.[2] The question arises, however, as to whether it is possible, or even sensible, to make a clear-cut distinction between procedure

1 Brown Weiss, 'The Evolution of International Water Law', 199–207; McCaffrey, *The Law of International Watercourses*, pp. 465–480; S. Salman, 'Evolution and Context of International Water Resources Law', in Boisson de Chazournes and Salman (eds.), *Les ressources en eau*, pp. 62ff. Tanzi and Arcari refer to 'procedural applications' of the obligation to cooperate, in Tanzi and Arcari, *The United Nations Convention*, pp. 181ff.

2 R. Higgins, 'General Course on Public International Law' (1991) 230 *RCADI* 183.

and substance with respect to the rules of cooperation in transboundary water law.

For practitioners or trial lawyers, the terms 'procedural' and 'procedural law' generally evoke the idea that norms falling under this category are essentially rules that describe methods for enforcing substantive rights in court or before government authorities (e.g., the obligation to file a complaint or the response to a complaint by a given date in order not to forfeit the right to claim enforcement of a substantive right). Leading law dictionaries define 'substantive law' as the law 'which creates duties, rights and obligations, while "procedural or remedial law" prescribes methods of enforcement of rights or obtaining redress for their invasion'.[3] Substantive and procedural laws are presented as separate sets of norms, in which the latter ensures the enforcement of the former.

The distinction between these two possible characteristics of norms is less evident in international law. A number of authors point to the close interrelationship between 'procedural law' and 'substantive law' in this context. Jenks clarified that:

> [i]n every legal system, law and procedure constantly react upon each other. Changes in the substantive law call for new procedures and remedies; new procedures and remedies make possible changes in the substantive law. So it is in international law; if we wish so as to develop the law to respond to the challenge of our times, our remedies and procedure must be sufficiently varied and flexible for the purpose.[4]

Bourne argued that in situations in which substantive law offers insufficient guidance to States on how they may proceed to give effect to these norms – such as in the case of the principle of equitable and reasonable utilisation – these shortcomings 'may be offset to some extent by a body of procedural law'.[5] McCaffrey noted that 'the "substantive" obligation of equitable and reasonable utilisation may itself be thought of as a

[3] *West's Law and Commercial Dictionary in Five Languages* (St. Paul, MN: West Publishing Company, 1985). The *PONS Fachwörterbuch Recht English-Deutsch/Deutsch-Englisch* (München: Ernst Klett Verlag, 1997) defines 'procedural' as 'referring to legal procedure', 'procedural law' as 'rules governing how the civil or criminal law is administered by the courts', and 'substantive law' as 'all laws including common law and statutory law which deal with legal principles' (as opposed to procedural law, which refers to the procedure for putting law into practice).

[4] C. W. Jenks, *The Prospects of International Adjudication*, cited in C. Bourne, 'Procedures in the Development of International Drainage Basins: Notice and Exchange of Information' (1972) 22 *University of Toronto Law Journal* 172.

[5] *Ibid.*

process' and can only be effective through cooperation.[6] In practice, it is true that cooperation duties can be used to facilitate observance of other rights as well as the creation of new rights; however, they also comprise a substantive obligation in and of itself.

That the line between the substantive and procedural rules is not as clearcut as it may seem was also confirmed by the ICJ in its decision on the *North Sea Continental Shelf Cases*:

> ... [i]n accordance with the ideas which have always underlain the development of a legal régime of the continental shelf in this field; namely (a) the parties are under an obligation to enter into negotiations with a view to arriving at an agreement and not merely to go through a formal process of negotiation ...; they are under an obligation so to conduct themselves that the negotiations *are meaningful*, which will not be the case when either of them insists upon its own position without contemplating any modification of it ... [emphasis added].[7]

The ICJ emphasises that 'procedural obligations' under international law are not satisfied by the conduct of a formal process and therefore are not merely 'procedural'; in fact, they are qualified by 'substantive' requirements that must be complied with during their implementation. Similarly, the general duty to cooperate and the rights and obligations emanating from it consist not merely in performing purely procedurally cooperative acts but also in performing procedural acts that meet substantive requirements. The obligation to carry out these acts in good faith and with a view to achieving mutual benefits, for example, is such a substantive requirement. The distinction between substantive and procedural law is difficult to draw in the context of international law because procedural obligations are interlaced with substantive content.[8] The doctrinal distinction between procedural and substantive rules of international water law that is observed by many authors facilitates schematisation for analytical purposes, yet it is a simplified conception of the nature of the rules of cooperation.

[6] McCaffrey, *The Law of International Watercourses*, p. 464.

[7] *North Sea Continental Shelf Case (Federal Republic of Germany/Denmark, Federal Republic of Germany/Netherlands)* (1969), ICJ Reports, p. 47, para. 85.

[8] As an exception, treaty articles outlining the procedures of adoption of protocols and annexes, as well as those stipulating the deposit of the instrument with a certain (international) organisation, are purely procedural rules.

4.1.2 *The procedural and substantive content of specific cooperation obligations*

Indeed, many authors acknowledge that the line dividing substantive and procedural rules is blurred in international water law.[9] Some argue that the general duty to cooperate 'serves as a bridge between substantive and procedural rules';[10] because specific cooperation obligations include both substantive and procedural content, they are substantive as well as procedural obligations. Referring to the obligation of notification for planned measures that risk causing appreciable harm to other riparian States, for example, Okowa argued that 'a number of substantive principles must be implied if the obligation is not to be deprived of practical significance'.[11] The objective of the obligation to notify is to give affected States the opportunity to assess the risk of harm with respect to their own interests and rights. Hence, notification consists not only of the procedural act of announcing to potentially affected States that a project is planned on a shared water system. The notification also must be accompanied by adequate technical data that will allow potentially affected States to carry out their assessments relative to the impact of the planned measure.[12] The obligation to notify outlines the procedure needed to achieve the goal that is equally inherent in the very same obligation. Thus, in the context of notification of planned measures, the obligation to notify and the act of notification are qualified by substantive requirements. Substance and procedure do not exist without one another but instead are intertwined in the obligation.

The dual character of obligations that contribute to the implementation of the general duty to cooperate was also highlighted with respect to the obligation to carry out an environmental impact assessment (EIA) when there is a risk of significant adverse transboundary impact of planned measures.[13] Okowa argued that this is a procedural obligation, which is

9 Brown Weiss, 'The Evolution of International Water Law', p. 207; P. Okowa, 'Procedural Obligations in International Environmental Agreements' (1996) 67 *BYIL* 277.

10 P. Wouters *et al.*, *Sharing Transboundary Waters – An Integrated Assessment of Equitable Entitlement: The Legal Assessment Model*, IHP-VI, Technical Documents in Hydrology, No. 74 (Paris: UNESCO, 2005), p. 24.

11 Okowa, 'Procedural Obligations', p. 291.

12 *Case Concerning Pulp Mills on the River Uruguay (Argentina v. Uruguay)* (2010), ICJ Reports, para. 104; ILC, *Third Report, Schwebel*, p. 103; Bourne, 'Procedures: Notice and Exchange of Information', p. 178.

13 *Case Concerning Pulp Mills on the River Uruguay (Argentina v. Uruguay)* (2010) ICJ Reports, para. 204.

closely linked to notification.[14] In its decision on the *Pulp Mills on the River Uruguay* case, the ICJ discussed this obligation primarily in the section that addresses substantive obligations and as one that is linked to the obligation to prevent transboundary environmental harm.[15] At the same time, the ICJ acknowledged the close link between the EIA obligation and the obligation to notify of planned measures, which the Court considers to be procedural. It acknowledges that the findings contained in an EIA comprisse a component of the notification obligation.[16]

The obligation to carry out an EIA is closely linked not only to notification obligations in this context but also contains procedural and substantive elements. To meet the obligation, it is not sufficient to go through the procedure of carrying out an EIA study of some kind.[17] There are content requirements to ensure that the inherent substantive goal can be achieved: the study must assess the potential environmental effects of planned measures, including potential transboundary effects. The ICJ observed that 'due diligence, and the duty of vigilance and prevention which it implies, would not be considered to have been exercised, if a party planning works liable to affect the régime of the river or the quality of its waters did not undertake an environmental impact assessment on the potential effects of such works'.[18] National regulations generally impose more specific components, procedural and content requirements,[19] and – in some instances – international treaties also define minimum content and procedural requirements for EIAs.[20] As with notification, the obligation includes both procedural and substantive elements. If transboundary environmental harm occurs and the State implementing the project has failed to carry out an EIA, the degree of due diligence required from the State has not been fulfilled.[21] The process of conducting an EIA is required

[14] Okowa, 'Procedural Obligations', p. 279.

[15] *Case Concerning Pulp Mills on the River Uruguay (Argentina v. Uruguay)* (2010) ICJ Reports, paras. 203–219.

[16] *Ibid.*, paras. 119f.

[17] Okowa citing Bowett: '[t]here was a risk that a State might appear to have fulfilled its obligations by carrying out a study of some kind, whereas in reality, it had totally failed to have the potential risk properly assessed'. Okowa, 'Procedural Obligations', p. 281.

[18] *Case Concerning Pulp Mills on the River Uruguay (Argentina v. Uruguay)* (2010) ICJ Reports, para. 204.

[19] The *UNEP Goals and Principles of Environmental Impact Assessment*, UNEP Governing Council Decision 14/25, 16 January 1987, UN Doc. UNEP/GC/Dec/14/25, provides guidelines to States and promotes harmonisation of national standards on the international level.

[20] See *Convention on Environmental Impact Assessment in a Transboundary Context (Espoo Convention)*, 26 February 1991, 1989 *UNTS* 309.

[21] Okowa, 'Procedural Obligations', p. 333.

to determine the likely risk of transboundary harm and to prevent it from occurring.

4.1.3 Importance of comprehensive reading

In the analysis of the close link between procedural and substantive elements of rules, it is worth examining the decision of the ICJ on the *Case Concerning Pulp Mills on the River Uruguay* in detail because cooperation obligations are at the heart of this dispute. Argentina claimed that Uruguay had breached obligations established by the 1975 *Statute of the River Uruguay*, a treaty that has as its objective the creation of 'a joint machinery necessary for the optimum and rational utilization of the River Uruguay'.[22] Argentina argued that in the process of authorising, constructing and planning the future commissioning of two pulp mills close to the riverbanks of the contiguous stretch of the river, Uruguay failed to adequately respect information obligations as well as those to prevent transboundary harm. Although the dispute is based largely on the application and interpretation of treaty instruments, the issues analysed by the Court are also pertinent to international water law in general, as well as to the legal nature of cooperation obligations more specifically.

In the structure of its decision, the ICJ followed the doctrinal distinction between procedural and substantive obligations. The decision contains two main parts: (1) the 'alleged breach of procedural obligations', and (2) 'substantive obligations'.[23] The Court recognised that there is a functional link between these two categories and that they complement one another in achieving the overall objective of the treaty in question.[24] Yet, the Court did not clarify the true nature of the relationship between 'procedure' and 'substance';[25] instead, it elaborated on how the functional link can be broken. According to the ICJ, the breach of a procedural obligation does not automatically entail the breach of a substantive obligation; compliance with procedural obligations is not a *sine qua non* for compliance with substantive obligations.[26] States must 'answer for these obligations separately'.[27]

However, in the process of analysing the two categories separately, the ICJ does not and cannot always keep the two completely apart. In two

[22] Art. 1, 1975 *Statute of the River Uruguay*.

[23] *Case Concerning Pulp Mills on the River Uruguay (Argentina v. Uruguay)* (2010) ICJ Reports, paras. 67–158 and 159–266.

[24] *Ibid.*, paras. 77–79.

[25] *Ibid.*, paras. 71–79; and *Joint Dissenting Opinion of Judges Al-Khasawneh and Simma*, paras. 26f.

[26] *Ibid.*, para. 78. [27] *Ibid.*, para. 79.

instances in its decision in the *Pulp Mills* case, the Court recognised the causality of procedure on the achievement of substantive obligations. With respect to EIA requirements, the Court argued that the obligation exists to 'enable the notified party to participate in the process of ensuring that the assessment is complete, so that it can then consider the plan and its effects with full knowledge of the facts'.[28] At another point, the Court stated that 'the obligation to inform CARU allows for the initiation of cooperation between the Parties which is necessary in order to fulfill the obligation of prevention'.[29] With this, the ICJ acknowledged the importance of procedure for the achievement of substantive goals and it recognised the tight links between cooperation and adherence to procedural and substantive obligations.

The approach of separating substantive and procedural obligations, which was adopted by the ICJ in its analysis, therefore seems somewhat artificial. And the Court itself considers:

> . . . that the procedural obligations of informing, notifying and negotiating constitute an appropriate means, accepted by the Parties, of achieving the objective which they set themselves in Article 1 of the 1975 Statute. These obligations are all the more vital when a shared resource is at issue, as in the case of the River Uruguay, which can only be protected through close and continuous cooperation between the riparian States.[30]

> . . . that it is by cooperating that the States concerned can jointly manage the risks of damage to the environment that might be created by the plans initiated by one or other of them . . . .[31]

Making a distinction between the two overlooks the possibilities and opportunities to improve substantive results through adequate procedures. For example, without notification of EIA findings, States are likely to miss opportunities to improve the design of a project in order to accommodate the interests of all parties concerned. As the Court observed, this link of procedure and substance is of particular importance regarding the complexity of managing and protecting transboundary water resources. Judges Simma and Al-Khasawneh eloquently summarised the argument against drawing a distinction between the two in their joint dissenting opinion by stating that in the absence of procedure, substantive results obtained are 'obviously no different from a *fait accompli*'.[32] The role of

[28] *Ibid.*, paras. 119f. [29] *Ibid.*, para. 102. [30] *Ibid.*, para. 81. [31] *Ibid.*, para. 77.

[32] *Case Concerning Pulp Mills on the River Uruguay, Joint Dissenting Opinion of Judges Al-Khasawneh and Simma*, para. 26.

procedure in this context is not to 'enforce' substantive rights and obligations; rather, its role is to facilitate their achievement.

In most cases, the same rule contains both procedural and substantive elements. Even if one component might be more prominent than another within one norm, drawing a distinction between procedural and substantive obligations can be misleading because it overlooks the important interrelationship of procedure and substance in achieving the objectives of the general duty to cooperate. It is therefore more sensible to discuss the substantive and procedural content of norms together rather than addressing substantive and procedural law or obligations separately.[33]

4.2 Informing the process

A number of specific cooperation obligations aim to enhance the cooperation process and the quality of its outcomes through promotion of information exchange and interaction among States. Information is provided for various reasons: (1) to inform about the general status of a water system; (2) to improve development and management planning capacity; and (3) to prevent harm by notifying other States of imminent danger or of planned activities that might negatively impact the water system or another State's territory.[34] Specific obligations concerning information exchange have crystallised in international water law. Depending on whether they are mutual or unilateral information-sharing duties, and based on their regular or ad hoc nature and their content, they can be grouped into three categories: (1) regular exchange of data and information, (2) notification of natural emergencies and those caused by human activity and (3) notification of planned measures.

33 McCaffrey, *The Law of International Watercourses*, p. 464.

34 See, for example, Art. 5, ILA *Montreal Rules on Water Pollution in an International Drainage Basin*, which differentiates among these three reasons to provide information: 'Basin States shall: (a) inform the other States concerned regularly of all relevant and reasonably available data, both qualitative and quantitative, on the pollution of waters of the basin, its causes, its nature, the damage resulting from it, and the preventive procedures; (b) notify the other States concerned in due time of any activities envisaged in their own territories that may involve a significant threat of, or increase in, water pollution in the territories of those other States; and (c) promptly inform States that might be affected, of any sudden change of circumstances that may cause or increase water pollution in the territories of those other States'.

4.2.1 *Regular data and information exchange*

Regular exchange of data and information is a duty closely linked to the effectiveness of cooperation on transboundary water systems. Information is essential for the administration and sustainable development of rivers. It is also of crucial importance to achieve equitable and reasonable utilisation and to avoid causing harm.[35] When only partial information is available on a shared water system, a meaningful assessment on whether a specific water use is equitable, reasonable or optimal cannot be carried out. Information sharing, therefore, is a fundamental requirement for the achievement of water governance objectives aimed towards efficient, equitable and sustainable solutions.[36]

Evolution of the duty

Acknowledging the importance of regular data exchange for successful cooperation on and effective management of transboundary waters, the *UN Watercourses Convention* placed this obligation in Part II on general principles. It affirmed that the obligation constitutes a specific application of the general duty to cooperate: '*Pursuant to Article 8 [the general duty to cooperate]*, watercourse States shall on a regular basis exchange readily available data and information on the condition of the watercourse . . . ' [emphasis added].[37]

Appreciation of the necessity of information exchange in instruments with intended universal scope crystallised over time and hand in hand with the general duty to cooperate. Whereas the 1961 IIL *Salzburg Resolution*[38] does not yet mention such a duty at all, Article XXIX of the 1966 ILA *Helsinki Rules*[39] included a recommendation to exchange readily available information first to prevent disputes. A recommendation to

[35] ILC, 'Third Report on the Law of the Non-navigational Uses of International Watercourses, by Mr. Stephen McCaffrey, Special Rapporteur' (1987) II(1) *YBILC*, para. 94.

[36] R. McDonnell, 'Challenges for Integrated Water Resources Management: How Do We Provide the Knowledge to Support Truly Integrated Thinking?' (2008) 24 *International Journal of Water Resources Management* 132f.

[37] ILC, 'Draft Articles . . . and Commentaries', p. 107.

[38] IIL, *Annuaire – Session de Salzbourg*, pp. 381–384.

[39] '1. With a view to preventing disputes from arising between basin States as to their legal rights or other interests, it is recommended that each basin State furnish relevant and reasonably available information to the other basin States concerning the waters of a drainage basin within its territory and its use of, and activities with respect to such waters'. ILA, *Report of the Fifty-Second Conference*, p. 518.

water system States to 'as soon as possible' collect and exchange 'physical and economic data essential for the planning and realization of the rational use of waters' is also included in Article V of the *Declaration of Buenos Aires of the Inter-American Bar Association*.[40] The ILA and IIL subsequently introduced regular data exchange as an obligation in their rules on pollution of international water systems in the 1976 IIL *Athens Resolution*[41] and in the 1982 ILA *Montreal Rules*.[42] The ILC recognised the need for the availability of information and its exchange for effective basin planning early in its discussions.[43] In his first report, Special Rapporteur Schwebel introduced Articles on data collection, exchange and allocation of costs of these activities,[44] which were later refined by the ILC.[45] Data exchange related to water quality was not yet considered in Article 9 of the 1994 *Draft Articles*. To correct this omission, it was included in the text of the *UN Watercourses Convention* during negotiations in the Drafting Committee.[46] With respect to the duty's location within the structure of the *Draft Articles*, it is noteworthy that the duty was first included among the obligations implementing the general principle of cooperation.[47] Only during the elaboration of the final draft in 1994 was it moved to the section on general principles governing non-navigational uses of international watercourses, in which it now also appears as Article 9 of the *UN Watercourses Convention*. The 2008 ILC *Draft Articles* also include an

40 Inter-American Bar Association, *Proceedings of the Tenth Conference, Buenos Aires, 14–21 November 1957* (Buenos Aires: Inter-American Bar Association, 1958), pp. 246–248.

41 Article VI: '*Dans l'exercice de leur devoir de coopération, les Etats faisant partie d'un même basin hydrographique doivent, dans toute la mesure du possible, notamment par la voie d'accords, recourir aux modalités suivantes de coopération: a) informer régulièrement les autres Etats riverains sur toutes les données appropriés relatives à la pollution du bassin, ses causes, sa nature, les dommages qu'elle provoque et les modalités de prévention*', in IIL, *Annuaire – Session d'Athènes*, p. 360.

42 Article 5: 'Basin states shall; (a) inform the other states concerned regularly of all relevant and reasonably available data, both qualitative and quantitative, on the pollution of waters of the basin, its causes, its nature, the damage resulting from it, and the preventive procedures'; in ILA, *Report of the Sixtieth Conference*, p. 540.

43 ILC, 'First Report on the Law of the Non-navigational Uses of International Watercourses, by Mr. Stephen Schwebel, Special Rapporteur' (1979) II(1) *YBILC*, para. 111.

44 *Ibid.*, paras. 123, 131 and 134.

45 ILC, 'First Report, Evensen', paras. 137–151; and ILC, 'Fourth Report on the Law of the Non-navigational Uses of International Watercourses, by Mr. Stephen C. McCaffrey, Special Rapporteur' (1988) II(1) *YBILC*, paras. 11–27.

46 Sixth Committee, *Summary Record of the 24th Meeting of the Sixth Committee*, 25 October 1996, UN Doc. A/C.6/51/SR.24, para. 21; Tanzi and Arcari, *The United Nations Convention*, p. 196.

47 ILC, 'First Report, Evensen', paras. 102–151.

obligation to regularly exchange data and information among the principles applicable to the law of transboundary aquifers.[48] The two provisions of the *UN Watercourses Convention* and the 2008 ILC *Draft Articles* are almost identical yet not completely congruent; the latter replaces the words 'and related to the water quality' of Article 9 of the *UN Watercourses Convention* with 'and related to the hydrochemistry' in the first paragraph and adds two new categories to be provided: geological and hydrogeological data. It also includes an additional paragraph that elaborates on data collection.[49]

States frequently acknowledge the necessity of regular information exchange for effective management and protection of shared water systems in international water treaties. An analysis of water treaties since 1900 showed that about 39 per cent of the agreements evaluated include a clause on regular information sharing.[50] The content of these provisions, however, is not identical: some are general, lacking a description of what kind of data should be shared and when; others, such as Article VI of the *Indus Treaty*[51] – which defines both the content and frequency of data exchange – are very precise. The basic obligation common to all of them is the mutual and regular exchange of data with the objective of informing the process of State (inter)-action on shared water resources.

The *Indus Water Treaty* is an example that is frequently cited as exemplary in this context. According to Article VI, the treaty parties are obligated to exchange, on a monthly basis, data that is related to the gauge and discharge to Indus tributaries, daily extraction and release from reservoirs, and daily flow volume in canals. Most other treaties are less precise than the *Indus Water Treaty* relative to the data to be shared. Numerous boundary-water treaties that the former Soviet Republic concluded with neighbouring States included provisions on regular sharing of data and information regarding water levels and hydrometeorological measurements, but they do not define data in more detail.[52] Other

48 Art. 8, 2008 ILC *Draft Articles*.

49 Art. 8 (2), 2008 ILC *Draft Articles*: '2. Where knowledge about the nature and extent of a transboundary aquifer or aquifer system is inadequate, aquifer States concerned shall employ their best efforts to collect and generate more complete data and information relating to such aquifer or aquifer system, taking into account current practices and standards. They shall take such action individually or jointly and, where appropriate, together with or through international organizations'.

50 See Annex in this volume.

51 *Indus Water Treaty*, 19 September 1960, 419 *UNTS* 125.

52 Art. 3 (10), *Agreement Concerning the Use of Water Resources in Frontier Waters*, 17 July 1964, 552 *UNTS* 186; Art. 14, *Treaty between the Soviet Union and Iran Concerning the Regime of the Soviet–Iranian Frontier and the Procedure for the Settlement of Frontier Disputes and Incidents*,

treaties that established obligations for regular data exchange include the 1992 *UNECE Water Convention*,[53] the 2000 *SADC Revised Protocol on Shared Watercourses*,[54] and the 2002 *Senegal Water Charter*.[55]

Although it is now established that the notification obligations relative to emergencies and planned measures (to be discussed herein) form part of customary international law,[56] regular exchange of data and information does not constitute universal practice, and its nature as a customary obligation is debated. The ILA concluded that although an obligation to share available information is generally accepted when there is an immediate problem, opinions remain divided about the obligation with respect to regular exchange.[57] The ILC does not discuss the nature of this obligation in its commentaries to the 1994 *Draft Articles*. In State practice, regular data exchange takes place only on the basis of international agreements or other arrangements. For the European continent, the existence of a regional custom of information exchange can be derived from wide State participation in the 1992 *UNECE Water Convention*[58] and the 2000 *European WFD*.

Quality and content of data provided

The texts of Article 9 of the *UN Watercourses Convention* and Article 8 of the 2008 ILC *Draft Articles* shed light on the content of the obligation to exchange information. These provisions explain the substance and quality of information, as well as the standard of diligence to be applied in order to make data available. They describe minimum requirements needed to assess what constitutes an equitable and reasonable use[59] and

14 May 1957, 457 *UNTS* 161; Art. 5, *Agreement between the Soviet Union and China on Joint Research Operations to Determine the Natural Resources of the Amur River Basin and the Prospects for Development of its Productive Potentialities and on Planning and Survey Operations to Prepare a Scheme for the Multi-Purpose Exploitation of the Argun River and the Upper Amur River*, 18 August 1956, 18 *SDD* 323; Art. 19, *Treaty between the Governments of the Soviet Union and Hungary Concerning the Regime of the Soviet–Hungarian State Frontier and Final Protocol*, signed 24 February 1950, 14 *SDD* 29; Art. 19, *Treaty between the Soviet Union and Romania Concerning the Regime of the Soviet–Romanian State Frontier and Final Protocol*, signed 25 November 1949, 14 *SDD* 157; Art. 19, *Agreement between the Soviet Union and Poland Concerning the Regime of the Soviet–Polish State Frontier*, 8 July 1948, 37 *UNTS* 25.

53 Arts. 6, 11 and 13, 1992 *UNECE Water Convention*.

54 Art. 3 (6), 2000 *SADC Revised Protocol*.

55 Art. 7, 2002 *Senegal Water Charter*.

56 Bourne, 'Procedures: Notice and Exchange of Information', pp. 204–206.

57 Finnish Branch of the ILA, *The Work of the International Law Association*, p. 289.

58 Ratified by thirty-six UNECE member States and the EU.

59 ILC, 'Draft Articles . . . and Commentaries', p. 107.

to prevent potential harm.[60] The list of required data provided in the first paragraph of both Articles is non-exhaustive. Provision of geological and hydrogeological information for aquifer systems in particular, as well as hydrological, meteorological, ecological and quality information for all water systems, fulfills basic information needs for effective management of transboundary water resources. The final paragraphs of the provisions specify that States shall 'employ their best efforts' in collecting and processing data and shall do this in a manner that facilitates utilisation of the data and information by other States.[61] At the very least, information should meet adequate quality standards so that the receiving State knows more than it did before about the water system and the interdependence created by the shared waters.[62] This is required based on the assumption that, provided availability of means, the systematic and ongoing process of information exchange that follows from the requirement of 'regular' data exchange would lead to standardisation of information and data-collection systems because the process gives States the opportunity to realise new possibilities and ways to coordinate their activities.[63] Standardisation of data collection on a basin-wide scale will give States the benefit of receiving comparable data that they can then easily process within their own information management systems.

At the same time, requirements regarding the content and scope of data and information remain limited by considerations of national defence and security. Unless they agree to it, States are not obliged to provide confidential material relating to national defence.[64] The *UN Watercourses Convention*[65] and the 2008 ILC *Draft Articles*[66] therefore include provisions that establish an exception to the obligation of regular data and information exchange.[67] This limitation may be problematic in its practical

60 Tanzi and Arcari, *The United Nations Convention*, p. 195.

61 Art. 9 (3), 1997 *UN Watercourses Convention*, and Art. 8 (4), 2008 ILC *Draft Articles* differ slightly in their wording. Article 9 (3): 'Watercourse States shall employ their best efforts to collect and, where appropriate, to process data and information in a manner which facilitates its utilization by other watercourse States to which it is communicated'. Article 8 (4): 'Aquifer States shall, where appropriate, employ their best efforts to collect and process data and information in a manner that facilitates their utilization by the other aquifer States to which such data and information are communicated'.

62 Bourne, 'Procedures: Notice and Exchange of Information', pp. 199f.

63 Ely and Wolman, cited in McCaffrey, *The Law of International Watercourses*, p. 479.

64 ILC, 'Third Report, McCaffrey', para. 109.

65 Art. 31, 1997 *UN Watercourses Convention*.

66 Art. 19, 2008 ILC *Draft Articles*.

67 This exception stipulated in the *UN Watercourses Convention* and the 2008 ILC *Draft Articles* also applies to obligations for *ad hoc* information obligations discussed herein.

application: there is a risk that it may be used as a protective shield to limit information exchange relative to hydrologic data that is not confidential or vital for national defence but which may nevertheless be withheld for political reasons.

Following a different line of argument, Article 12 (5) of the 1994 *Danube River Protection Convention*[68] established limits regarding information 'related to personal data [and] intellectual property including industrial and commercial secrecy', in addition to national security. In case a party to the *Convention* nevertheless decides to provide protected data, the receiving party is obliged to handle it in a manner that gives due respect to its confidentiality.[69] The *Convention*, at the same time, obliges States to provide 'as much information as possible under the [given] circumstances'.[70] A breach of this obligation (as well as of the obligation to regularly exchange readily available data and information more generally), however, may be difficult to establish because the recipient does not know which information is available to the provider. Individual States may choose to share only minimal information.[71]

Availability of data and the duty to collect

The initial obligation consists of the exchange of readily available data on the condition of the shared system. This criterion of 'readily available', which was the cause of extensive discussion during negotiations of the *UN Watercourses Convention*,[72] was inserted in the *Convention* to reflect that the obligation should not become too onerous for individual States to carry out and that it should correspond to the respective capacities of States to produce data.[73] A riparian State must provide only such information as is 'readily at its disposal';[74] this includes data that is collected for the State's own purposes and hence is readily available to the authorities.[75] The criterion gives due consideration to feasibility of data generation and reasonableness[76] of the effort to be employed. It is common in

68 *Convention on Cooperation for the Protection and Sustainable Use of the Danube River (Danube Protection Convention)*, 29 June 1994, OJ 1997 L342/19.
69 Art. 12 (6), 1994 *Danube Protection Convention*.
70 Art. 31, 1997 *UN Watercourses Convention*, and Art. 19, 2008 ILC *Draft Articles*.
71 Okowa, 'Procedural Obligations', p. 301.
72 For detail, see Tanzi and Arcari, *The United Nations Convention*, p. 198.
73 ILC, 'Third Report, Schwebel', paras. 191f.
74 ILC, 'Draft Articles . . . and Commentaries', p. 108.
75 Tanzi and Arcari, *The United Nations Convention*, p. 198.
76 The standard of 'reasonably available data' is employed by Art. 13, 1992 *UNECE Water Convention*.

multi-riparian water systems for individual system States to vary in their technical and financial capacity to generate water-related data. Paragraph (2) of Article 8 of the 2008 ILC *Draft Articles* therefore offers a remedy to imbalances in data collection capacity by stipulating that States shall collect data 'individually or jointly and, where appropriate, together with or through international organizations'. This encourages institutionalisation among system States as well as the promotion of cooperation with international organisations. However, the insertion of the phrase 'where appropriate' also provides some flexibility relative to the design of State cooperation. There is no obligation to institutionalise the process of data collection or to involve international organisations.

The obligation to collect data is inherent in the obligation of regular data and information exchange. The fact that the first obligation is implied in the latter is made explicit in the 2008 ILC *Draft Articles*. Article 8 (2) establishes a due diligence obligation to generate data about an aquifer system when current knowledge is inadequate. Article 9 of the *UN Watercourses Convention* lacks this more specific provision. Paragraph (3) refers to an obligation to collect data; yet, from the location of the provision within the Article – namely, after the paragraph establishing a due diligence obligation to provide data upon request by other States – and from the text of the provision itself, it is not clear whether this duty to collect usable data applies only to cases in which the information is requested by another system State or more generally.

A rule implicit in the obligation to collect data is the obligation to monitor system conditions, which was included in a number of treaties. Monitoring of water quality and water system conditions also necessarily implies obligations to generate and collect data. In this context, Bourne pointed to legal precedents in the US federal states systems; the US Supreme Court confirmed the existence of monitoring obligations in cases in which federal states hold water quantity rights on a shared water system. In the 1945 case of *Nebraska v. Wyoming*, the Supreme Court established that measuring devices and proper recordkeeping are required to know whether other obligations relative to quantity of water use and provision are being met.[77] The Court further decided that these records must be made available to co-basin States. Article 11 of the 1992 *UNECE Water Convention* also established obligations regarding joint monitoring of water quality and joint assessments on floods and ice drifts. The detailed

[77] Bourne, 'Procedures: Notice and Exchange of Information', pp. 198f.

description in Article VI of the 1960 *Indus Waters Treaty* on the topic of which data should be provided equally implies that this data must be generated to make possible regular assessments of system conditions by the other party.[78]

Timing and frequency of data exchange

Neither the relevant provisions of the *UN Watercourses Convention* nor the 2008 ILC *Draft Articles*, nor the ILC comments thereto, gives any indication regarding the frequency of data exchange. Both indicate that the term 'regular' 'provides for an ongoing and systematic process'.[79] Many treaties that refer to the obligation of regular data exchange are non-specific with regard to time intervals.[80] Provisions that specify time intervals for the information transfer – such as Article VI of the *Indus Waters Treaty*, which stipulates that available daily flow data must be transmitted on a monthly basis between parties – are rare. Yet, the treaties point towards a general duty to establish a continuous process of State cooperation over shared water resources. Time intervals and regularity must be considered based on what is reasonable for individual water systems and the factors to be monitored. At the least, data transfer must take place with such regularity that States are in possession of the information that is necessary to comply with obligations related to equitable and reasonable use, including environmental protection and prevention of significant harm.[81]

[78] Art. VI: '(1) The following data with respect to the flow in, and utilization of the waters of, the Rivers shall be exchanged regularly between the Parties: – (a) Daily (or as observed or estimated less frequently) gauge and discharge data relating to flow of the Rivers at all observation sites. – (b) Daily extractions for or releases from reservoirs. – (c) Daily withdrawals at the heads of all canals operated by government . . . – (d) daily escapages from all canals, including link canals. – (e) Daily deliveries from link canals. – These data shall be transmitted monthly by each Party to the other as soon as the data for a calendar month have been collected and tabulated, but not later than three months after the end of the month of which they relate: . . . '.

[79] ILC, 'Draft Articles on the Law of Transboundary Aquifers', 51; ILC, 'Draft Articles . . . and Commentaries', p. 108.

[80] Art. 3, 1964 *Agreement Concerning the Use of Water Resources in Frontier Waters*; Art. 14, 1957 *Treaty between the Soviet Union and Iran Concerning the Regime of the Soviet–Iranian Frontier and the Procedure for the Settlement of Frontier Disputes and Incidents*; Art. 19, 1950 *Treaty between the Governments of the Soviet Union and Hungary Concerning the Regime of the Soviet–Hungarian State Frontier and Final Protocol*; Art. 19, 1949 *Treaty between the Soviet Union and Romania Concerning the Regime of the Soviet–Romanian State Frontier and Final Protocol*; Art. 19, 1948 *Agreement between the Soviet Union and Poland Concerning the Regime of the Soviet–Polish State Frontier*.

[81] ILC, 'Draft Articles . . . and Commentaries', p. 107.

In general, treaties also make no clear reference to the form in and the means by which information transmission shall take place. The ILC comments that States are 'free to utilize for this purpose any mutually acceptable method'.[82] Given the nature of the information (e.g., hydrological and hydrochemical data), it seems plausible that it must be put in written form or transmitted electronically, when adequate technology exists.[83] The transfer can happen directly among States or via joint institutions that were established *inter alia* to coordinate information exchange.[84]

Indeed, in many cases, system States have entrusted joint bodies with tasks of joint data collection and information transfer among them. For example, the International Boundary and Water Commission, established in 1889 between the United States and Mexico, is mandated to keep records of water use by both countries.[85] To this end, the Commission is charged with the installation of gauging stations and systems necessary to collect the data. At the same time, States are under obligation to transfer information related to their diversions and consumptive uses to the Commission. Other examples of joint mechanisms charged with the collection of data and making it available to member States include the Niger Basin Authority[86] and the Mekong River Commission.[87] The pooling of data through joint institutions can improve data assessment and processing, and it increases transparency – and, thus, potentially trust – among member States.

Information exchange upon request

If a water system State is requested by another system State to provide data that is not readily available, the former is under a due diligence obligation to comply with the request. The State from which information is sought has the right to ask the requesting State for a contribution to

82 *Ibid.*, 107f.

83 In evolution of the argument made by Bourne in 1972. Bourne, 'Procedures: Notice and Exchange of Information', p. 200.

84 ILC, 'Draft Articles . . . and Commentaries', p. 107.

85 Arts. 2 and 9 (j), *Treaty between the United States of America and Mexico Relating to the Utilization of the Waters of the Colorado and Tijuana Rivers and of the Rio Grande*, 3 February 1944, *USTS* 994.

86 Art. (2), *Convention Creating the Niger Basin Authority*, 21 November 1980, 1346 *UNTS* 208.

87 Art. 24, *Agreement on the Cooperation for the Sustainable Development of the Mekong River Basin* (1995 Mekong Cooperation Agreement), 5 April 1995, 2069 *UNTS* 3.

or coverage of all reasonable costs of data collection.[88] A question arises concerning whether nationals of the requesting State can enter the territory of the State from which they are requesting information either to inspect information available there or to collect their own data – for instance, in a case in which the requested State does not comply. Numerous treaties confer such a right with respect to joint projects and joint studies.[89] For example, the 1949 exchange of notes by which Great Britain and Egypt agreed to the construction of the Owen Falls Dam in Uganda[90] provides that compliance with downstream interests in sustained water discharge during the construction phase and operation of the dam will be monitored by a resident Egyptian engineer at the site. The 1933 *Montevideo Declaration by the Pan American Union Concerning the Industrial and Agricultural Use of International Rivers*[91] called for the right of States interested in developing hydropower potential to collect information on the usability of shared waters for this purpose in the territory of other States. If those States are unwilling to provide the data directly, they are obligated to facilitate access to their territory. Treaty practice does not support the existence of a customary right for an interested State to access the territory of another State that is not willing to provide such information. Such a right would be contrary to the principle of inviolability of State territory[92] and is therefore only possible if explicitly provided for in a treaty.

In cases in which direct data exchange is impossible due to hostilities or other serious obstacles to direct contact between States, the *UN Watercourse Convention* provides that the States are obligated to find mutually acceptable indirect procedures for data exchange.[93] This provision highlights the importance of the systematic process of information exchange, affirming that it should not be interrupted even in times of armed conflict.

[88] Art. 4, *Agreement between the People's Republic of Bulgaria and the Republic of Turkey Concerning Cooperation in the Use of Waters of Rivers Flowing through the Territory of Both Countries*, 23 October 1968, 807 *UNTS* 117.

[89] Bourne, 'Procedures: Notice and Exchange of Information', p. 105.

[90] Para. 4, *Exchange of Notes Constituting an Agreement between Great Britain and Egypt Regarding the Construction of the Owen Falls Dam, Uganda*, 30 and 31 May 1949, 226 *UNTS* 274.

[91] Pan American Union, *Seventh International Conference of American States, Plenary Sessions, Minutes and Antecedents* (Montevideo: Pan American Union, 1933), p. 114.

[92] Bourne, 'Procedures: Notice and Exchange of Information', 203; *Corfu Channel Case* (1949) ICJ Reports, pp. 33–35.

[93] The obligation to use indirect procedures of Art. 30, 1997 *UN Watercourses Convention*, also applies to obligations to notify, consult and negotiate.

4.2.2 *Notification of emergencies*

Inherent in the necessity to regularly inform on conditions of water systems and changes therein for effective water resources management is the need to exchange information on 'one-off' events that affect the overall condition of hydrologic and related ecosystems. These ad hoc notification obligations are intended primarily to prevent or mitigate harm; their nature as obligations of customary law is widely accepted.[94] Numerous treaties focus on or are limited to ad hoc information obligations regarding the notification of events that might cause adverse impact in the territory of another system State.[95] The events regularly referred to in these treaties fall into two broad categories: (1) emergencies due to natural causes or human conduct, and (2) measures planned by individual water system States. Notification of these events informs the process of cooperation with respect to individual or joint response to emergencies, on the one hand, and aims to ensure that interests and rights of States that are potentially affected by planned measures are taken into account, on the other hand. The notification of emergencies is addressed first.

Emergencies occur due to natural causes (e.g., floods and ice conditions) and/or human activity on or in proximity to a transboundary water system. This happened in October 2010 when toxic sludge spilled from a reservoir at an alumina plant in western Hungary and polluted several villages, eventually reaching the Danube River through a tributary.[96]

[94] Okowa, 'Procedural Obligations', pp. 281 and 319; C. Bourne, 'Procedure in the Development of International Drainage Basins: The Duty to Consult and to Negotiate' (1972) 10 *Canadian Yearbook of International Law*, 218; Finnish Branch of the ILA, *The Work of the International Law Association*, pp. 211f and 231ff; McCaffrey, *The Law of International Watercourses*, p. 473. For jurisdiction of the ICJ on the customary character of the obligation to notify, see *Pulp Mills on the River Uruguay Case (Argentina v. Uruguay)* (2010) ICJ Reports, paras. 101f; *Case Concerning Military and Paramilitary Activities in and against Nicaragua (Nicaragua v. United States of America), Merits* (1986) ICJ Reports, para. 215; *Corfu Channel Case (United Kingdom of Great Britain and Northern Ireland v. Albania)* (1949) ICJ Reports, p. 22; in support, see GA *Res. 2995 (XXVII)* on *Cooperation between States in the Field of Environment*, 15 December 1972, UN Doc. A/RES/2995(XXVII).

[95] *Convention Concerning Water Economy Questions Relating to the Drava*, 25 May 1954, 227 *UNTS* 128; *Agreement Concerning Water Economy Questions*, signed 4 April 1958, 367 *UNTS* 104; *Agreement between Greece and Yugoslavia Concerning Hydro-economic Questions*, 18 June 1959, 363 *UNTS* 135; slightly more limited: 1944 *Treaty Relating to Utilization of Waters of the Colorado and Tijuana Rivers, and of the Rio Grande*; 1960 *Indus Waters Treaty*; 1976 *Convention on the Protection of the Rhine against Chemical Pollution*; 1995 *Mekong Cooperation Agreement*.

[96] BBC, 'Danube Toxic Sludge Operation Reaches Crucial Day', *BBC Online*, 8 October 2010, www.bbc.co.uk/news/world-europe-11498884.

Emergencies are defined as 'situation[s] that cause, or pose an imminent threat of causing, serious harm'.[97] The principle of good neighbourliness forms the legal basis for the obligation of States in whose territory an emergency occurs to notify States that might be adversely affected by it.

When and whom to notify

The purpose of the obligation of notification of emergencies is to give affected States the opportunity to react with a view to preventing or mitigating damage in their territory. For this reason, notification must be timely and must provide all necessary information. This is reflected in Principle 18 of the 1992 *Rio Declaration on Environment and Development*, which requires 'immediate' notification to potentially affected States. Article 28 of the *UN Watercourses Convention* demands notification 'without delay and by the most expeditious means available'.[98]

Some agreements stipulate that notification shall be provided not only to affected States but also to joint mechanisms. Article 10 of the1995 *Mekong Agreement* provides that the Joint Committee shall be informed so that immediate remedial action can be taken not only on a bilateral but also a sub-basin level. Furthermore, the obligation of notification in the case of emergencies is not restricted to other water system States but extends also to potentially affected non-riparian States.[99] The obligation to notify of emergencies is not specific to international water law but is an obligation appertaining to international environmental law in general.[100] As expressed in Principle 18 of the 1992 *Rio Declaration*:

> States shall immediately notify other States of any natural disasters or other emergencies that are likely to produce sudden harmful effects on the environment

[97] Art. 28 (1), 1997 *UN Watercourses Convention*; Art. 16 (1), 2008 ILC *Draft Articles* adds that emergency situations 'result suddenly'.

[98] Also required, for example, in Art. 10, 1995 *Mekong Cooperation Agreement*, and Art. 4 (5), 2000 *SADC Revised Protocol*.

[99] For example, Art. 28 (2), 1997 *UN Watercourses Convention*; Art. 17 (2), 2008 ILC *Draft Articles*. In contrast, according to Art. 14, 1992 *UNECE Water Convention*, Art. 10, 1992 *UNECE Convention on the Transboundary Effects of Industrial Accidents*, 18 March 1992, 2105 *UNTS* 457, and Art. 10, 1995 *Mekong Cooperation Agreement*, the notification obligation applies between parties.

[100] M. Farrajota, 'Notification and Consultation in the Law Applicable to International Watercourses', in Boisson de Chazournes and Salman (eds.), *Les ressources en eau*, pp. 319, 322; Okowa, 'Procedural Obligations', 330–332; P. Birnie and A. Boyle, *International Law and the Environment*, 3rd edition (New York: Oxford University Press, 2009), pp. 322f.; Bourne, 'Procedures: Notice and Exchange of Information', p. 183.

of those States. Every effort shall be made by the international community to help States so afflicted.

Duty to minimise damage

This obligation must be seen in the wider context of the obligation to cooperate in the protection of transboundary water systems. It is linked to the corollary duty to adopt both individually and jointly all necessary measures to prevent and mitigate harm. The State in whose territory an emergency occurs and the State that is affected share a duty to take all necessary measures to minimise damage.[101] This corollary duty is expressed in Paragraph 3 of Article 28 in the *UN Watercourses Convention* and Article 17 of the 2008 ILC *Draft Articles* on the law of transboundary aquifers. Water system States shall 'in cooperation with potentially affected States and, where appropriate, competent international organizations, immediately take all measures necessitated by the circumstances to prevent and mitigate and eliminate harmful effects of the emergency'.[102]

This duty also applies to the time before the actual occurrence of an emergency.[103] Water system States are already under obligation to prevent the occurrence of emergencies. The risk of occurrence of natural emergencies (e.g., floods and dangerous ice conditions) can be mitigated, for example, by cooperation on flow regulation in watercourses[104] or through regular exchange of information on water levels.[105] The risk of the spread of dangerous waterborne diseases can best be mitigated through national and joint monitoring and information exchange systems.[106]

Damage to water systems caused by human activity (e.g., the risk of pollution through industrial activities on or in proximity to water systems) can be averted or mitigated by ensuring that industrial facilities comply with state-of-the-art technical and security standards. The 1992 *UNECE Convention on the Transboundary Effects of Industrial Accidents* is an example of a

101 Bourne, *ibid.*, 186; with respect to pollution, see Lammers, *Pollution of International Watercourses*, pp. 342, 352.

102 Art. 28 (3), 1997 *UN Watercourses Convention*; Art. 17 (2) (b), 2008 ILC *Draft Articles*.

103 For example, Art. 27, 1997 *UN Watercourses Convention*.

104 Art. 25, 1997 *UN Watercourses Convention*.

105 Art. 19, 1950 *Treaty between the Governments of the Soviet Union and Hungary Concerning the Regime of the Soviet–Hungarian State Frontier and Final Protocol*.

106 Farrajota, 'Notification and Consultation', p. 321; Tanzi and Arcari, *The United Nations Convention*, p. 223.

treaty that aims to prevent emergencies with transboundary impact that are due to human conduct through close cooperation of State parties; the *Convention* also applies to transboundary waters. It includes the obligation of notification of industrial accidents[107] and embeds it in a comprehensive legal framework that aims to avoid the occurrence of such accidents and provide for cooperation in the event of one.[108]

The case of pollution

The prevention of damage by pollution is a generally well-regulated issue in international water law. Many legal instruments have been adopted to address this question of water management,[109] which reflects the fact that cooperation in this matter tends to have little controversy because riparian States generally share an interest in good water quality. Notification obligations regarding pollution concern both types of ad hoc information obligations: (1) emergencies in the case of accidents, and (2) planned measures that will or could pollute waters.[110] The IIL 1979 *Athens Resolution* concerning the pollution of international rivers and lakes,[111] for example, stipulates that '[in] carrying out their duty to cooperate, States bordering the same hydrographic basin shall, as far as practicable, especially through agreements, resort to' *inter alia* (1) regular data exchange regarding the pollution of a basin, (2) notification of planned activities that may involve a significant threat of transboundary pollution and (3) prompt notification of sudden increases in pollution levels.[112] The 1979 *Athens Resolution* also provided for other cooperation obligations, including consultations

[107] Art. 10 (2), 1992 *Convention on the Transboundary Effects of Industrial Accidents*; the *Convention* applies to impacts of industrial accidents not only related to water. The *Convention* does not apply to 'dam failures, with the exception of the effects of industrial accidents caused by such failures' (Art. 2 (2)).

[108] Arts. 6–17, *ibid.*

[109] For example, *Framework Agreement on the Sava River*, 3 December 2002, 2366 *UNTS* 479; 2000 *EU WFD*; 1994 *Danube Protection Convention*; *Treaty Relating to Boundary Waters, and Questions Arising between the United States and Canada (Boundary Waters Treaty)*, 11 January 1909, *British Treaty Series 1910*, No. 3; a list of eighty-eight other treaties relating to water pollution and dating from 1868 to 1979 is provided in Lammers, *Pollution of International Watercourses*, pp. 124–147.

[110] IIL, *Yearbook: Session of Athens 1979 – Preparatory Works* (Basel: S. Karger, 1980), Part I, pp. 236–239.

[111] IIL, *Yearbook: Athens Session 1979 – Deliberations of the Institute during Plenary Meetings* (Basel: S. Karger, 1980), Part II, pp. 196–203.

[112] Art. VII, *Resolution on the Pollution of Rivers and Lakes and International Law*, *ibid*. See also Art. 5, 1982 ILA *Rules on Water Pollution in an International Drainage Basin* in ILA, *Report of the Sixtieth Conference*, pp. 531–548.

on actual or potential pollution problems, establishment of joint research and monitoring programs, adoption of relevant legislation and common quality standards and establishment of joint mechanisms. A subsequent ILA Resolution on the same topic, the 1982 *Montreal Rules on Water Pollution in an International Drainage Basin*, and the 1979 IIL text both emphasise the importance of State cooperation relative to water pollution, pointing out that this problem can be addressed most effectively by joint action. The issue of cooperation relative to pollution and water quality control measures is addressed in more detail in Chapter 5.

4.2.3 Notification of planned measures

Pollution is one type of potential harm that can result from planned measures, triggering a notification obligation. In the specific context of international water law, the obligation of notification of planned measures exists not only for situations in which other riparian States are under threat of significant damage to the environment of their territory but also in which other rights and interests are at risk of being significantly affected. The threshold of harm that triggers different notification obligations is not always the same; treaty texts establish nuanced differences and the ILC applies different thresholds concerning emergency situations and planned measures. Regarding the former, States are obligated to notify potentially affected States about emergencies that cause or pose a risk of causing 'serious harm'.[113] When planning new measures, States must notify other watercourse States in case of risk for 'significant adverse effects'.[114] According to the ILC, the threshold of 'significant adverse effects' is lower than that of 'serious' or 'significant' harm.[115] This lower threshold for planned measures was introduced in order to require notification even before legally significant harm might occur.[116]

Potential harm or adverse impact of planned measures can occur in different ways and manifestations. It can be in the form of physical harm – for instance, through change in flow patterns and/or inundation of parts of a territory – or it can occur in the form of non-physical harm, such as regarding current or future rights. Upstream plans for consumptive water use may conflict with the obligation to prevent significant harm in instances in which the plans reduce flow volumes downstream to an

[113] Art. 28, 1997 *UN Watercourses Convention*; Art. 17, 2008 ILC *Draft Articles*.
[114] Art. 12, 1997 *UN Watercourses Convention*; Art. 15, 2008 ILC *Draft Articles*.
[115] ILC, 'Draft Articles . . . and Commentaries', p. 111.
[116] McCaffrey, *The Law of International Watercourses*, p. 473.

extent that impedes existing downstream uses. At the same time, any new use by one riparian State creates a new existing use that will be assessed against any future use planned by another riparian State. The concept is interesting in two respects: first, because it contradicts an often-held assumption that harm travels downstream in transboundary basins; and, second, because it highlights that harm is not always physical and can be intertemporal – that is, it can travel across generations, from present to future. This type of harm can travel downstream as well as upstream.[117] In discussing the two-way phenomenon of harm in the context of transboundary water systems, McCaffrey and Salman pointed to the fact that by implementing planned measures, States create 'facts on the ground' that may foreclose or restrict future uses of other water system States.[118] This foreclosure of future use rights may be considered as constituting significant harm.[119]

The obligation to notify those water system States that are at risk of significant damage caused by planned measures forms part of customary international law.[120] GA *Resolution 2995 (XXVII)*[121] and Principle 19 of the *Rio Declaration*[122] formulate the obligation more generally for the environment as an obligation that aims to prevent significant adverse transboundary environmental impact. *Resolution 2995* clarifies that notification 'will be given and received in the best spirit of cooperation and good neighborliness, without this being construed as enabling each State to delay or impede the programs and projects of' the notifying State. In the spirit of good neighbourliness, the *UN Watercourses Convention* also

[117] Upstream harm caused by downstream utilisation also occurs when a dam is constructed close to the border on a consecutive river and where the reservoir inundates the territory of an upstream riparian.

[118] S. Salman, 'Downstream Riparians Can also Harm Upstream Riparians: The Concept of Foreclosure of Future Uses' (2010) 34 *Water International* 4, 350–364; McCaffrey, *The Law of International Watercourses*, pp. 474f.; S. McCaffrey, 'Some Developments in the Law of International Watercourses', in M. Kohen (ed.), *Promoting Justice, Human Rights and Conflict Resolution through International Law – Liber Amicorum Lucius Caflisch* (Leiden: Koninlijke Brill NV, 2007), pp. 785–791.

[119] McCaffrey, 'Some Developments', p. 788.

[120] McCaffrey, *The Law of International Watercourses*, p. 473; G. Sauser-Hall, '*L'utilisation industrielle des fleuves internationaux*' (1953) 83 *RCADI* 568; Okowa, 'Procedural Obligations', 289; McIntyre, *Environmental Protection*, p. 340.

[121] 15 December 1972, UN Doc. A/RES/2995(XXVII).

[122] Principle 19, *Rio Declaration on Environment and Development*: 'States shall provide prior and timely notification and relevant information to potentially affected States on activities that may have a significant adverse transboundary environmental effect and shall consult with those States at an early stage and in good faith'.

stipulates more general information obligations related to planned measures, in addition to the ad hoc notification obligation.[123] Article 11 provides the following:

> Watercourse States shall exchange information and consult each other and, if necessary, negotiate on the possible effects of planned measures on the condition of an international watercourse.

In stipulating that information shall be exchanged on all potential effects – adverse as well as beneficial – the obligation follows the imperatives of sound water resources management.[124]

Timing of notification

The timing of notification is an important aspect for the success of meaningful cooperation. Notification only after design plans have been finalised and after authorisation for construction (when necessary) has been obtained makes it more difficult to adapt plans with a view to accommodating interests raised by other riparian States to mitigate risks of significant harm. If it is found that interests raised by affected riparian States are legitimate and must be considered, early notification avoids the risk of additional redesign costs to the planning States. Notification after implementation steps have already been taken (e.g., the beginning of infrastructure construction) is particularly problematic. Notification delayed to this point in time infringes on customary law and puts notified States into a situation in which they are faced with established realities; the room for manoeuvring and the space for cooperation are thereby severely limited. When there are legitimate reasons to object to such a project, the notified State has a dilemma; a request for reversal of project implementation faces the unlikelihood of being respected by the planning State. If the planning State continues implementation without adjusting to the legitimate objection, the notified State must wait until significant harm actually materialises to be able to claim compensation.[125]

123 Art. 12, 1997 *UN Watercourses Convention.*

124 ILC, 'Draft Articles . . . and Commentaries . . . ', p. 111; similar information obligations with respect to planned measures are in Art. XIV, *Treaty Relating to Cooperative Development of the Water Resources of the Columbia River Basin*, 17 January 1961, 542 *UNTS* 246; Art. 5, 1999 *Convention on the Protection of the Rhine*; and Art. 12, 2002 *Tripartite Interim Agreement for Cooperation on the Protection and Sustainable Utilization of the Water Resources on the Incomati and Maputo Watercourses*, 29 August 2002, LEX-FAOC034943.

125 Art. 7 (2), 1997 *UN Watercourses Convention.*

It is up to the States to agree on a precise point in time when notification must take place. This was done, for example, by Great Britain and Belgium in an agreement regarding water rights on the border of their former colonies, Tanzania and Rwanda-Burundi.[126] They agreed that in the event that either Government desired to use waters of a boundary river, they would have to notify the other '6 months before commencing operations for the utilization'.[127] The two colonial powers further agreed that such utilisation rights would be granted on the condition that gauging stations – open to inspection by both Governments – would be installed to verify the correct implementation of these rights.[128]

If no time period for prior notification was specified by treaty, reference to more general rules is necessary. Article 12 of the *UN Watercourses Convention* speaks of 'timely' notification. Provisions in other documents use the same expression[129] or other similarly vague terms, such as 'in good time',[130] 'promptly'[131] and 'as early as possible'[132] – or they contain no provision regarding the point in time when notification must be provided.[133] The ILC defined the term 'timely' as 'intended to require notification sufficiently early in the planning stages to permit meaningful consultations and negotiations . . . if such prove necessary'.[134] This definition should also be applied to those cases in which no other more specific provision was made;[135] this is necessary to allow for the reasonable conduct of cooperation. The ICJ also confirmed this in the *Pulp Mills on the River Uruguay* case and clarified that notification should happen at

[126] *Agreement between the United Kingdom and Belgium Regarding Water Rights on the Boundary between Tanganyika and Ruanda-Urundi*, 22 November 1934, 190 *LNTS* 95; other provisions stipulating specific points in time for notification are found in Annexures D, paras. 6, 7, 9, 10, 12, 13, 20 and E, paras. 4–6, 12, 13, 15, 16, 25, 1960 *Indus Waters Treaty*; and in the Third Article, second para., *Agreement for the Full Utilization of the Nile Waters*, 8 November 1959, 453 *UNTS* 51.

[127] Para. 6, 1934 *Agreement Regarding Water Rights on the Boundary between Tanganyika and Ruanda-Urundi*.

[128] Para. 7, *ibid*.

[129] Art. 20, 2008 *Niger Basin Water Charter*; and Art. 24, 2002 *Senegal Water Charter*.

[130] *Agreement with Final Protocol Regulating the Withdrawal of Water from Lake Constance*, 30 April 1966, 620 *UNTS* 191.

[131] Art. 57, 2004 ILA *Berlin Rules* in ILA, *Report of the Seventy-First Conference*, pp. 340–411.

[132] Art. 3, 1991 *Espoo Convention*.

[133] Art. 6, 1966 *Agreement Regulating the Withdrawal of Water from Lake Constance*; 1975 *Statute of the River Uruguay*.

[134] ILC, 'Draft Articles . . . and Commentaries', p. 111.

[135] Tanzi and Arcari, *The United Nations Convention*, p. 203.

such a point in time as is required relative to the intended objective of notification.[136]

Whom to notify?

The customary obligation to notify exists with respect to all States that are potentially affected by significant adverse effects of planned measures.[137] Following from the previous discussion, this includes both downstream and upstream riparian States of international watercourses. Some legal instruments establish an obligation of notification of planned measures that is independent of the question of whether they cause significant harm. The potentiality of transboundary impact alone – regardless of whether it is positive or negative – triggers the treaty-based obligation. Article 9 of the 1992 *UNECE Water Convention*, for example, creates an obligation to establish joint bodies, which – among other duties – shall 'serve as a forum for the exchange of information on existing and planned uses of water and related installations that are likely to cause transboundary impact'.[138] If a planned measure benefits from World Bank funding, the organisation's safeguard policy for projects on international waterways requires that the planning State notify all other riparian States, independent of risk of harm to these States.[139]

Article 14 of the *Lake Tanganyika Agreement* requires States under whose jurisdiction or control certain activities are proposed that might 'give rise to transboundary adverse impacts' to notify the Secretariat of the joint-cooperation mechanism – that is, the Lake Tanganyika Authority.[140]

136 *Case Concerning Pulp Mills on the River Uruguay (Argentina v. Uruguay)* (2010) ICJ Reports, paras. 104f.

137 Bourne, 'Procedures: Notice and Exchange of Information', p. 175, citing legal instruments such as Art. 1 (3), 1957 *Resolution of the Inter-American Bar Association*; Arts. 4 and 5, 1961 IIL *Salzburg Resolution*; Art. XXIX, 1966 *Helsinki Rules*; 1954 *Recommendation of the Committee on Electric Power of the Economic Commission for Europe*, UN Doc. E/ECE/EP/14. See also Art. 12, 1997 *UN Watercourses Convention*; Art. 4 (1), 2000 *SADC Revised Protocol*; Art. 3, 1991 *Espoo Convention*; Art. 20, 2008 *Niger Basin Water Charter*.

138 Art. 9 (1) (h), 1992 *UNECE Water Convention*.

139 Operational Policy 7.50 – *Projects on International Waterways*, June 2001, as revised in March 2012, www.worldbank.org/safeguards; Bourne cites further the 1933 *Montevideo Declaration of the 7th International Conference of American States* (Arts. 6–7), and the 1965 *Draft Convention on the Industrial and Agricultural Use of International Rivers and Lakes of the Inter-American Juridical Committee* (Art. 8) as instruments that require notification irrespective of whether there is a risk of harm. Bourne, 'Procedures: Notice and Exchange of Information', p. 174.

140 *Convention on the Sustainable Management of Lake Tanganyika*, 12 June 2003, www.ltbp.org.

This provision shows that notification does not necessarily have to take place directly between two States; agreements on alternative notification procedures are possible. In fact, treaties establishing joint management mechanisms frequently mandate these mechanisms to take on the role of intermediary in the notification process. By these means, the joint mechanism remains the main platform through which information is collected, stored and shared. When more than two States are involved in the joint mechanism, the latter intervenes as a Secretariat and makes the notification process easier for the planning State. The Senegal River framework is a good example of this; other States are notified of projects via the High Commissioner's office of the OMVS,[141] which then distributes the information to the three other riparian States.

With regard to implementation of notification through joint mechanisms, the decision of the ICJ in the *Pulp Mills on the River Uruguay* case exemplifies that when States have agreed to notification via an intermediary, the agreed-upon procedure must be followed. According to Article 7 of the *Statute of the River Uruguay*, the party planning a project first must notify the CARU, which, in a preliminary assessment, will determine whether the planned work 'might cause significant damage to the other party'. If this is deemed to be the case, the planning party 'shall notify the other Party of the plan through the said Commission'.[142] In its decision in the *Pulp Mills* case, the ICJ found that because Uruguay had not adequately informed the CARU but instead notified Argentina directly of the results of the EIAs of the plants (and only after environmental authorisation for the plants in question had been granted), the State had failed to comply 'with its obligation to notify the plans to Argentina through CARU'.[143]

In the absence of notification, which might occur when the planning State estimates that no other State would be materially affected, a riparian State that nevertheless believes it is being subjected to a risk of significant damage may demand notification and provision of data from the planning State.[144]

[141] Art. 24, 2002 *Senegal Water Charter*. [142] Art. 7, para. 2, 1975 *Statute of the River Uruguay*.

[143] *Case Concerning Pulp Mills on the River Uruguay (Argentina v. Uruguay)* (2010) ICJ Reports, paras. 120–122.

[144] Art. 18, 1997 *UN Watercourses Convention*; *Lake Lanoux Arbitration (France v. Spain)*, 16 November 1957 (1958) XXIX *RGDIP* 115.

Content of notification

The obligation to notify interested States is not accomplished by effectuating a simple act of notification that a certain measure is being planned; it carries with it special requirements relating to the substance of the notification. Notification must contain sufficient information so that notified States are in a position to evaluate potential effects of the project.[145] As the decision of the Arbitration Tribunal in the *Lac Lanoux* case confirms, the potentially affected State is best placed to judge its own interests.[146] The precise content of notification naturally varies depending on the nature of the planned measure.[147] In some cases, States have agreed on specific content requirements for notification. For example, in Annexures D and E of the 1960 *Indus Water Treaty*,[148] India and Pakistan annexed several lists outlining data items to be provided for certain planned hydroelectric power stations and storage reservoirs; information was required about plant location, hydraulic and hydrologic data, design specifications and timing of implementation. In cases in which the treaties do not provide such detailed content requirements, Article 12 of the *UN Watercourses Convention* provides general guidance: it requires provision of 'available technical data and information, including the result of any environmental impact assessment' that would allow reasonable assessment by the notified State.[149] According to this rule, the notifying State must provide only information that has been collected in the planning of the measure and that is readily available.[150] In case a notified State is of the opinion that it needs additional information, and gathering the information would require further research and data collection by the planning State, the latter could ask for payment of reasonable expenses that arise in carrying out the request.[151]

145 Art. 12, 1997 *UN Watercourses Convention*; see also Okowa, 'Procedural Obligations', 299f.

146 *Lake Lanoux Arbitration (France v. Spain)* (1958) XXIX *RGDIP* 115: '*L'Etat exposé a subir les répercussions des travaux entrepris par un Etat limitrophe est seul juge des ses intérêts . . .*'

147 Okowa, 'Procedural Obligations', pp. 299f.

148 419 *UNTS* 125.

149 Art. 4 (1) (b), 2000 *SADC Revised Protocol*, and Art. 20, 2008 *Niger Basin Water Charter*, repeat these requirements, whereas Art. 13 (2), 2003 *Protocol for Sustainable Development of Lake Victoria Basin*, requires notification of 'technical data and information concerning the planned project to enable the notified . . . ' without specific reference to available EIAs. Art. 13 (1), 1992 *UNECE Water Convention* requires transmission of 'reasonably available data'.

150 ILC, 'Draft Articles . . . and Commentaries', p. 112.

151 Art. 14, 1997 *UN Watercourses Convention*; the ILC commentary of the 1994 *Draft Articles* states that it would be appropriate of the State that requests information to offer to

The requirement to transmit 'the results of any environmental impact assessment'[152] was strengthened by the ICJ decision in the *Pulp Mills on the River Uruguay* case. Whereas the *UN Watercourses Convention* stops short of prescribing that an impact assessment must be carried out, the *Pulp Mills* decision confirms that 'it may now be considered a requirement under general international law to undertake an environmental impact assessment where there is a risk that the proposed industrial activity may have a significant adverse impact in a transboundary context, in particular, on a shared resource'.[153] Restated in other terms, it was confirmed that the obligation to notify interested States of planned measures also implies a duty to assess the transboundary consequences of planned measures when significant harm is likely and to forward the results of this assessment to concerned States as part of the notification.

It follows from the previous discussion that the intended purpose of notification obligations is not only to advise potentially affected riparian States about what is going to happen but also to 'open a door' to determine how the interests and rights of notified States might be accommodated and which mitigation strategies might need to be adopted.[154] Notification is part of a wider process; it aims to lead to consultations to facilitate joint assessment of the risks of the proposed measures and, if necessary, to negotiations with a view to reaching agreement on changes to the plan.[155] The duty to notify is thereby linked to the obligation to consult.[156] In the case of planned measures with significant transboundary impact, the duty to consult is implied in the obligation to notify[157] and vice versa. When the obligation of notification serves merely to transmit information, such as in the case of emergencies or relative to adopted legislation, an obligation

indemnify the State for carrying out this request. Art. 13 (3), 1992 *UNECE Water Convention*, allows the planning State to request payment for the collection of additional data.

152 The fact that the 1997 *UN Watercourses Convention* uses the term 'environmental impact assessment' rather than the more general notion of 'impact assessment' was criticised for being too limited. It overlooks other potential impacts, such as social impacts of projects. McCaffrey, *The Law of International Watercourses*, p. 475.

153 *Pulp Mills on the River Uruguay (Argentina v. Uruguay)* (2010) ICJ Reports, para. 204.

154 Okowa, 'Procedural Obligations', p. 302.

155 *Pulp Mills on the River Uruguay (Argentina v. Uruguay)* (2010) ICJ Reports, para. 115.

156 L. Boisson de Chazournes and M. Tignino, '*Droit international et eau douce*' (2007) *Jurisclasseur – Environnement*, Fasc. 2900.

157 McCaffrey, *The Law of International Watercourses*, p. 476; A. Nollkaemper, *The Legal Regime for Transboundary Water Pollution: Between Discretion and Constraint* (Dordrecht: Martinus Nijhoff/ Graham & Trotman, 1993), p. 176.

to consult does not follow from the former. The duty to consult – as well as the duty to negotiate[158] – also implies an obligation to notify because consultation is unlikely to take place in the absence of notification.[159] Consultation is the process that ensues in case of a response by the notified State; it is a process of bilateral – or multilateral, when more than two States are concerned – engagement.

Related obligation to suspend implementation

To create a facilitative environment for the consultation process,[160] the *UN Watercourses Convention* and other instruments provide for an obligation to suspend implementation of the planned measure in question after notification.[161] Notification of planned measures generally entails a period that follows during which the notified State can respond. During this time, the notified State can study potential effects of the project on rights and interests that might be affected by it.[162] The length of the period accorded for response differs among legal instruments. Many of those instruments that define time limits grant a period of about six months;[163] others provide for a shorter period, with the possibility of extension if needed.[164] Stated more simply, other instruments provide for 'a reasonable period of time to make an assessment'.[165] According to Article 14 of the *UN Watercourse Convention*, the planning State must

[158] Negotiation is distinguished from consultation by an additional characteristic: negotiation is qualified by having the aim of the achievement of a mutual accord.

[159] Bourne, 'Procedure: The Duty to Consult', 175.

[160] *Ibid.*, 181.

[161] For example, Arts. 14, 17 and 18, 1997 *UN Watercourses Convention*; Art. 5.4.3, *Mekong Procedures for Notification, Prior Consultation and Agreement*, 30 November 2003, www.mrcmekong.org/agreement_95/Agreement-procedures-guidelines.htm; Art. 4, 2000 *SADC Revised Protocol*; Art. 20, 2008 *The Niger Basin Water Charter*; Art. 58 (4), 2004 ILA *Berlin Rules*; Art. 9, 1975 *Statute of the River Uruguay*; recommended in Art. XXIX (3), 1966 ILA *Helsinki Rules*; Art. 7, 1961 IIL *Salzburg Resolution* provides for an obligation to suspend in case of negotiations on a planned measure; Art. 24, 2002 *Senegal Water Charter* requires prior consent by other contracting States before implementation of certain measures.

[162] Art. 13, 1997 *UN Watercourses Convention*.

[163] *Ibid.*; Art. 8, 1975 *Statute of the River Uruguay*; Art. 5.5.1, 2003 *Mekong Procedures for Notification, Prior Consultation and Agreement*; Art. 4, 2000 *SADC Revised Protocol*; Art. 18, *Treaty Concerning the Rio de la Plata and the Corresponding Maritime Boundary*, 19 November 1973, 1295 *UNTS* 307.

[164] For example, Art. 20, 2008 *Niger Basin Water Charter* provides for a period of three months, with a possibility for extension by one additional month; Annexure D, paras. 7 and 10, 1960 *Indus Waters Treaty* provide for a period of three months.

[165] Art. XXIX, 1966 *Helsinki Rules*.

await a reaction from the notified State for the duration of the period allowed for response before proceeding or permitting implementation of the planned measure.

The ILC, which discussed whether to provide for a 'reasonable period of time' or to specify a given time limit,[166] opted for a specified period of six months, relative to the duty to suspend implementation. The obligation to suspend implementation is a logical corollary to the duty to notify of planned measures. The goal of notification is to prompt a response that would provide the notifying State with additional information in order to assess whether (1) its measure will cause significant harm or impact any interests or rights of other riparian States; and (2) it falls within or outside of equitable and reasonable use of the shared resources.[167] It therefore seems logical that the planning State shall not proceed until it has received such information.[168] If the period for response was only vaguely defined, implementation could be delayed until the moment that the responding State decides what a 'reasonable time for response' might be. This would not only provide the notified State with a virtual veto over implementation;[169] it also bears considerable potential for dispute between the States.[170]

The duty to suspend implementation extends to the period after a response has been received, if the responding State so requests, to allow for meaningful consultation and/or negotiations.[171] The *UN Watercourses Convention* again limits this time period to six months, with a view to ensuring that implementation is not suspended for longer periods than would be reasonable.[172] When the notifying State does not receive a response within the given time limits, it can proceed with implementation in accordance with the technical specifications provided in the notification and within the limits provided by the principle of equitable and reasonable utilisation and the obligation not to cause significant harm.[173] In a case in

166 ILC, 'Third Report, McCaffrey', 37; ILC, 'Draft Articles . . . and Commentaries', p. 114.

167 ILC, 'Draft Articles . . . and Commentaries', p. 114.

168 *Ibid.*; Bourne, 'Procedure: The Duty to Consult', 180f.; *Lake Lanoux Arbitration (France v. Spain)*, 111.

169 Nollkaemper, *The Legal Regime*, p. 175.

170 ILC, 'Draft Articles . . . and Commentaries', p. 114.

171 Art. 17, 1997 *UN Watercourses Convention*; Art. 58, 2004 ILA *Berlin Rules*.

172 ILC, 'Draft Articles . . . and Commentaries', p. 116.

173 For example, Art. 16, 1997 *UN Watercourses Convention*; Art. 4, 2000 *SADC Revised Protocol*; Art. 9, 1975 *Statute of the River Uruguay*; Art. 21, 2008 *Niger Basin Water Charter*.

which ensuing consultations (and negotiations) do not result in an agreement, the planning State may carry out the planned measure according to the same limitations. The ICJ speaks of implementing the measure at the planning State's 'own risk' in the *Case Concerning Pulp Mills on the River Uruguay (Argentina v. Uruguay)*.[174] According to the *UN Watercourses Convention* and agreements modelled on this instrument, a planning State may immediately proceed with planned measures in case of urgency, if these serve 'to protect public health or safety or equally important interests'.[175] Such a situation, however, does not abrogate the obligation to notify or any subsequent consultation obligations.

Hence, except for cases of public emergency, the planning State is obligated to suspend implementation during the period accorded for reply after notification and remains under this obligation, if the responding State so requests,[176] to provide space for meaningful consultations. As a consequence, and based on the principle of good faith, the notified State has the duty to respond as early as is convenient.[177] Any undue delay with the response would lead to an obstruction of activities of the planning State.

4.3 Identifying mutual benefits and competing interests and rights

One-way communication through notification is insufficient to identify possible mutual gains and competing interests and rights that may stand in their way: two-way (or multi-way) communication among involved parties is necessary.

4.3.1 Consultation

Consultation in the context of notification of planned measures is the process that ensues in case of a response by the notified State claiming significant adverse effect; the response to notification is already a part of the consultation process. If notification can be seen as the act that invites consultation, a response would be the act that opens the mutual process of consultation. Consultation is straightforward in a bilateral

[174] *Case Concerning Pulp Mills on the River Uruguay (Argentina v. Uruguay)*, para. 154.
[175] Art. 19, 1997 *UN Watercourses Convention*; Art. 24, 2008 *Niger Basin Water Charter*.
[176] The ICJ makes the obligation to suspend independent of the request from the responding State in the *Case Concerning Pulp Mills on the River Uruguay (Argentina v. Uruguay)*, para. 144.
[177] Okowa, 'Procedural Obligations', p. 285.

context.[178] When several States are involved, a joint commission or any other dialogue platform can facilitate multilateral consultations.

The obligation to consult highlights the role of notification in the process of cooperation. There would be no need for notification if there were no consequences attached to it.[179] Regarding an ensuing obligation to consult, an underlying objective of notification comes to light: that interests addressed in the response of the notified State are to be considered.[180] Consultation in this context is a process of information exchange that carries with it a legal consequence: the duty to take into account the information obtained throughout this process. The legal framework applying to cooperation in the lower Mekong River Basin specifies that 'prior negotiation [and consultation for that matter] is neither a right to veto the use nor [a] unilateral right to use water by any riparian without taking into account other riparians' rights'.[181] Nollkaemper defined consultation as 'a process in which a State planning an activity which may present transboundary environmental harm engages in an exchange of views and interests with potentially affected States so as to make the consideration of their interests a component in its final determination'.[182] As to be discussed herein, the obligation to consult exists not only relative to planned measures;[183] several treaties also establish consultation obligations relative to coordination in managing shared water resources.[184] Whereas the obligation of prior consultation resulting from notification of planned measures that

178 *Ibid.*, p. 290.

179 *Lake Lanoux Arbitration (France v. Spain)*, p. 115.

180 Okowa, 'Procedural Obligations', 302; Bourne, 'Procedure: The Duty to Consult', p. 227.

181 2003 *Mekong Procedures for Notification, Prior Consultation and Agreement*; Chapter II, 1995 *Mekong Cooperation Agreement*.

182 Nollkaemper, *The Legal Regime*, p. 176.

183 For example, Art. 17, 1997 *UN Watercourses Convention*; Art. XII (5), XIII (6), and Annex A, 1961 *Columbia Basin Treaty*; Art. 11, 1994 *Danube Protection Convention*; Art. 11, *Agreement on the Guaraní Aquifer (Acuerdo sobre el Acuífero Guaraní)*, 2 August 2010, http://internationalwaterlaw.org/documents/regionaldocs/Guarani_Aquifer_Agreement-Spanish.pdf.

184 For example, Art. 5, *Agreement on the Protection and Utilization of Transboundary Waters Concluded between China and Mongolia*, 29 April 1994, http://ocid.nacse.org/tfdd/tfdddocs/528ENG.pdf, envisages obligatory State consultations regarding fishing in these waters. Art. 16, 1994 *Danube Protection Convention*, stipulates a consultation obligation regarding harmonisation of emergency warning systems. Art. 5 (2), 1997 *UN Watercourses Convention*, provides for consultations regarding the application of factors that are relevant to determine equitable and reasonable utilisation. Art. 3 (10), 2000 *SADC Revised Protocol*, provides for consultations regarding the elimination or mitigation of harm.

might cause significant harm emerged as a norm of customary law,[185] other consultation obligations exist when they are constituted by treaty instruments.

Consultation obligations apply for different forms of planned measures, which can be of various natures, ranging from physical structures to the adoption of legal measures. Physical infrastructure, particularly of a large scale, might seem to carry a more obvious and immediate risk of transboundary harm. Yet, the adoption of legislative acts or the conclusion of treaties that have as their objective activities that then in turn might cause significant harm also fall under the category of planned measures that require notification and consultation. In a case filed by Costa Rica against Nicaragua in 1916, the complainant alleged that Nicaragua violated its obligations of prior consultation by entering into a treaty with the United States – the so-called *Bryan–Chamorro Treaty* of 1914 – and by conceding preferential rights to build an inter-oceanic channel on Nicaraguan territory through the San Juan River and the Great Nicaragua Lake or in other parts of Nicaraguan territory. Although the San Juan River is located entirely in the territory of Nicaragua, Costa Rica holds navigation rights where the river forms the boundary between the two countries.[186] The decision of the Central American Court of Justice confirmed[187] that, based on the 1858 *Treaty of Limits*, Nicaragua has an obligation of prior consultation before concluding agreements with third parties that might damage the adjacent territory of Costa Rica or that would affect its rights on the San Juan River. The Court elaborated that where 'the concession is one that violates [Costa Rica's] "natural rights", it would seem that [Costa Rica's] consent is necessary . . . and in the event . . . that the concession does not affect such rights, her voice must be purely consultative'.[188]

[185] McIntyre, *Environmental Protection*, p. 340; J. Bruhács, *The Law of Non-Navigational Uses of International Watercourses* (Dordrecht/Boston: M. Nijhoff, 1993), pp. 176f.

[186] According to the 1858 *Treaty of Limits*, the San Juan River forms part of the boundary between Costa Rica and Nicaragua. The boundary line runs along the Costa Rican riverbank, placing the river in the territory of Nicaragua. Costa Rica holds navigation rights for purposes of commerce. *Tratado de Limites Territoriales con Nicaragua*, 15 April 1858, *Dispute Regarding Navigational and Related Rights (Costa Rica v. Nicaragua)*, application instituting proceedings filed in the Registry of the Court, 29 September 2005, pp. 31–33.

[187] The ICJ relied on an earlier arbitration decision by President Cleveland for its decision ('Cleveland Award'), 24 March 1888, published as 'The Costa Rica Boundary – President Cleveland's Decision as Arbitrator', *New York Times*, 25 March 1888.

[188] Case of *Costa Rica v. Nicaragua*, Decision of the Central American Court of Justice, San José, Costa Rica, 30 September 1916, in (1917) 11 *AJIL* 222.

Related duty: Taking interests of others into account

This decision illustrates the fact that the scope of consultations extends beyond situations in which the rights of another State might be affected. The planning State not only must respect the potentially affected rights of other States; prior consultation also requires that it seek out the interests and views of affected States. As discussed in the case of notification, the consultation process consists not only of taking note of the information provided but also of considering the interests and views of potentially affected States.[189] The planning State must make an effort to accommodate the interests of potentially affected States.[190] This duty results from the principle of good faith.[191] Okowa confirms that 'the duty to enter into consultations is not a mere formality, but it is in fact a procedural rule embodying a substantive value'.[192] Substantive and procedural aspects of the consultation obligation are inextricably intertwined.

The arbitration tribunal presiding over the decision in the *Lac Lanoux* case[193] elaborated on elements that would provide evidence of good faith efforts made by the planning State. These elements include consultations with regard to different implementation options, continued efforts to engage affected countries in a dialogue and to not break off consultations without justification, and avoiding unnecessary delays.[194] At the same time, the planning State is not required to put its own interests completely behind other States' interests. Unless it is otherwise provided by treaties, the right to be consulted does not include a right to veto the planned measure. The interests of others must be taken into account only as far as reasonably possible. If the interests and views of others can be accommodated with minimal additional cost, the planning State should act accordingly.[195]

In the *Pulp Mills on the River Uruguay* case, the ICJ further pointed out that when notification and consultation obligations exist relative to utilisations that might affect water quality and/or the régime of a watercourse,

189 *Lake Lanoux Arbitration (France v. Spain)*, 116; Okowa, 'Procedural Obligations', p. 306.

190 *Ibid.* 191 *Lake Lanoux Arbitration (France v. Spain)*, p. 116.

192 Okowa, 'Procedural Obligations', p. 306.

193 This dispute concerned a plan by France to utilise the waters of Lac Lanoux, located entirely in France, for the production of hydroelectricity. The scheme involved an interbasin transfer of the Lake's waters into another French river; the abstracted waters would have been substituted via a second interbasin transfer into the original river upstream of the point where the river entered Spanish territory. *Lake Lanoux Arbitration (France v. Spain)*, pp. 79–119.

194 *Ibid.*, p. 106. 195 Okowa, 'Procedural Obligations', p. 312.

'such utilizations could not be considered to be equitable and reasonable if interests of the other riparian State in the shared resources and the environmental protection of the latter were not taken into account'.[196] This dictum illustrates that consultation and information exchange comprise one part of the implementation process of the principle of equitable and reasonable utilisation; the obligation to consider accordingly the additional information that emerges from this process is another element of this cooperative process.

4.3.2 Negotiation

Whereas the role of consultation concerning cooperation is to inform a mutual process that enables consideration of the interests and rights of those involved, the intention of negotiation is to take this process one step further. Negotiation is like consultation but with the additional goal of the subsequent conclusion of an agreement.[197] Although the intended objective of consultations and negotiations is different, the process and engagement required of States are very similar. What was stated previously regarding requirements of meaningful conduct of consultations also applies with respect to negotiations.[198] Negotiating in good faith requires that States are not limited to reiterating their own positions; they must be ready to consider modification of their position so that compromise and agreement become a realistic possibility.[199]

In addition, whereas the objective of negotiation is to achieve an agreement that creates new legal circumstances, States are not under obligation to reach agreement. The agreement remains an objective but is not a result required by law.[200] However, if negotiations do not end in agreement, participating States are nonetheless not absolved from the duty to take into account the other States' interests.[201] For example, when a

196 *Case Concerning Pulp Mills on the River Uruguay (Argentina v. Uruguay)*, p. 54.

197 For example, Art. 4, 1923 *Geneva Convention Relating to the Development of Hydraulic Power Affecting More than One State*; and Art. 17, 1997 *UN Watercourses Convention*; *North Sea Continental Shelf Case (Federal Republic of Germany/Denmark, Federal Republic of Germany/Netherlands)* (1969) ICJ Reports, p. 3, para. 85.

198 Indeed, many authors wrote about consultation and negotiation simultaneously when analysing cooperation obligations of States; for example, Bourne, 'Procedure: The Duty to Consult', 212–234; McIntyre, *Environmental Protection*, pp. 337–344.

199 *North Sea Continental Shelf Case (Federal Republic of Germany/Denmark, Federal Republic of Germany/Netherlands)* (1969) ICJ Reports, p. 3, para. 85.

200 *AO Railway Traffic between Lithuania and Poland*, PCIJ (1931), Ser. A/B, No. 42, p. 116.

201 *Lake Lanoux Arbitration (France v. Spain)*, pp. 108 and 118.

State planning a new measure is under a treaty-based obligation to negotiate and the negotiations do not result in an agreement, the planning State still has the duty to take into account the rights and interests of affected States. Thus, when negotiations are not successful, the objective of consultations nonetheless remains a legal obligation. Hence, the factual outcome of the two obligations – consultation and negotiation – ultimately might be the same.

Therefore, to an outsider who cannot discern the intentions of the involved parties, negotiation and consultation might appear to be the same; the difference between the two is in the subjective element of intent at the time of engagement in a mutual process of exchange. Although this element is not visible to an outside observer, technically, consultation and negotiation are not the same. The difference is subtle: (1) it is the States' intention to exchange information concerning respective rights and interests at stake with a view to consider them; or (2) States engage with the intention of achieving mutual agreement on a change in legal circumstances. When a planned measure is adapted to the interests of other States based on consultation, the planning State makes changes to the initial design based on its judgement of whether and how the interests of others are taken into account. When adaptation of planned measures occurs based on negotiation, it is the result of mutual agreement between concerned States.

Consultation and negotiation can also take place in sequence. This was captured in Article 3 (5) of the *UN Watercourses Convention*, which stipulates that '[w]here a watercourse State considers that adjustment and application of the provisions of the present Convention is required because of the characteristics and use of a particular international watercourse, watercourse States shall consult with a view to negotiating in good faith for the purpose of concluding a watercourse agreement or agreements'. Based on this provision, some commentators argue that the *Convention* 'envisages consultation as a step that precedes formal negotiation'.[202]

Negotiation as a customary obligation

Because of the requirement for the subjective element of intent, attempts to establish from State practice whether the obligation to negotiate has passed into customary law turn out to be more difficult than for consultation. Identification *ex post facto* of State intention and *opinio iuris* about

202 McCaffrey, *The Law of International Watercourses*, p. 477; McIntyre, *Environmental Protection*, p. 342.

a result that is not required to materialise (i.e., an agreement) is intricate. At the same time, the existence of numerous legal instruments that concern transboundary water systems indicates that basin States frequently refer to the tool of negotiation and do so successfully.[203]

Earlier documents by regional and international expert bodies were not conclusive on the point of whether the obligation to negotiate was considered a customary norm. A number of resolutions adopted by expert bodies in the Americas[204] and in the 1961 IIL *Salzburg Resolution*[205] stipulate an obligation to negotiate (i.e., concerning planned measures). This obligation is implied in the instruments adopted in the Americas through an even stronger requirement of prior consent, which led Argentina to argue the existence of a singular regional practice to seek prior agreement in its Memorial in the *Pulp Mills Case*.[206] The 1966 *Helsinki Rules* do not provide for such an obligation to find a settlement with affected States, except when pollution has resulted.[207] Article XXX of the *Helsinki Rules* merely recommends negotiations for dispute settlement. The ILA *Berlin Rules* do not distinguish between consultation and negotiation, stipulating only an obligation to consult with respect to planned measures, thereby leaving it to the States to decide whether to negotiate.[208]

The text of the *UN Watercourses Convention* makes an interesting distinction. When there is a response to a notification of planned measures, States 'shall enter into consultations, and *if necessary*, negotiations' [emphasis added].[209] Article 17 contrasts in this respect with the two

[203] See also C. Bourne, 'Mediation, Conciliation and Adjudication in the Settlement of International Drainage Basin Disputes' (1971) 9 *Canadian Yearbook of International Law* 114.

[204] Bourne, 'Procedure: The Duty to Consult', 213f., citing the 1933 *Declaration of the Seventh Pan-American Conference on the Industrial and Agricultural Use of International Rivers*, 24 December 1933 in (1934) 28 *AJIL* Suppl., 59–60; 1957 *Resolution Adopted by the Inter-American Bar Association at Buenos Aires*; 1959 *Resolution by the Council of the Section of International and Comparative Law of the American Bar Association*; 1965 *Draft Convention on the Industrial and Agricultural Use of International Rivers and Lakes Prepared by the Inter-American Juridical Committee of the OAS, OAS Official Records*, OEA/Ser.I/VI.2 (English).

[205] Art. 6, 1961 IIL *Salzburg Resolution*: 'In case objection is made, the States will enter into negotiations with a view to reaching an agreement within a reasonable time . . .'

[206] *Case Concerning Pulp Mills on the River Uruguay (Argentina v. Uruguay)*, Memorial at pp. 99f., para. 3.85.

[207] Art. XI (2), 1966 *Helsinki Rules*: ' . . . if a State fails to take reasonable measures [to abate existing water pollution], it shall be required promptly to enter into negotiations with the injured State with a view towards reaching a settlement equitable under the circumstances'.

[208] Arts. 58 and 59, 2004 ILA *Berlin Rules*.

[209] Art. 17, 1997 *UN Watercourses Convention*.

articles that follow. Article 18 addresses the case in which notification did not take place and a potentially affected State demands that the planning State review whether it nevertheless is obligated to notify because of likely adverse impact. If the planning State maintains its position that it is not under an obligation to notify and the potentially affected State disagrees with this position, 'the two States shall, at the request of the [latter], promptly enter into consultations and negotiations'.[210] This obligation to consult *and* negotiate also exists when implementation of planned measures cannot be delayed for reasons of urgency 'in order to protect public health, public safety or other equally important interests'.[211] In its commentary to the 1994 *Draft Articles on the Law of Non-navigational Uses of International Watercourses*, the ILC did not clarify why this distinction is made.[212] In this matter, it limits the commentary to recognising that there is a difference between consultation and negotiation.[213] The rule that States are to hold negotiations *if necessary*, which is repeated in Article 14 of the ILC 2008 *Groundwater Articles* on planned measures, and the explicit obligation to negotiate in Articles 18 and 19 of the *UN Watercourses Convention* lead to the conclusion that in situations when States are in obvious disagreement with one another or there is a risk thereof, they are under obligation to negotiate to avoid disputes.

Regarding the nature of the obligation to negotiate as a customary norm, it can therefore be said that when the interests of the planning State and affected States are not divergent, consultation will suffice to resolve the issues.[214] Bourne concludes with respect to planned measures that '[i]t is possible, then, to assert that international law imposes on a basin state the obligation to consult and to negotiate in good faith with co-basin States that object to a proposed work or utilization of waters on the ground that it might cause them serious injury'.[215] In addition, it can be said that when consultations reveal or when it is obvious from the start that interests are conflicting and there is a risk of dispute, States are obligated to negotiate to find a peaceful solution under customary law.[216] Treaty instruments that include rules on dispute settlement almost invariably

210 Art. 18 (2), 1997 *UN Watercourses Convention*.
211 Art. 19, 1997 *UN Watercourses Convention*.
212 ILC, 'Draft Articles . . . and Commentaries', pp. 115–118.
213 *Ibid.*, p. 116. 214 *Ibid.* 215 Bourne, 'Procedure: The Duty to Consult', p. 233.
216 Also suggested in Art. 13 (2) and (3), 1997 *UN Watercourses Convention* concerning the reply to notification.

require States to first attempt to find a solution through negotiations.[217] An alternative legal basis for an obligation to negotiate is in Article 33 of the UN Charter. This Article, which obliges States to 'seek a solution by negotiation' and other peaceful means, would be applicable when a dispute about the utilisation of transboundary waters endangers the maintenance of international peace and security.[218]

The right to participate in negotiations

Along this same line of reasoning, Article 4 of the *UN Watercourses Convention* stipulates a right to participate in negotiations, which is complementary to the duty to negotiate.[219] Watercourse States 'whose use of an international watercourse may be affected to a significant extent by the implementation of a proposed watercourse agreement that applies to only part of the watercourse or to a particular project, program or use' have a right to participate, when appropriate, in the negotiation of such an agreement.[220] When an agreement applies to the entire watercourse, States have a right not only to participate in negotiations but also to become a party thereto.[221] This right seems self-evident given that basin agreements are negotiated with the objective of developing legally binding effects relative to the entire basin. Insofar as agreements should apply to resources within the territory of an individual State, it seems obvious that the State must have the right to participate in negotiations because, even if it does occur on occasion, it cannot be expected that a State would automatically accept terms negotiated by others. At the same time, an obligation to negotiate agreements that concern parts of or an entire water system does not exist.[222] Questions regarding water use may not always require settlement by way of agreement among all basin States; they may be settled locally or by other more informal arrangements.[223]

[217] For example, Art. 33, 1997 *UN Watercourses Convention*; Art. 2, *Agreement Concerning the Regulations Governing Lake Saimaa and the Vuoksi River*, 26 October 1989, 1663 *UNTS* 337; Art. IX, 1960 *Indus Waters Treaty*; Art. 4, 1954 *Convention Concerning Water Economy Questions Relating to the Drava*.

[218] Bourne cites instances in which water disputes have taken this dimension, such as the Indus in 1948, in Bourne, 'Mediation, Conciliation and Adjudication', 148; also ILC, 'Second Report on the Law of the Non-navigational Uses of International Watercourses, by Mr. Stephen Schwebel, Special Rapporteur' (1980) II *YBILC* (1), paras. 71f.

[219] ILC, 'First Report, Evensen', para. 30.

[220] Art. 4 (2), 1997 *UN Watercourses Convention*.

[221] Art. 4 (1), 1997 *UN Watercourses Convention*.

[222] ILC, 'Second Report, Schwebel', para. 91.

[223] *Ibid*.

The ILC *Groundwater Articles* do not mention a right to participate in negotiation, yet Article 19 recommends the conclusion of agreements to manage shared aquifers or aquifer systems. For these agreements, it can be equally argued that those States that will be affected by such agreements would be entitled to participate in the negotiations thereof. At the regional level, the 2000 *SADC Revised Protocol on Shared Watercourses* repeats the provisions of Article 4 of the *UN Watercourses Convention* for the Southern African Region.[224] The 1992 *UNECE Water Convention* goes further than other comparable instruments: rather than stipulating a right to participate in negotiations, it establishes an obligation for States to negotiate and enter into 'bilateral or multilateral agreements or other arrangements, where these do not yet exist . . . to define their mutual relations and conduct regarding prevention, control, and reduction of transboundary impact'.[225] At the same time, this provision does not establish an explicit obligation to enter into basin-wide treaties; it leaves the choice regarding arrangements and instruments to achieve the *Convention*'s objectives to the State parties.

Obligation to reach agreement

Although there follows no obligation to achieve agreement from the customary duty to negotiate, in some cases, agreement is necessary to proceed with planned activities. In some instances, States have agreed on such an obligation in legal instruments. The 1972 *Convention on the Statute of the Senegal River* and the 2002 *Senegal Water Charter*[226] require prior approval by all signatory parties for any project that might perceptibly alter the river characteristics.[227] A similar provision is included in the 1978 *Convention Relating to the Status of the River Gambia*.[228] An agreement on the use of transboundary waters concluded by the German Democratic Republic

[224] Art. 6 (6) and (7), 2000 *SADC Revised Protocol*.

[225] Art. 9 (1), 1992 *UNECE Water Convention*.

[226] French version in L. Boisson de Chazournes *et al.*, *Protection internationale de l'environnement*, 2nd edition (Paris: Pedone, 2005), pp. 301–314.

[227] Art. 24, 2002 *Senegal Water Charter*: '*En tout état de cause, aucun projet susceptible de modifier d'une manière sensible les caractéristiques du régime du Fleuve, ses conditions de navigabilité, d'exploitation industrielle, l'état sanitaire des eaux, les caractéristiques biologiques de sa faune ou de sa flore, son plan d'eau, ne peut être exécuté sans avoir été au préalable approuvé par les Etats contractants*'. Art. 4, 1972 *Convention Relating to the Statute of the Senegal River*, requires also that the approval must be preceded by discussions and justification in case of opposition.

[228] 30 June 1978, ECOLEX TRE-153223.

and Poland provides that where the removal or reconstruction of existing water-engineering works (e.g., bridges, dams and embankments) may alter the water level in the other party's territory or impair navigability, the projects can be executed only upon the latter's consent. Consent is also required for the construction of new engineering works.[229]

According to Sauser-Hall, in the absence of a treaty-based obligation, prior agreement is also required for development of hydroelectric potential on contiguous stretches of transboundary rivers. On a contiguous river, where the border is drawn along a line in the riverbed – for example, along the *thalweg* or the median line rather than on one of the riverbanks – the water flows on the territory of two States and the hydroelectric-power potential constitutes an indivisible resource.[230] If one State wishes to exploit this resource, agreement is required from its neighbour.[231] This obligation to reach agreement is a concern of solely the planning State; if it wants to proceed, it must obtain approval. In other words, it is not a mutual obligation of the concerned States to enter into an agreement; the neighbouring State is under no obligation to accept the proposed project. State practice shows that hydropower stations on contiguous stretches have been built almost exclusively on the basis of international treaties.[232]

Separate from these obligations of prior agreement, Lauterpacht argued for an obligation to reach agreement more generally when States have concluded a *pactum in contrahendo*. When States have agreed to conclude an agreement, Lauterpacht argued, this obligation must be interpreted in good faith.[233] In particular, when the objective of the future agreement has been reasonably specified in a legal instrument, there would be 'an obligation to accept an agreement offered by the other party if its terms

229 Art. 18, *Agreement (with Final Protocol and Additional Protocol) Concerning Navigation in Frontier Waters and the Use and Maintenance of Frontier Waters*, 6 February 1952, 304 *UNTS* 160. Other treaties requiring prior agreement include the *Agreement Regulating the Use and Development of the Water Resources of the International Reaches of the Rivers Miño, Limia, Tajo, Guadiana and Chanza and of Their Tributaries*, 29 May 1968, 1206 *UNTS* 25; *Treaty Concerning Regulation of Water Management Questions Relating to Frontier Waters between Austria and Czechoslovakia*, 7 December 1967, 728 *UNTS* 313.

230 Sauser-Hall, '*L'Utilisation industrielle des fleuves internationaux*', p. 569. 231 *Ibid.*

232 1977 *Treaty Concerning the Construction and Operation of the Gabčíkovo–Nagymaros System of Locks*; *Treaty of Yacyretá*, 3 December 1973, 1380 *UNTS* 80; *Final Act Relating to the Establishment and Operation of the Iron Gates Water Power and Navigation System*, 30 November 1963, 513 *UNTS* 12.

233 *Report on the Law of Treaties by Mr. H. Lauterpacht, Special Rapporteur*, UN Doc. A/CN.4/63, cited in Bourne, 'Procedure: The Duty to Consult', p. 226.

are such that an impartial tribunal would consider them as taking into account the legitimate interests of both parties'.[234] Lauterpacht's theory has been criticised based on the fact that it gives the principle of good faith too much weight. The argument against Lauterpacht's proposal is that the principle of good faith cannot import an obligation to reach an accord because the agreement to conclude an agreement in the future merely stipulates an intention to negotiate.[235]

4.3.3 *Conclusion of treaties*

Independent of whether an obligation to conclude agreements exists, the abundance of treaties concerning transboundary freshwater systems highlights the fact that negotiations between States are frequently successful. States find it useful to establish a sound legal base for coordination and cooperation in the development and management of shared water resources.

The conclusion of a legal agreement represents yet another step in a continuous process of cooperation among States sharing water resources; it can also 'freeze' cooperation at a certain stage (and for a certain period). Treaties that define each party's rights with respect to specific uses of the shared water resources and the duties with respect to one another and do not define future cooperation obligations tend to 'lock in' cooperation at a certain level. The sample treaty analysis conducted for this book indicates that such treaties are relatively rare. Where they do exist, no further cooperation obligations (e.g., information exchange or consultation) are specified in the text because the treaty concerns either water uses relating to specific projects or establishes specific rules for identified water uses. Examples include the 1910 treaty[236] between the United Kingdom and the Sultan of Abdali (now in Yemen), which concerned the building of wells that would draw on the Wadi-As Saghir to supply water to the port of Aden then occupied by the British; and the 1925 agreement between the United Kingdom and Italy on the utilisation of the River Gash,[237] which quantifies water-discharge obligations and water use rights based on planned irrigation works in Eritrea. Neither instrument contains any

234 *Ibid.*

235 Bourne, 'Procedure: The Duty to Consult', 228f., agreeing with and citing Fawcett, 'The Legal Character of International Agreements' (1953) 30 *BYIL* 381, 396 and 398.

236 *Convention Regarding the Water Supply of Aden between Great Britain and the Sultan of Abdali*, 11 April 1910, http://ocid.nacse.org/tfdd/tfdddocs/47ENG.pdf.

237 *Exchange of Notes between the United Kingdom and Italy Respecting the Regulation of the Utilization of the Waters of the River Gash*, 12/15 June 1925, 38 *LNTS* 190.

provisions that specify obligations such as consultation, negotiation or information exchange duties.

In 1922, Finland and the Russian Federal Soviet Republic specified rules on the maintenance and minimum width of boundary river channels to provide adequate conditions for navigation, timber floating and migration of fish, as well as on rules regarding fishing practices, in such detail that the treaty envisages no obligation for further engagement of the two parties.[238] The treaty does not envisage later adjustment of its terms but establishes a status quo. Yet, although this treaty may have locked in cooperation between the two States regarding these uses, it did not stop them from concluding a treaty, almost forty years later, that covered *inter alia* the same water uses. In their 1964 *Agreement Concerning the Use of Water Resources in Frontier Waters*, the two States both reiterated and enhanced specific rules agreed to in 1922 and set additional cooperation obligations for joint management of transboundary waters.

This example highlights that the conclusion of a treaty need not be the end point or culmination of cooperation. Individual treaties can be temporal phenomena in evolving relationships among States as they are replaced or amended by other treaties. Treaties that outline cooperation obligations regarding specified areas can limit cooperation to those areas; at the same time, they can also form the legal bases for widening and ever-increasing cooperation among States. These aspects are discussed in Chapter 5, which explores the wide array of treaty-based cooperation obligations that build the legal foundation of cooperation beyond customary norms. The discussion illustrates that States have agreed on a variety of modalities of joint management of shared waters – which, in some cases, take on a life of their own and transform the cooperative relationships among States.

4.4 Conclusion

This chapter explores the character and content of specific cooperation obligations, which are referred to by many authors as the 'procedural' obligations of international water law. Procedure is crucial for the realisation of the substantive objectives inherent in the general principles of international water law, such as equitable and reasonable utilisation

[238] *Convention Concerning the Maintenance of River Channels and the Regulation of Fishing on Water Courses Forming Part of the Frontier between Finland and Russia*, 28 October 1922, 19 *LNTS* 184.

and the cooperative management of transboundary resources. Although it is true that the obligations described in this chapter largely prescribe procedures in which States shall engage, they also contain significant substantive elements that define the character and content of the cooperation process that they are there to facilitate. This is the case, for instance, with the qualitative content requirements of notification. Procedure is essential, yet the objective that is to be obtained by it also must be clear. Some cooperation duties have been qualified by different authors as either procedural or substantive obligations, most notably the obligation to carry out EIAs. This illustrates that in the context of international water law, it is difficult to draw a clear distinction between the two sets of norms. More often, specific cooperation obligations contain both substantive and procedural elements. These rights and obligations represent individual components of a much wider process of State cooperation – for example, the notification of planned measures initiates a mutual process of information sharing and consultation. The measures aim to prevent and mitigate harm as well as to ensure that rights and interests of all concerned States are taken into account and can be realised in an equitable and reasonable manner.

5 Adoption of joint measures

When their interests in the utilisation and management of transboundary water resources align, States frequently engage in cooperation to realise their common objectives. In this process, they may agree on a variety of measures: the adoption of common policy standards or common water management principles, the construction of infrastructure, or the creation of joint mechanisms that then control certain aspects of the development of their shared resources.[1] Similar to the multitude of possible measures of joint water management, the legal agreements that were adopted to regulate their implementation display a variety of different characteristics. Peace treaties and other instruments that establish international borders sometimes include single provisions concerning the management and utilisation of boundary waters.[2] Other treaties are more comprehensive in scope; they provide a general framework of management principles and may mandate the contracting parties to conclude subsequent and more detailed legal agreements, as needed.[3] Most agreements concerning the cooperative management and development of shared water resources, however, are dedicated directly to planned

[1] J. Lautze, M. Giordano, and M. Borghese, 'Driving Forces behind African Transboundary Water Law: Internal, External, and Implications', paper presented at the international workshop on *African Water Laws: Plural Legislative Frameworks for Rural Water Management in Africa*, Johannesburg, South Africa (2005), p. 2.

[2] For example, *Treaty Concerning the Regime of the Soviet–Iranian Frontier and the Procedure for the Settlement of Frontier Disputes and Incidents*, 14 May 1957, 457 *UNTS* 161; *Convention Regarding the Determination of the Legal Status of the Frontier between Brazil and Uruguay*, 20 December 1933, 181 *LNTS* 81.

[3] For example, 2000 *SADC Revised Protocol*; 1992 *UNECE Water Convention*; 1978 *Treaty for Amazonian Cooperation*.

infrastructure, other specific joint measures, or the allocation of rights over the shared resources.[4]

Many subject areas have become the object of cooperative management. This chapter is an overview of different aspects of water management and ways in which States pursue their common interests. Because water and the interests of States in its utilisation touch on all areas and aspects of life, this overview must remain limited to a sample set of topics. Those presented herein include pollution prevention and water quality control, ecosystem preservation, river regulation and joint infrastructure projects. These topics were selected because they are representative of the catalytic role that cooperation can play regarding the implementation of other general principles of international water law (e.g., environmental protection and equitable and reasonable utilisation), including the fair distribution of benefits.

5.1 Protection and development of shared waters

Water resource management measures that impact the quality and flow regime of water systems and, therefore, also the connected ecosystems are generally more effective when coordinated among all concerned parties. This holds true at the local, national and international levels. Consequently, these are issue areas that have led to riparian cooperation in international river basins and some transboundary aquifer systems.

5.1.1 Coordination for water quality control

In addition to the customary due diligence obligation to prevent pollution that causes or risks causing significant harm in the territory of others, States developed cooperative regimes to control pollution at levels below the legally relevant 'significant harm' threshold.[5] 'Pollution' is any detrimental alteration in the composition, content or quality of water that results from human conduct.[6] It occurs when human activity leads to a

4 For example, *Agreement Concerning the Cachuela Esperanza Hydroelectric Plant*, 8 February 1984, 1353 UNTS 375; *Exchange of Notes Concerning a Special Operating Program for the Duncan and Arrow Storages on the Columbia River System*, 30 December 1968/26 February 1969, p. 714 *UNTS* 298; *Final Act Relating to the Establishment and Operation of the Iron Gates Water Power and Navigation System*, 30 November 1963, 513 *UNTS* 12.

5 ILC, 'Draft Articles ... and Commentaries', p. 122; Nollkaemper, *The Legal Regime*, p. 85; Lammers, *Pollution of International Watercourses*, p. 538; ILA, *Report of the Sixtieth Conference*, p. 537; IIL, *Yearbook – Session of Athens*, p. 341.

6 Art. 21 (1), *UN Watercourses Convention*; Art. IX, 1966 *Helsinki Rules*.

change in the chemical composition and the purity of waters, as well as from a change in water temperature.[7] Water that is used to cool power stations, for example, is reintroduced into rivers at higher temperatures. This thermal pollution can impact riverine ecosystems in ways similar to chemical pollution.[8] States seek to prevent changes to or to improve water quality because of quality requirements of specific uses, including drinking water, irrigation and other economic applications. Furthermore, States have an interest in keeping water at a quality that supports healthy ecosystems in order to maintain the self-purification capacity of water.

It is only natural that downstream States demonstrate particular interest in cooperation on pollution prevention and control; they are the recipients of accumulated chemical and organic pollution from upstream. It is for this reason, for example, that the Netherlands was the main driving force in establishing cooperation on Rhine River pollution.[9] At its instigation, joint activities that initially focused merely on assessing the extent of pollution were converted to cooperation measures that eventually led to effective improvements in water quality. In the first twenty years of its existence, the International Commission for the Protection of the Rhine against Pollution – established in 1950 among Germany, France, Luxembourg, the Netherlands and Switzerland on the basis of an exchange of notes and later established on the basis of an international treaty[10] – conducted periodic investigations of pollution levels. Despite this work, it was not able to deliver on its mandate to propose successful mitigation measures.[11] Eventually, in 1976, the member countries agreed to more specific rules based on which toxic substances needed to be eliminated or reduced. Two conventions, on chemical pollution and on chloride pollution, were adopted, and the introduction of toxic substances slowly abated. The 1976 *Convention on the Protection of the Rhine against Chemical Pollution* is a framework that includes in Annex I a list of substances that must be eliminated and in Annex II a list of substances that must be reduced over time. The Commission was mandated to set emission standards for Annex I substances,[12] whereas emission standards for Annex II substances are subject to national regulation to be adopted in

[7] A treaty that addresses thermal pollution is the *Convention on the Establishment and Exploitation of Certain Industrial Works on the Moselle*, 12 March 1986 (1987) 91 *RGDIP* 705.

[8] Lammers, *Pollution of International Watercourses*, p. 173.

[9] For a detailed account, see *ibid.*, pp. 166–195.

[10] 1963 *Agreement on the International Commission for the Protection of the Rhine against Pollution*.

[11] Art. 2, *ibid.* [12] Art. 5, 1976 *Convention on the Protection of the Rhine against Chemical Pollution*.

coordination with other member States.[13] The second treaty, the *Convention on the Protection of the Rhine against Pollution by Chlorides*, sets water-quality standards regarding overall chloride levels, as measured at the German–Netherlands border.[14]

Quality and emission standards

As illustrated in the two cited *Conventions* concerning the pollution of the Rhine River, there are different approaches to pollution control and prevention that set either water quality or emission standards, or a combination of the two. In most cases, instruments adopt either one or the other; for example, the *Convention on the Protection of the Rhine against Chemical Pollution* follows the approach of emission standards, and the *Chloride Pollution Convention* sets water quality standards. Article 28 of the 2004 ILA *Berlin Rules* adopts a water quality standard approach. These standards focus on the water body receiving the pollution; they define the maximum permissible presence of certain substances in the waters.[15] Quality standards are most useful when pollution stems from a large variety of sources, including point source and non point source pollution (i.e., in cases in which some or all of the sources are difficult to identify). Water quality standards adopted by agreement between basin States can be drawn from in order to specify the 'significant harm' threshold and to limit discretion of acceptable pollution levels by States.[16] Yet, the water quality standard approach may be criticised for not attributing any responsibility to identified polluters regarding regulation of discharge and for providing a 'license to pollute' up to a certain level. In contrast, emission standards focus on limiting the discharge of pollutants by defining 'maximum admissible concentration of substances in effluents entering the waters'.[17] The disadvantage of emission standards, however, is that they may have no effect on overall water quality in situations in which a water body receives additional pollution inflow from a variety of unidentified pollution sources.

Regimes that combine the two approaches are more effective for pollution control than either of the two individually. The 2000 European WFD

[13] Art. 6, *ibid.* [14] 3 December 1976, 1404 *UNTS* 90.

[15] S. Boehmer-Christiansen, 'Environmental Quality Objectives versus Uniform Emission Standards', in D. Freeston and T. Ijlstra, *The North Sea: Perspectives on Regional Environmental Cooperation* (London: Graham and Trotman, 1990), p. 140.

[16] Nollkaemper, *The Legal Regime*, p. 107. Where quality standards have been adopted by national law, they can provide guidance but do not represent legal limits with respect to the 'significant harm' threshold at the international level.

[17] Boehmer-Christiansen, 'Environmental Quality Objectives', p. 140.

is one instrument that adopts a combined approach for emission and quality standards.[18] It is not an international treaty but rather forms part of the EU's body of norms of supra-national law and is binding on member states with respect to the results that must be achieved, whereas the choice of adequate methods, tools and form of implementation is left up to the member states. The primary concern of the WFD is the achievement of a 'good' (and non-deteriorating) quality status of surface and ground waters, as well as the protection of coastal waters from land-based pollution. Article 10 provides that member states must control discharge according to the best technologies available, relevant emission limits, and best environmental practices as set in EU legislation. In those cases in which quality standards imposed by EU legislation require stricter conditions than by national legislation, more stringent emission limits must be set accordingly.

The coordination process

Another instrument, which adopts a combined emission- and quality-standard approach, is the 1992 *UNECE Water Convention*, to which the EU is a contracting party. The *Convention* frames this approach through long-term objectives that outline policies rather than by prescribing specific rules and standards. It does not establish any numeric values and leaves the adoption of specific parameters to the contracting parties.[19] The *Protocol on Water and Health*, which aims to protect human health through improvements in water management, does not provide any numeric values either; it merely states that guidelines of the World Health Organization (WHO) shall be taken into account when adopting targets related to drinking water quality.[20]

The approach of prescribing the harmonisation of national measures was adopted in a number of other international instruments,

[18] This strategy had been adopted to mitigate deficiencies of earlier European water legislation that followed either one or the other approach; for example, *Council Directive 75/440/EEC Concerning Quality Required of Surface Water Intended for Abstraction of Drinking Water; Council Directive 76/160/EEC Concerning Quality of Bathing Water; Council Directive 76/464/EEC on Pollution Caused by Certain Dangerous Substances Discharged into the Aquatic Environment; Council Directive 78/659/EEC on Quality of Fresh Waters Needing Protection of Improvement in Order to Support Fish Life; Council Directive 79/923/EEC on Quality Required of Shellfish Waters; Council Directive 80/778/EEC on Quality of Water Intended for Human Consumption.*

[19] Arts. 11 and 13, 1992 *UNECE Water Convention.*

[20] Art. 6 (2), *Protocol on Water and Health to the 1992 UNECE Water Convention*, 17 June 1999, UN Doc. MP.WAT/2000/1/EUR/ICP/EHCO 020205/8Fin.

including the *UN Watercourses Convention*.[21] Article 21 captures this practice; Paragraph (1) defines pollution and Paragraphs (2) and (3) stipulate the following:

2. Watercourse States shall, individually and, where appropriate, jointly, prevent, reduce and control the pollution of an international watercourse that may cause significant harm to other watercourse States or to their environment, including harm to human health or safety, to the use of the waters for any beneficial purpose or to the living resources of the watercourse. Watercourse States *shall take steps to harmonize their policies in this connection.*

3. Watercourse States shall, at the request of any of them, consult with a view to arriving at mutually agreeable measures and methods to prevent, reduce and control pollution of an international watercourse, such as:

(a) Setting joint water quality objectives and criteria;
(b) Establishing techniques and practices to address pollution from point and non-point sources;
(c) Establishing lists of substances the introduction of which into the waters of an international watercourse is to be prohibited, limited, investigated or monitored [emphasis added].

Article 21 combines and applies fundamental rules of international water law. The first part of Paragraph 2 references the principle of equitable and reasonable utilisation and the obligation not to cause significant harm; the remainder of the Article provides for specific applications of the general duty to cooperate.[22] The last sentence of Paragraph 2 considers the inescapable interdependence among riparian States and the fact that pollution control and water quality management measures are most effective when they are of the same or a similar nature in all States sharing the resource. The provision also alludes to the character of cooperation as a process. As the ILC pointed out in commentary on the 1994 *Draft Articles*, 'initial achievement of harmonization will often involve several steps or stages; it is this aspect of the process that is addressed in paragraph 2, as indicated by the words "take steps" . . . even after policies have been successfully harmonized, continuing cooperative efforts will ordinarily be required to maintain their harmonization as conditions change'.[23] Paragraph 3 of Article 21 of the *UN Watercourses Convention* further stipulates

[21] ILC, 'Draft Articles . . . and Commentaries', p. 123, n367.
[22] *Ibid.*, pp. 122f. [23] *Ibid.*, p. 123.

a consultation obligation for this process and suggests measures that follow both the quality and the emission standard approaches for pollution control.

Expert commissions have adopted documents that equally codify universal norms on pollution of transboundary waters. The 1982 ILA *Montreal Rules on Water Pollution*[24] and the 2004 ILA *Berlin Rules on Water Resources*[25] mandate States to adopt national water quality standards.[26] These resolutions, however, lack the more specific cooperation obligations regarding harmonisation of standards at the basin level. This duty is merely implied in the general duty to cooperate that is stipulated in both documents.[27]

The advantage of a step-by-step approach – which limits initial cooperation obligations to adoption and harmonisation of national standards with a view to achieving long-term quality objectives rather than quantifying standards *ab initio* – is obviously in its flexibility. This approach allows States to adapt standards based on continuously evolving scientific knowledge on the harmful effects of different pollutants through more flexible instruments, such as regularly reviewed protocols and administrative agreements. Only a few international treaties quantify quality or emission standards; the *Rhine Chloride Pollution Convention* and the initial *Great Lakes Water Quality Agreement* of 1978 are two such instruments.[28] Later revisions of the *Great Lakes Water Quality Agreement*, including those from 1987 and 2012, introduced more flexibility to adjust to changes in scientific knowledge and pollution levels and to provide for regular review of substance objectives and loading targets.[29] Both treaties benefitted from and are the result of ongoing processes of cooperation on water quality management among riparian States. The case of the Rhine River was discussed previously. The *Great Lakes Water Quality Agreement* implements the 1909 *Boundary Waters Treaty* between the United States

[24] ILA, *Report of the Sixtieth Conference*, pp. 531–548.

[25] ILA, *Report of the Seventy-First Conference*, pp. 340–411.

[26] Art. 3, *Montreal Rules*; Art. 28, *Berlin Rules*.

[27] Art. 4, *Montreal Rules*; Art. 11, *Berlin Rules*.

[28] 1976 *Convention on the Protection of the Rhine against Pollution by Chlorides*; *Agreement between the United States and Canada on Great Lakes Water Quality*, 22 November 1978, 1153 *UNTS* 188; *Agreement on Great Lakes Water Quality*, 15 April 1972, 837 *UNTS* 214; others include the *Mexico–US Agreement on the Permanent and Definitive Solution to the Salinity of the Colorado River, Minute No. 242, International Boundary and Water Commission*, 30 August 1973, 12 *ILM* 1105.

[29] *Protocol Amending the Agreement between the United States of America and Canada on Great Lakes Water Quality*, 1978, as amended on 16 October 1983 and 18 November 1987, 7 September 2012, www.ec.gc.ca/Publications.

and Canada, which is historically one of the earliest instruments to refer to pollution control.[30]

Monitoring and warning systems

In a number of basins, States established specialised commissions with the mandate to develop, harmonise and monitor water quality management measures.[31] Joint commissions are also effective mechanisms for setting basin-wide warning and alarm procedures in the case of accidents.[32] Each riparian State can certainly monitor water quality independently with subsequent data exchanges, as recognised in Article 5 of the ILA *Montreal Rules*. Yet, direct cooperation and establishment of joint monitoring systems have the added advantage that same or comparable parameters are used in data assessment. Data is also pooled in one database and can be processed much faster than between non-integrated national systems.

The 1992 *UNECE Water Convention*, with its strong quality-management focus, established joint monitoring and assessment obligations, as well as an obligation to set up early warning systems, when appropriate.[33] Monitoring and warning systems were established on European rivers before the early 1990s;[34] yet, the recent expansion and enhancement of those systems must be evaluated in light of obligations established by the 1992 *UNECE Water Convention*, the WFD and related European legislation. The European and UNECE systems are probably the most elaborate and advanced systems of supra-national and international rules related to water quality regulation and pollution control. The advanced level of State cooperation in this respect, as compared internationally, can be attributed to the fact that because of extensive industrial and agricultural water use, water quality is a main concern in the transboundary basins of the UNECE region.

Navigation and pollution

Pollution management of surface waters is also an area in which navigational uses overlap with the 'law of non-navigational uses'. Article 1 (2) of

[30] Art. IV, 1909 *Boundary Waters Treaty*.

[31] For example, 1978 *Agreement between the United States and Canada on Great Lakes Water Quality*; *Convention Concerning the Protection of Italo-Swiss Waters against Pollution*, 20 April 1972, 957 UNTS 279; 1963 *Agreement on the International Commission for the Protection of the Rhine against Pollution*.

[32] Art. 9, 1992 *UNECE Water Convention*.

[33] Arts. 11 and 14, 1992 *UNECE Water Convention*.

[34] A. Kiss, 'The Protection of the Rhine against Pollution' (1985) 25 *Natural Resources Journal* 613–638.

the *UN Watercourses Convention* indicates that non-navigational uses might affect navigation, and vice versa, and clarifies that, indeed, international water law applies to both types of uses insofar as they affect protection, development and management of shared waters and their ecosystems.[35] The harmful impact of navigation on water quality was recognised in a number of treaty instruments. The 2003 *Protocol for Sustainable Development of Lake Victoria Basin* established an obligation for contracting parties to enact and harmonise national laws and policies to prevent pollution from vessels dumping waste into the lake and to regulate 'the movement of hazardous wastes in the Basin'.[36] The text of this provision can be interpreted as applying to both the shipment of hazardous material and hazardous effluents that move through the basin once they have been introduced to its waters. The 2003 *Convention on the Sustainable Management of Lake Tanganyika* includes similar obligations concerning pollution from installations and waste, as well as from shipping activities.[37] Additionally, the riparian States of the Lower Mekong Basin agreed to prevent pollution of the river resulting from any of the uses covered by their 1995 *Cooperation Agreement*;[38] these include navigation, irrigation, hydropower plants, fisheries, recreation and tourism.

The joint commissions, which have respective mandates for quality management and navigation in the Danube River Basin, coordinate their programs with a view to addressing problems that stem from pollution caused by navigation. The Danube Commission, which was established in 1948 with a mandate to regulate navigation on the Danube River between Ulm and the Black Sea;[39] the International Commission for the Protection of the Danube River (ICPDR);[40] and the International Commission for the Protection of the Sava River Basin[41] issued the *Joint Statement on Inland Navigation and Environmental Sustainability in the Danube River Basin*[42] in October 2007. It marked the initiation of a comprehensive program of institutional cooperation to implement the WFD with the objective of

35 ILC, 'Draft Articles . . . and Commentaries', pp. 89f.

36 Art. 32, 2003 *Protocol for Sustainable Development of Lake Victoria Basin*.

37 Art. 8, 2003 *Convention on the Sustainable Management of Lake Tanganyika*.

38 Art. 3, 1995 *Mekong Cooperation Agreement*.

39 *Convention Regarding the Regime of Navigation on the Danube*, 18 August 1948, 33 *UNTS* 180.

40 Established by the 1994 *Danube River Protection Convention*.

41 Established by the 2002 *Framework Agreement on the Sava River*. The Sava Commission has a dual mandate regarding navigational and non-navigational uses of this Danube River tributary.

42 www.icpdr.org/icpdr-pages/navigation_and_ecology_process.htm.

achieving a good ecological status of the waterway. The document contains principles and criteria for environmentally sustainable navigation, and it is regularly reviewed by the parties concerning progress made and possible enhancements in its application. A similar process was initiated on other European waterways – for example, on the Rhine River by the ICPR and the Central Commission for Navigation on the Rhine. Thus, cooperation with respect to pollution abatement in these basins is carried out not only directly among States but also by joint mechanisms.

Groundwater

Pollution control related to groundwater resources is of particular importance because of the vulnerability of these waters and aquifers, which demand a higher level of pollution protection. Groundwater experts consider quality monitoring to be a requisite for proper aquifer management.[43] In developing the 2008 *Draft Articles*, the ILC therefore modified and enhanced the rules of pollution prevention in comparison to the rules established in the *UN Watercourses Convention*.[44] The 2008 *Draft Articles* elevate the threshold of pollution prevention; they refer explicitly to the application of 'a precautionary approach in view of uncertainty about the nature and extent of a transboundary aquifer or aquifer system and of its vulnerability to pollution'.[45] They equally provide special protection to recharge and discharge zones[46] and an obligation for joint monitoring and harmonised standards, allowing national monitoring with subsequent data exchange only as a fall-back option if joint monitoring is not possible.[47] Given the particular vulnerability of aquifer systems to pollution, it is interesting that neither the reference to the precautionary approach nor the special monitoring obligations were included in the first aquifer treaty adopted since the adoption of the *Draft Articles*: that is, the 2010 *Guaraní Aquifer Treaty*.

5.1.2 Ecosystem management and protection

Water quality control is only one aspect of conservation, protection, and management of ecosystems linked to water bodies; when pollution

[43] ILC, 'Draft Articles on the Law of Transboundary Aquifers', p. 60.

[44] The provisions of the *UN Watercourses Convention* were criticised as being too weak and requiring modification. ILC, *Second Report on Shared Natural Resources: Transboundary Groundwaters by Mr. Chusei Yamada, Special Rapporteur*, 9 March 2004, UN Doc. A/CN.4/539, p. 3, para. 7.

[45] Art. 12, 2008 ILC *Draft Articles*; the precautionary approach is considered implicit in Article 20 of the *UN Watercourses Convention*.

[46] Art. 11, 2008 ILC *Draft Articles*. [47] Art. 13, *ibid*.

reaches unsustainably high levels, serious or irreversible ecological damage can occur. Organic pollution and alteration of water temperature can also lead to the extinction of species. It is for these reasons that many treaties concerned with pollution control simultaneously pursue wider overall objectives of protection of aquatic ecosystems and the environment and of conservation of biological diversity in transboundary basins.[48]

In most cases, the objective of sustainable ecosystem management and protection is framed in relationship with or directly linked to a general obligation to cooperate. Article 3 on the Scope of Cooperation of the *2003 Lake Victoria Protocol*, for instance, stipulates that 'Partner States have agreed to cooperate in the areas as they relate to the conservation and sustainable utilization of the resources of the Basin including . . . ' and then continues to list management activities linked to individual elements of the basin's ecosystem, including water resources, fishery resources, wetlands, forests and connected lands. Article 1 on the Basic Objectives of the *2004 Lake Ohrid Agreement*[49] between Albania and Macedonia begins with the following: '[i]ndividually and in cooperation, the Parties shall work in order to: 1. Assure equal and integrated protection as well as sustainable development for the Lake Ohrid and its watershed . . . ' Article 20, which introduces Chapter IV of the 1997 *UN Watercourses Convention*, stipulates an obligation for individual or joint action, when appropriate, with respect to preservation and protection of the ecosystem.[50] Thus, similar to the *Lake Ohrid Agreement*, it recognises that not all measures of ecosystem conservation must be taken jointly; basin States may opt to adopt stricter or more comprehensive measures on the national level than their co-riparian States. An obligation that always requires joint action would delay such initiatives.

More precise definitions of what general statements regarding the duty to cooperate for ecosystem protection mean in the context of a given basin or treaty are usually clarified through the establishment of specific cooperation obligations. These include *inter alia* the obligations on data exchange, consultations, notification, monitoring and harmonisation of national laws already discussed in this and the previous chapter. More specifically, they also include conservation and management obligations

[48] For example, Art. 3, 2003 *Lake Victoria Protocol*; Art. 2, 2003 *Convention on the Sustainable Management of Lake Tanganyika*; Art. 3, 1999 *Convention on the Protection of the Rhine*; Art. 2 (3), 1994 *Danube Protection Convention*; Art. 2, 1992 *UNECE Water Convention*.

[49] *Agreement for the Protection and Sustainable Development of Lake Ohrid and its Watershed*, 17 June 2004, ECOLEX TRE-151137.

[50] ILC, 'Draft Articles . . . and Commentaries', p. 118.

with respect to other natural resources that depend on or whose existence is intertwined with the water resources of shared basins (e.g., fish and other aquatic fauna and flora).

Management of fish resources

Some treaties governing the use of transboundary surface water resources include rules on fishing activities and sustainable management of those resources. Insofar as fish are a migrating species or live in contiguous water bodies, they constitute a separate resource that is shared by two or more States.[51] Consequently, legal principles applicable to shared natural resources also apply to these fish, including the principle of equitable and reasonable utilisation, the obligation not to cause significant harm, and the general duty to cooperate.[52]

A distinction is made between the rules managing fishery activities[53] as an economic exploitation of natural resources and the rules governing protection and conservation of these resources in a wider sense and inasmuch as they form an integrative part of basin ecosystems. Many general documents on fishery management do not distinguish between saltwater and freshwater species,[54] and they usually reference international rules adopted within the framework of the *United Nations Convention on the Law of the Sea* (UNCLOS).[55] Specific rules regulating freshwater fisheries are more commonly adopted at the national level through national legislation. At the same time, a small number of international treaties exist that regulate protection as well as economic exploitation of fish resources at the basin level. Some of these agreements include fishery management as their primary focus: (1) they specify geographic areas where fishing is prohibited or restricted in order to protect migrating routes and spawning grounds; and (2) they include rules regarding hydraulic infrastructure and pollution of transboundary waters as far as they relate to impacts on fish stock and migrating routes.[56] These treaties often define a minimum length for catch of specific species, prohibited fishing methods (e.g., the

51 Migratory fish form part of the category of 'shared resources'.

52 J. Barberis, *Los Recursos Naturales Compartidos entre Estados y el Derecho Internacional* (Madrid: Editorial Tecnos, 1979), pp. 143–166.

53 The term 'fisheries' is used here in its broad meaning and also includes shellfish.

54 For example, FAO, *Code of Conduct for Responsible Fishers* (Rome: FAO, 1959).

55 10 December 1982, 1833 *UNTS* 397.

56 For example, *Convention for the Establishment of the Lake Victoria Fisheries Organization*, 30 June 1994, 1930 *UNTS* 127; *Convention Concerning Fisheries in Swiss-Italian Waters*, 19 March 1986, ECOLEX 0.923.51, RO 1989 539; *Convention Concerning Fishing on the Waters of the Danube*, 29 January 1958, 339 *UNTS* 23.

use of certain nets or explosives) and time periods during which fishing for certain species is prohibited.

Treaties regarding the development of works sometimes include provisions to ensure that planned infrastructure does not impact or cut off migrating routes, such as the 1959 *Agreement between Greece and Yugoslavia Concerning Hydro-economic Questions*.[57] Agreements with the goals of sustainable development and management of basin ecosystems may equally include provisions on fisheries as one of multiple elements that are linked to protection and conservation of aquatic biodiversity.[58] Cooperation regarding fisheries and prevention of overexploitation of certain species is of vital importance to maintain the ecological balance in aquatic ecosystems. Riparian States have established specialised commissions to manage fish resources in various international lakes. The Great Lakes Fishery Organization,[59] for example, was established with a mandate that is limited to fisheries. In contrast, the Lake Victoria Fisheries Organization (LVFO),[60] established in 1999 as a specialised institution of the East African Community, has a wider mandate: it regulates fisheries and coordinates riparian State actions relative to conservation of the ecosystem.

Protection of ecosystems from invasive species

Experience with fish management in the Lake Victoria Basin also serves as a case study for another aspect of ecosystem protection: the problem of invasive species. The 2004 documentary, 'Darwin's Nightmare', which discusses the accidental introduction of the Nile perch into Lake Victoria and its subsequent economic and environmental impacts, pointed out the fragility of ecosystems. Although the Nile perch has a higher nutritional and financial value to riparian populations than endemic species because of its larger size and suitability for export, it also contributed to

[57] 18 June 1959, 363 *UNTS* 135.

[58] For example, 2003 *Convention on the Sustainable Management of Lake Tanganyika*; 2003 *Lake Victoria Protocol*; 1994 *Agreement on the Protection and Utilization of Transboundary Waters Concluded between China and Mongolia*; 1975 *Statute of the River Uruguay*; *Agreement (with Final Protocol and Additional Protocol) Concerning Navigation in Frontier Waters and the Use and Maintenance of Frontier Waters*, 6 February 1952, 304 *UNTS* 160; *Agreement on the Transfer from the Course of the Näätämo (Neiden) River to the Course of the Gandvik River of Water from the Garsjoen, Kjerringvatn and Forstevannene Lakes*, 25 April 1951, in Finland, *Finlands Författningssamlings. Fördagasserie. Överenskommelser Med Främmande Makter*, 1951 (Helsinki: Finland, 1952), No. 8, p. 24.

[59] Established by the *Convention of Great Lakes Fisheries*, 10 September 1954, 238 *UNTS* 97.

[60] 1994 *Convention for the Establishment of the Lake Victoria Fisheries Organization*.

the disappearance of a number of indigenous fish species and upset the ecological balance of the lake.

To avoid this type of harm to the natural environment, a number of treaties provide a general cooperation obligation with a view to preventing the introduction of alien species and to eliminating them when they are present.[61] How exactly this introduction is to be prevented by cooperative measures is usually not regulated. Article 10 (2) of the *Convention on the Sustainable Management of Lake Tanganyika* is a rare exception in providing one such measure. It establishes obligations to consult with other Lake Basin States and to obtain their prior consent before a State can grant permission to introduce any exotic species or organisms into the lake. These obligations are far stricter than the obligation established in Article 22 of the 1997 *UN Watercourses Convention* on preventing the introduction of alien and new species.[62] This provision establishes a due diligence obligation to prevent the introduction of only those species 'that may have effects detrimental to the ecosystem of the watercourse resulting in significant harm to other watercourse States'. It thus puts the threshold – in accordance with the general principles stipulated in Part II of the *Convention* – at the 'significant harm' level. However, in its 1994 commentary, the ILC clarified that the use of the term 'may' is to be understood as an obligation to adopt a precautionary approach regarding the risk of harm.[63] Because the potential harm that can be inflicted by alien species is often not known, the precautionary approach should be applied. Watercourse States must take all necessary measures to prevent harm, including joint and cooperative measures.

Obligations with respect to the introduction of alien species naturally apply not only to fauna but also to flora. The *Lake Tanganyika Convention* applies to any type of organism that could affect the basin's biodiversity. The case study of the Lake Victoria Basin again illustrates the reasons why all alien species are to be considered. As in many other basins where the plant is not endemic, the bloom of the water hyacinth in Lake Victoria and its tributaries directly causes damage to the ecosystem and indirectly to human health and navigational uses. Unless it is controlled, the water hyacinth can cover large surface areas of water bodies, blocking sunlight

[61] Art. 22, 1997 *UN Watercourses Convention*; Art. 12, 2008 *Niger Basin Water Charter*; Art. 10, 2003 *Convention on the Sustainable Management of Lake Tanganyika*.

[62] New species, as opposed to alien species that occur naturally, are species that have been genetically modified.

[63] ILC, 'Draft Articles . . . and Commentaries', p. 124.

and depleting water of oxygen and thereby killing more sensitive organisms and living species. The plant also blocks waterways and locks and creates a convenient habitat for mosquitoes and snails that are vectors for disease (e.g., malaria and schistosomiasis).

Water supply and flow control benefit from healthy ecosystems

Treaties that stipulate a general obligation to protect the basin environment recognise the multiple benefits that healthy ecosystems provide to humans, including health, water purification for safe drinking and irrigation water, flood prevention and other benefits that facilitate economic activity. The 1997 *UN Watercourses Convention* does not explicitly reference these human benefits in Part IV but instead focuses on aspects of protection and preservation of the ecosystem. Article 20 of the *Convention* stipulates a general obligation to 'protect and preserve the ecosystem of international watercourses'. The ILC defines the concept of 'watercourse ecosystem' as the 'ecological unit consisting of living and non-living components that are interdependent and function as a community'; all components depend on one another, are interrelated, and can benefit one another.[64] Humans very much form part of the living components of an ecosystem; their activities impact ecosystems and they can benefit from them – for example, the ecosystem's contribution to water purification.[65] Water percolates through watersheds and groundwater recharge zones; microorganisms, minerals and plants function as filters, eliminating or reducing toxic substances in freshwater.

This is one reason why the 2008 ILC *Draft Articles* accorded special protection to groundwater recharge and discharge zones. Groundwater exploitation meets 50 per cent of drinking water demand globally.[66] According to Article 11 of the 2008 ILC *Draft Articles*, aquifer States are under obligation to prevent and minimise detrimental impacts on the recharge processes.[67]

64 *Ibid.*, 118. This definition is similar but more restricted and, at the same time, more precise than the drainage basin approach promoted by the ILA. Whereas the drainage basin approach focuses on the geographic space that is determined by the watershed limits and everything within that space, the ecosystem approach focuses on the functional interlinkages among living and non-living components (i.e., among water, flora, fauna, minerals and soil).

65 *Legality of the Threat or Use of Nuclear Weapons*, Advisory Opinion (1996) ICJ Reports, p. 241, para. 29.

66 I. S. Zekster and L. G. Everett, *Groundwater Resources of the World and their Use*, IHP-VI, Series on Groundwater No. 6 (Paris: UNESCO, 2004), p. 22.

67 Art. 11, 2008 ILC *Draft Articles*.

This provision provides an elevated level of protection because it is not limited to a 'significant harm' threshold. It also obligates non-aquifer States, in whose territory these zones are located, to cooperate in protecting aquifer ecosystems.[68] The 1994 *Convention on Cooperation for the Protection and Sustainable Use of the Danube River* includes a similar provision, in particular with respect to considering the importance of groundwater for drinking water supply. Article 6 provides the following:

> Contracting Parties shall take appropriate measures aiming at the prevention or reduction of transboundary impacts and at a sustainable and equitable use of water resources as well as at the conservation of ecological resources, especially: (a) enumerate groundwater resources subject to a long-term protection as well as protection zones valuable for existing or future drinking water supply purposes; (b) prevent the pollution of groundwater resources, especially those in a long-term perspective reserved for drinking water supply....

The 1999 *Convention on the Protection of the Rhine* similarly draws an implicit link between sustainable development of ecosystems and their contribution to drinking water production. Article 3 lists a series of activities that aim to protect the health of the ecosystem as part of the actions to be carried out under its first objective: 'sustainable development of the Rhine ecosystem'; 'production of drinking water from the waters of the Rhine' is the second objective of the *Convention*. For further implementation of these goals, the *Rhine River Basin Management Plan*, agreed to by all riparian States in 2009, makes several references to the protection of groundwater and related ecosystems to guarantee drinking water supply.[69]

Additional ecosystem services are provided by wetlands and forest cover. Wetlands fulfill an important water-purification function. The wetlands north of New York City provide natural filtering for pollution runoff and facilitate drinking water availability for the city's inhabitants.[70] In addition, wetlands fulfill important flood and drought control and flow management functions, as do plant cover and forests in the watershed area. Plant cover retains both soil and humidity and can mitigate the impact of higher-than-normal surface runoff by absorbing excess water

[68] Additional protection is provided through strict provisions concerning pollution control of aquifers and aquifer systems in Article 12 of the 2008 ILC *Draft Articles*.

[69] ICPR, *Internationally Coordinated Management Plan for the International River Basin District of the Rhine, Part A* (Koblenz: International Commission for the Protection of the Rhine [ICPR], 2009), pp. 14, 27, 45 and 50.

[70] Committee on Public Health, *Statement on Preservation of New York City's Drinking Water Quality* (New York: The New York Academy of Medicine, 1989), p. 903.

into the ground. Studies of the Amazon Basin indicate that deforestation leads to persistent drought conditions and increased sedimentation, which blocks waterways and reduces hydropower production.[71] This is one reason why the Amazon Cooperation Treaty Organization (ACTO), which was established in 1995 based on the Amazon Cooperation Treaty,[72] is engaged in an integrated basin management project, including forest management.[73] Natural wetlands are also being restored under the Rhine Flood Prevention Program, which implements Article 3 (1) of the *Convention on the Protection of the Rhine*.[74] One reason identified for an increase in floods in the Rhine Basin was the fact that river-regulation projects had cut off more than 85 per cent of the former natural inundation areas from the river.[75]

To achieve the benefits that ecosystems can provide for human use, many conventions include comprehensive basin management plans to be developed by riparian States. In most cases, joint commissions or mechanisms are mandated to develop these plans. One instrument stipulating such an obligation is the *WFD*, which has an international dimension that reaches beyond the area of its direct applicability concerning comprehensive basin management planning.[76] For basins that cross into the territory of non-member States, Article 13 mandates that member States endeavour to establish appropriate coordination with the non-member States, with a view towards achieving *WFD* objectives. Other treaties that mandate basin commissions to establish comprehensive, integrated basin development plans include the *Senegal Water Charter*,[77] the *Framework Agreement on the Sava River Basin*,[78] and the *Protocol for Sustainable Development of the Lake Victoria Basin*.[79]

5.1.3 *Flow regulation and joint infrastructure*

Another aspect of ecosystem management, which applies primarily to surface water bodies, is flow regulation through human intervention. Changes to the natural flow of water systems invariably impact the

71 UNEP/Global Environmental Facility (GEF) Project Document, *GEF Amazonas PDF-B Project Brief*, 2005 p. 6, www.otca.info/gefam/index.php?page=HomePage&cat=82 .
72 1978 *Treaty for Amazonian Cooperation*.
73 UNEP/GEF Project Document, *GEF Amazonas PDF-B Project Brief*.
74 Conference of Rhine Ministers, *Rhine 2020: Program on the sustainable development of the Rhine* (Koblenz: ICPR, 2001), p. 14.
75 *Ibid.* 76 Art. 13 (3), WFD. 77 Art. 17, 2002 *Senegal Water Charter*.
78 Art. 12, 2002 *Framework Agreement on the Sava Basin in Implementation of the WFD*.
79 Art. 27, 2003 *Lake Victoria Protocol*.

often delicate equilibrium of connected ecosystems. These interferences may need to be balanced against other obligations to protect the environment.

Intervention in natural flows

In contrast to flow management through conservation and renaturalisation of the ecosystem – as the Rhine River riparian States are doing for flood management – 'regulation' 'means the use of hydraulic works or any other continuing measure to alter, vary or otherwise control the flow of the waters of an international watercourse'.[80] Typical measures of regulation include dams, reservoirs, embankments, and canals.[81] In this respect, it is interesting that the provisions for flow regulation are included in Part IV of the *UN Watercourses Convention*, titled 'Protection, Preservation and Management.' It begins by stipulating a general duty to preserve and protect ecosystems in Article 20. The ILC clarifies in its commentary that it 'was of the view that [this obligation] should precede the other more specific articles in part four'.[82] The last two Articles of Part IV, however, concern measures that upset the natural state of the ecosystem: Article 25 mandates that States cooperate in responding to needs or opportunities 'for regulation of the flow of waters of an international watercourse'; and Article 26 addresses the safe operation of installations and works related to watercourses. The inclusion of these two Articles in Part IV is interesting in that hydraulic works do anything but preserve the natural ecosystem. Interventions for flow regulation and installations inevitably lead to alterations of the ecosystem and disturb its natural balance.[83] Conversely, flow regulation facilitates navigation,

[80] Art. 25 (3), 1997 *UN Watercourses Convention*. The 1980 ILA *Belgrade Articles on Regulation of the Flow of Water of International Watercourses* defines 'regulation' as 'continuing measures intended for controlling, moderating, increasing or otherwise modifying the flow of the waters in an international watercourse for any purpose; such measures may include storing, releasing and diverting of water by means such as dams, reservoirs, barrages and canals'. See ILA, *Report of the Fifty-Ninth Conference, Belgrade 1980* (London: ILA, 1982), p. 362.

[81] ILC, 'Draft Articles . . . and Commentaries', p. 127. [82] *Ibid.*, p. 118.

[83] It could be argued that in an interpretation in which the human being is understood as integral part of the ecosystem and thus also constitutes a natural element of it, alterations of the ecosystem carried out by humans are natural alterations, just as alterations that occur due to change in precipitation patterns or temperature. However, even under such a broad interpretation of the ecosystem concept, in which humans use material for hydraulic works that were not initially present in the concerned area (e.g., cement brought in from outside to build a dam), we certainly can no longer speak of natural causes for ecosystem change.

flood control, hydropower production and irrigation activities. Articles 25 and 26 can therefore be understood in one of two ways: (1) as restricting the more general obligation of Article 20 to preserve and protect ecosystems; or (2) as being ancillary to this obligation in that they must be implemented with a view to preserving ecosystems and by striking a balance between environmental protection and the economic interests of riparian States.

Literature does not provide clarity on this issue of dissonance between protection and preservation of ecosystems, on the one hand, and flow regulation by human intervention on the other.[84] Tanzi and Arcari, who addressed Article 25 in more detail than other authors who analysed the 1997 *UN Watercourses Convention*, are of the opinion that placing Article 25 in Part IV of the *Convention* is due to requirements that States must maintain the minimum flow necessary for the 'self-depuration of the watercourse'.[85] Yet, they did not discuss ecosystem impacts or potential damage caused by flow regulation measures and the positioning of the Article in the context of 'environmental protection of international watercourses'. Rather, they analysed it together with the obligation to cooperate and its procedural applications. Linking the discussion of flow regulation obligations to the cooperation obligation resonates with an argument made by Special Rapporteur Evensen in his second report – namely, that this provision would fit better in the chapter that addresses cooperation and management (now Part III on Planned Measures).[86] This seems reasonable because minimum flow requirements for nature and for human activity are difficult to observe in the absence of cooperation, particularly if various States engage in consumptive water uses in different parts of a transboundary basin.[87] The concept of minimum flow includes considerations of minimum environmental flow requirements – that is, flow required by nature to maintain environmental functions of a watercourse, as well as minimum flow requirements that are defined by States to guarantee water needs for their respective economic activities. Therefore, given the impact of flow regulation on riverine ecosystems, there is a solid argument for why Articles 25 and 26 are well placed in Part IV of the *Convention*.

[84] The ILC commentary on the 1994 *Draft Articles* is also silent on this point.

[85] Tanzi and Arcari, *The United Nations Convention*, p. 218.

[86] ILC, 'Second Report, Evensen', p. 121. Evensen resigned from the ILC after this report and his argument was not continued by subsequent Special Rapporteurs.

[87] Tanzi and Arcari, *The United Nations Convention*, p. 218.

Utton and Utton perceived the minimum-flow concept as implied in the *UN Watercourses Convention* and considered the obligation to maintain minimum flow requirements to be a rule of customary law.[88] This argument seems plausible – for instance, when thinking about situations in which one riparian State reduces flow (below minimum requirements) to a level where it causes irreversible damage to the ecosystems of other States. In such a case, the obligation of no significant harm and the principle of equitable and reasonable utilisation would be infringed.[89] The ICJ made a similar argument with respect to environmental flow requirements in its decision on the *Gabčíkovo–Nagymaros* project. It ruled that 'with the continuing effects of the diversion of... waters on the ecology of the riparian area of the Szigetköz' – the only continental delta in Europe – Czechoslovakia deprived Hungary 'of its right to an equitable and reasonable share of the natural resources of the Danube'.[90]

Treaty practice with respect to minimum flow

Article 25 of the *UN Watercourses Convention* addresses an obligation to cooperate, when appropriate, to respond not only to needs but also to opportunities for regulation.[91] The needs aspect addresses flow regulation activities to prevent harm (e.g., through damage caused by floods, ice or erosion) and to meet environmental flow requirements. The opportunities aspect concerns benefits that can be derived from a watercourse through, for example, fisheries, hydropower production, tourism and navigation. When States decide on joint flow regulation measures, they have the right and the obligation to participate equitably in construction, maintenance and defrayal of costs.[92] There are numerous international treaties in which States agree to joint projects for flow regulation to decrease harm and to increase benefits. The ways by which States aim to achieve the benefits from flow regulation and by which they manage cooperation in this respect differ among agreements.

88 A. E. Utton and J. Utton, 'The International Law of Minimum Stream Flows' (1999) 10 *CJIELP* 7–38.

89 Tanzi and Arcari, *The United Nations Convention*, pp. 218f.

90 *Gabčíkovo–Nagymaros Project (Hungary/Slovakia)* (1997) ICJ Reports, para. 85.

91 Article 2 of the ILA *Belgrade Rules* speaks of an obligation to 'cooperate in a spirit of good faith and neighborliness in assessing the needs and possibilities and preparing plans', and to jointly undertake regulation when appropriate.

92 Art. 25 (2), *UN Watercourses Convention*.

Minimum environmental flow requirements were recognised, for example, in the 1995 *Mekong Cooperation Agreement*. In Article 6, the lower riparian States agreed to cooperate in order to maintain 'not less than the acceptable monthly natural flow' during the dry season and to 'enable the acceptable natural reverse flow to the Tonlé Sap' during the wet season. The Tonlé Sap is Southeast Asia's largest water body, and the lake is connected through a river branch to the Mekong River. In the dry season, the lake shrinks as water drains into the river; this flow is reversed during the wet season and water from the river's main stem backs up into the lake. The lake increases in size more than twentyfold, from 2,700 km^2 to about 60,000 km^2, and provides habitat for more than 75 per cent of Cambodia's freshwater fish catch.[93] As a consequence, the Mekong River Commission (MRC) is tasked with the adoption of guidelines for flow level monitoring and regulation. Another treaty, which recognises minimum flow requirements for ecosystem purposes, is the 1996 *Treaty between India and Nepal on the Mahakali River*.[94] Article 1 (2) states that 'India shall maintain a flow of not less than 10 m^3/sec downstream of the Sarada Barrage in the Mahakali River to maintain and preserve the river ecosystem'.

Minimum flow requirements to satisfy human needs were addressed in the 1996 *Treaty on Sharing of the Ganges Waters at Farakka*.[95] In this treaty, the two parties agreed to a flow schedule and minimum flow provisions by India to Bangladesh. The Farakka barrage was built primarily to divert water from the Ganges's main stem to the Hooghly River. This river branches off the River Ganges southward before the Ganges reaches Bangladesh. The extra water provided to the Hooghly River through Farakka helps to flush sediments out of the Indian port of Calcutta.[96] The treaty aims to ensure that both the environmental and economic water needs of Bangladesh, as well as the minimum needed for maintenance of port operations at Calcutta, are met.

Minimum flow requirements with the aim of ensuring sufficient water for economic and human use downstream are also implicit in the 1929 and 1959 *Treaties on the Nile*, which were concluded between

93 Available at http://tonlesap.net.

94 12 February 1996, reproduced in S. Salman and K. Uprety, *Conflict and Cooperation on South Asia's International Rivers* (The Hague: Kluwer International Law, 2002), pp. 373–378.

95 *Treaty between the Government of the Republic of India and the Government of the People's Republic of Bangladesh on Sharing of the Ganga/Ganges Waters at Farakka*, 12 December 1996, reproduced in Salman and Uprety, *Conflict and Cooperation*, pp. 387–391.

96 Salman and Uprety, *Conflict and Cooperation*, pp. 144f.

Egypt and the former colonial power Great Britain, and between Egypt and Sudan, respectively. Both treaties aim to guarantee that a certain flow volume reaches Egypt, which is almost fully dependent on Nile flows for its national water needs. The *Niger Basin Water Charter* established an obligation to maintain 'essential minimum flows to protect human health against pollutants and water-borne diseases'.[97] This provision takes into account minimum water quantity requirements for pollution dispersion and minimum flow velocity requirements; slow-flowing or standing waters provide optimal breeding grounds for mosquitoes and snails that transmit waterborne diseases. Therefore, the practice of establishing minimum flow requirements is important to maintain the health of ecosystems, human health and economically beneficial uses of watercourses.

Flow regulation through joint infrastructure projects

The majority of legal instruments concerning joint infrastructure are concluded on a bilateral basis; however, the *Convention Concerning the Legal Statute of Common Works*, concluded among Senegal, Mali and Mauritania on 21 December 1978, constitutes a rare exception. The reason that bilateral treaties are more common is easily explained by the fact that infrastructure projects are complex undertakings and require a set of detailed regulations regarding technological specification, materials, access to construction sites and cost-sharing rules. In basins with more than two riparian States, however, it is important that bilateral treaties on joint infrastructure projects take into account the basin-wide impact of these works in accordance with the principle of equitable and reasonable utilisation. Some basin-wide framework agreements aim to ensure just that; for example, the 1978 *Amazon Cooperation Treaty* was established to promote regional economic integration as well as environmental protection and sustainable use of the basin, and it outlines overall basin management principles. The treaty mandates that new bilateral or multilateral treaties signed between the parties cannot be 'contrary to the achievement of the common aims for cooperation in the Amazonian region stated in [the] instrument'.[98] Similar rules for new planned measures are included in the *Treaty of the River Plate (Río de La Plata) Basin*[99] and in the 1972 *Convention Concerning the Statute of the Senegal River*.

[97] Art. 10, 2008 *Niger Basin Water Charter*.

[98] Art. XVIII, 1978 *Treaty for Amazonian Cooperation*.

[99] 23 April 1969, 875 *UNTS* 11.

Implementation of the Yacyreta Hydro-Power Project[100] was adopted in the *River Plate Treaty*. Navigation improvements and hydroelectricity production that are envisaged by this treaty also offer opportunities for flow regulation on the Paraná River. The works are co-owned by the two participating States, Argentina and Paraguay, and managed by a public bi-national entity established with equal participation in initial capital investment.[101] The treaty specifies that co-ownership of the joint infrastructure neither confers in any way a right of property or jurisdiction over any part of the territory of the other party nor affects State sovereignty.[102] Yet, these arrangements give both riparian States an equal voice in the joint management of the contiguous river stretch. The same cost-sharing and management model was adopted by Brazil and Paraguay for another hydropower project in the River Plate Basin – namely, the Itaipú hydropower development at Foz do Iguaçu.[103]

Joint flow regulation projects often serve the twin purposes of enhancing economic activities and protection from natural hazards. On the Danube River, the Iron Gates Water Power and Navigation System serves these economic activities and controls ice and flood hazards. The project is composed of two mirror-image electrical power plants on opposite sides of a contiguous river stretch between Romania and Serbia. It includes an overflow-spillway dam, two locks with the same parameters and technical characteristics, a road and works required for construction of a railway on the dam.[104] Whereas the System constitutes a single operating unit,[105] ownership of the infrastructure rests with the party in whose territory it is located. The frontier line, which crosses at the centre of the dam, separates the parts owned by the two States.[106] Management, operation and maintenance of the hydraulic works, however, are carried out by a complex cooperation and coordination mechanism between the two States, including both public national enterprises and joint government mechanisms.[107]

Not all hydraulic works are located on contiguous stretches of rivers or adjacent territories of States. Infrastructure and flow regulation projects on the Senegal River are unique in the way that the riparian States agreed on ownership and management. According to the 1978

100 Project agreed to by Argentina and Paraguay in the *Treaty of Yacyretá*, 3 December 1973, 1380 *UNTS* 80.

101 Art. III (2), 1973 *Treaty of Yacyretá*. 102 Arts. V and VII, 1973 *Treaty of Yacyretá*.

103 *Treaty of Itaipú*, 26 April 1973, 923 *UNTS* 92.

104 Art. 2, 1963 *Final Act Relating to the Establishment and Operation of the Iron Gates Water Power and Navigation System*.

105 Art. 10, *ibid*. 106 Art. 9, *ibid*. 107 Art. 13, *ibid*.

Convention Concerning the Legal Statute of Common Works, any hydraulic works that are carried out based on a legal instrument that declares joint property are 'joint infrastructure'.[108] It is not necessary that those works make contact with the territory of more than one riparian State; they can be located entirely in the territory of only one Member State of the OMVS. Article 4 of this *Convention* stipulates that – without prejudice to the sovereignty of the State on whose territory the infrastructure is located – the co-owning States have an individual right to an indivisible share and a collective right of use, enjoyment of benefits and administration of the joint infrastructure.[109] Flow regulation infrastructure that is located in the territory of one State but operated by another also was agreed to by India and Nepal. On both the Kosi and Gandak tributaries of the Ganges River, India leases land and has the right to carry out works for flood control, hydropower production and irrigation.[110]

The cooperation cases outlined herein highlight that the methods used by States to structure joint management of river regulation take many forms. The possibilities for designing cooperative processes on flow regulation and infrastructure development for economic and/or environmental benefits are endless. When States agree to joint infrastructure, they also generally agree to the establishment of joint mechanisms to manage the complex administration and distribution of benefits of the joint works. Riparian States also employ joint structures not only for infrastructure management; they also establish joint mechanisms for the more general management of transboundary water resources. The pooling of both financial resources and expertise in joint mechanisms is an effective means to reduce costs and enhance the management of shared resources.

5.2 Institutional frameworks for cooperation

The treaty analysis presented in Annex of this volume highlights that States more often refer to some type of joint mechanism for the

[108] Art. 1, 1978 *Convention Concerning the Legal Statute of Common Works*. A 'legal instrument' is understood to be any convention or resolution adopted by a decision-making organ of the OMVS.

[109] '*Sans préjudice de la souveraineté de l'Etat sur le territoire duquel se trouve tout ou partie d'un ouvrage commun, d'un ouvrage accessoire ou annexe, les Etats copropriétaires ont un droit individuel à une quote-part indivisible et un droit collectif d'usage, de jouissance et d'administration de l'ouvrage commun, de ses annexes et accessoires*'.

[110] *Revised Agreement on the Kosi Project*, 19 December 1966, and the *Agreement on the Gandak Irrigation and Power Project*, 4 December 1959, both reproduced in Salman and Uprety, *Conflict and Cooperation*, pp. 349–372.

management of shared water resources. Of the treaties reviewed, 70 per cent established joint mechanisms for their implementation or referred specific tasks to already existing institutions. Regarding the African continent, these results are even more marked: treaty practice after 1960 shows a striking rate of 93 per cent of agreements that established or referred to common institutions, which indicates a regional custom. Joint mechanisms represent the preferred choice of management structure because it is generally considered a more appropriate method to steer utilisation of shared waters than an agreement on rules that each State implements individually in the national context and with little or no coordination.[111]

Institutionalisation of cooperation can overcome State-centric water management practices. Joint mechanisms provide for regular interaction and information exchange among basin States. They provide opportunities to seek clarification about activities from those participating, and they perform an important function by ensuring that agreed-to principles or measures of water management and utilisation are implemented. In addition, joint mechanisms contribute to dispute prevention and settlement.

Structure, design and functional role with respect to cooperation

The design of individual institutions is of key importance in determining their potential and effectiveness for fulfilling their role in international cooperation.[112] Joint mechanisms perform numerous different functions; consequently, there are many varieties of institutional and functional designs for these mechanisms, each adapted to individual characteristics of a basin and the purposes they serve. Some mechanisms cover entire basins, others only parts of them. Their geographic scope is not always limited to individual basins; several bilateral commissions oversee multiple basins that form boundary waters between two States (e.g., the Indo-Bangladesh Joint Rivers Commission and the International Joint

[111] UNECE, *Legal Aspects of Hydroelectric Development of Rivers and Lakes of Common Interest* (Geneva: UNECE, 1952), para. 31.

[112] B. Hettne and F. Söderbaum, *Regional Cooperation* (Göteborg: International Task Force on Public Goods, 2004), pp. 33, 40; other factors that play a role with respect to effectiveness include the human resources and skills of individuals appointed to those mechanisms and the availability of technical and financial resources.

Commission of Canada and the United States);[113] some regional mechanisms address all basins in an area.[114]

Caflisch provided a useful categorisation of different manifestations of joint water management mechanisms.[115] He classified them based on (1) structure and composition, (2) mandates and functions, (3) decision-making procedures and (4) degree of independence from member States. Regarding the first criterion, he distinguished between mechanisms that consist of one sole organ and those composed of multiple organs united under a hierarchical organisational structure. The Permanent Indus Commission, which consists of two high-ranking engineers – one from India and one from Pakistan – who meet regularly and at least once a year, is an example of the first category.[116] Mechanisms with a hierarchical multiple-organ structure are common for management institutions with multiple technical and planning functions. The ICPDR, for example, consists of national delegations that oversee a Standing Working Group, as well as permanent or *ad hoc* Expert Groups for specific fields of work. The Commission is supported by a Permanent Secretariat headquartered in Vienna.[117] Mechanisms consisting of only one organ that meets in regular intervals (e.g., annually or biannually) often do not benefit from the existence of a permanent international Secretariat or another lasting structure that supports them in executing their functions. The majority of mechanisms exist solely as international structures; however, some also include national chapters that interact with the respective body established at the international level. The International Boundary and Water Commission of the United States and Mexico, for example, is composed of two national sections that together have the status of an international body.[118]

The second category, which entails the mandates and functions of mechanisms, concerns the large variety of activities with which joint mechanisms can be tasked. These include the nature of the water

[113] For example, the Indo-Bangladesh Joint Rivers Commission established by the *Statute of the Indo-Bangladesh Joint Rivers Commission*, 24 November 1972, reproduced in Salman and Uprety, *Conflict and Cooperation*, pp. 383–385; and the International Joint Commission established by the 1909 *Boundary Waters Treaty*.

[114] For example, the Secretariat and the Meeting of the Parties of the 1992 *UNECE Water Convention*, and the Water Division of the SADC Secretariat.

[115] Caflisch, '*Règles générales*', pp. 197–207.

[116] Art. VIII, 1960 *Indus Waters Treaty*.

[117] Annex IV, 1994 *Danube Protection Convention*.

[118] Art. 2, 1944 *Treaty Relating to the Utilization of the Waters of the Colorado and Tijuana Rivers and of the Rio Grande*.

utilisations they regulate (e.g., navigation, irrigation and environmental protection); the nature of their function as political or operational mechanisms; and their involvement relative to data generation, information exchange, planning, operational decision making and dispute settlement. The third classification, which relies on decision-making procedures, is straightforward: it distinguishes among consensus, unanimity, opting-out and majority procedures.[119] As will be discussed herein, the fourth classification criterion – that is, integration with or independence from member States regarding decision making – can be a 'mixed blessing' depending on the role of the mechanism in the cooperation process among basin States.

Rather than categorising joint mechanisms based on structure and design, here they are analysed with respect to the functional roles they have in shaping the process of cooperation. One key function they serve is that they form (1) platforms for continuing dialogue and information exchange as well as for planning. When cooperation proves effective, the joint mechanisms can also become (2) important vehicles for enhancing and widening cooperation. At the same time, they have a significant role (3) in ensuring compliance with agreed-to principles and measures and (4) in the prevention and settlement of disputes.

5.2.1 *Platforms for dialogue, planning and operation*

Aquifers, surface waters and their connected environment are complex and sensitive systems that are subject to continuous hydrologic changes. Sound use of these systems requires the regular interaction of experts from a large variety of technical sectors.[120] At the national level, management of these systems regularly falls under the competence of various ministries (e.g., agriculture, national planning, environment and energy). In the context of transboundary water systems, political boundaries add a further element to the complexity of integrated water resources

[119] For example, Article 5 of Annex IV, 1994 *Danube Protection Convention*, provides for decision by a four-fifths majority if consensus if not achievable; Article 7, 1977 *Agreement for the Establishment of the Organization for the Management and Development of the Kagera River Basin*, provided for unanimity; and the IJC established by the United States and Canada makes decisions by simple majority; see Article VIII, 1909 *Boundary Waters Treaty*.

[120] For example, the *Water Resources Management Act of Tanzania*, 12 May 2009, *Tanzania Official Gazette of 15 May 2009, Act Supplement No. 11*, recognises this complexity in providing for a multi-disciplinary composition of the National Water Board.

management (IWRM). As a consequence, early codification efforts of the rules of international water law included recommendations to address the need for regular interaction among those involved in managing shared water systems by establishing joint management mechanisms. As early as 1911, the IIL had recommended the establishment of permanent joint commissions for international watercourses in which management decisions might result in serious consequences.[121] Subsequent resolutions by both the IIL and the ILA reiterated the recommendation,[122] which also was addressed in Article 24 of the *UN Watercourses Convention*. Article 14 of the 2008 ILC *Draft Articles* and Article 64 of the 2004 ILA *Berlin Rules* included an obligation to establish basin commissions or other appropriate machinery for cooperation among interested parties 'wherever appropriate'. This reflects the fact that customary law does not specifically require basin States to set up such mechanisms; therefore, it is merely but strongly recommended.[123] In the absence of a customary rule, an obligation to set up a joint mechanism results from legal instruments, such as in the case of the 1992 *UNECE Water Convention*.

'Joint bodies' in the 1992 *UNECE Water Convention* are defined as 'any bilateral or multilateral commission or other appropriate institutional arrangements for cooperation between riparian States'. Article 9 stipulated not only the obligation to establish joint mechanisms but also provided a non-exhaustive list of tasks that the joint bodies should perform: data collection, exchange of information and drawing up of inventories related to pollution sources, establishment of monitoring programs for water quality and quantity, elaboration of emission limits and quality objectives, development of concerted action programs for pollution reduction and establishment of warning procedures. These joint bodies are expected to serve more generally as a forum for information exchange about uses and installations that are likely to cause transboundary impact. They shall promote cooperation on the best available technology and

[121] Art. II (7) of the IIL 1911 *Resolution on International Regulations Regarding the Use of International Watercourses*: 'It is recommended that the interested States appoint permanent joint commissions, which shall render decisions, or at least shall give their opinion, when from the building of new establishments or the making of alterations in existing establishments, serious consequences might result in that part of the stream situated in the territory of the other States'. (1974) II(2) *YBILC* 200.

[122] For example, Art. 9, 1961 IIL *Salzburg Resolution*; and Arts. XXV and XXXI, 1966 *Helsinki Rules*.

[123] ILA, *Report of the Seventy-First Conference*, p. 404; Tanzi and Arcari, *The United Nations Convention*, p. 191; ILA, *Report of the Fifty-Sixth Conference*, p. 121; Sauser-Hall, '*L'utilisation industrielle des fleuves internationaux*', p. 533.

scientific research programs and they shall participate in EIAs. The list of tasks provided in Article 9 must be viewed in light of the *Convention*'s objective, which is primarily the protection of water systems and mitigation of transboundary impacts. Another document that outlines minimum mandates for joint mechanisms is the non-binding 2004 ILA *Berlin Rules*. Article 65 lists three items over which joint mechanisms should have authority : '(a) coordination and pooling of their scientific and technical research programs, (b) establishment of harmonized, coordinated, or unified networks for permanent observation and control; and (c) establishment of joint or harmonized water quality objectives and standards for the whole of or each significant part of a basin'. Paragraph 2 of Article 65 further provides a useful checklist of elements that should be included in any agreement that establishes a joint mechanism: objective and purpose, nature and composition, form and duration, legal status, area of operation, functions and powers and financial arrangements.

One function that was not explicitly mentioned in the 1992 *UNECE Water Convention* and the 2004 *Berlin Rules*, but which is implicit to all of these mechanisms, is the more general role as a platform for dialogue. Here, the focus is not so much on their role relative to the political character of dialogue but rather on their contribution relative to information exchange and sound water resources management.[124] An increase in mutual knowledge by basin States about their shared hydrologic system provides opportunities for more informed decision making with respect to equitable and reasonable utilisation.[125] Depending on their institutional design, joint mechanisms can (1) serve as simple messengers for the transmission of data from one State to another; or (2) provide formidable mechanisms to pool, process and evaluate this data with a view to applying it towards more effective resource management.

Joint mechanisms that lack a permanent organ and are constituted merely as joint mechanisms during meetings of appointed officials or experts are less likely to be effective in joint data evaluation. In these cases, the time for interaction among riparian States – and, therefore, the time for joint assessment – is limited. Thus, exchanged data is most often submitted for assessment by each State at the national level rather than for

[124] The role that the Permanent Indus Commission and the Joint Israeli–Palestinian Water Committee, for example, play as channels for political dialogue during periods when other diplomatic channels are closed has been hailed by some authors and downplayed by others. The actual record of these institutions remains to be judged by history.

[125] McCaffrey, *The Law of International Watercourses*, p. 478.

joint analysis. Permanent mechanisms and, in particular, those that have a mandate to or can set up joint investigations or research commissions are better suited for joint evaluation of data and joint planning. Their institutional design ensures that unexpected questions can be addressed immediately and cooperatively.[126] The ICPDR is again illustrative: by 2011, its Permanent Secretariat had worked with expert committees on, among others, the issues of river basin management, monitoring and assessment, flood protection, accident prevention and public participation.[127] These committees assist in preparing proposals and recommendations for the ICPDR on respective issues with which they are dealing.[128]

In addition to basin organisations, which exhibit a similar type of comprehensive and integrated structure,[129] numerous entities focus on only one or a few issue areas, including navigation and fisheries, such as the Central Commission for Navigation on the Rhine and the LVFO, respectively. Others have been set up as purely operational mechanisms to carry out specific projects, such as the bi-national entities YACYRETA and ITAIPU. The OMVS is one of the rare commissions that fulfills all of these functions: information exchange, planning and operations. It consists of the Conference of Heads of State and Government, the Council of Ministers, the *Haut Commissariat* and the Permanent Water Commission.[130] The OMVS defines the basin management plan and makes decisions on interventions in the basin. When the Council is not in session, the High Commissioner represents the OMVS with support from the executive organ, the *Haut Commissariat*; he or she also oversees data exchange and the elaboration of joint basin management activities and infrastructure planning.[131] In addition, the OMVS supervises the activities of the two agencies (i.e., international public enterprises) that manage operations and electricity sales of the Diama and Manantali Dams: the *Agence de Gestion et d'Exploitation du barrage de Diama* (SOGED) and the *Société de Gestion de l'Energie de Manantali* (SOGEM), respectively.[132]

126 ILC, 'Sixth Report, McCaffrey', p. 43.

127 ICPDR, *Shared Waters – Joint Responsibilities, ICPDR Annual Report 2009* (Vienna: ICPDR, 2009), p. 4.

128 Art. 8 and Annex IV, 1994 *Danube Protection Convention*.

129 For example, the ICPR, the Lake Victoria Basin Commission and the OMVS.

130 1972 *Convention Establishing the Senegal River Basin Organization*.

131 Arts. 12 and 13, *ibid*.

132 Arts. 15–19, 1978 *Convention Concerning the Legal Statute of Common Works*; and *Convention for the Establishment of the Management Agency for the Operation of the Diama Dam* and *Convention Concerning the Establishment of the Manantali Energy Management Agency*, both

Joint mechanisms, which implement joint projects, are usually vested with a legal 'personality' that allows them to perform their functions independently from State authorities. For example, ITAIPU, the bi-national entity that exploits the hydropower potential at the Iugazu Falls (Foz do Iguazu) on the Paraná River, has been accorded all powers and rights necessary for implementing the infrastructure project and its operation. The *Statute of ITAIPU* specifies that the entity 'shall . . . have the juridical, financial and administrative capacity and the technical responsibility to study, plan, direct and execute the works for which it was established, bring them into service and operate them, for which purposes it may require rights and undertake obligations'.[133] According to the *Statute*, ITAIPU was constituted by the two public electricity companies, the *Centrais Eléctricas Brasileiras* of Brazil and the *Administración Nacional de Electricitdad* of Paraguay.[134] The Supervisory Board is composed of twelve members: six, including a representative of the Ministry of Foreign Affairs, are appointed by each of the two participating States, plus the ITAIPU Director-General and Deputy Director-General participating as non-voting members. The Executive Directorate consists of the Director-General and the Technical, Legal, Administrative, Financial and Coordinating Directors. Similar to other joint mechanisms that have a technical mandate (e.g., the Lake Victoria Basin Commission),[135] this mechanism is characterised by a mixed composition of technical staff and political representatives.

Technical staff and political representatives serve different functions with respect to State cooperation and dialogue about shared water resources. The presence of qualified technical staff ensures that basin management is in line with the highest scientific standards. Equitable and reasonable use becomes subject to scientific rather than political considerations. Institutional designs that bring together primarily technical experts might appear favourable from the point of view of scientifically efficient and sustainable resources management. However, given that the majority of joint mechanisms do not have decision-making authority regarding basin planning and adoption of new uses, this only seems to be

7 January 1997; see A. Samboly Ba and M. Mbengue, '*Le régime juridique du fleuve Sénégal: aspects du droit des cours d'eau dans un context regional*' (2004) 12 *African Yearbook of International Law* 345.

133 Art. 4 of Annex A, *Treaty of Itaipú*, 26 April 1973, 923 *UNTS* 92.

134 Art. 1 of Annex A, *ibid.*

135 See Art. 37, 2003 *Lake Victoria Protocol Concerning the Composition of Sectoral Committees.*

so. In institutions that are limited to providing recommendations rather than having the authority to make decisions, it is important that at least some of the individuals who comprise the institution have a close relationship with or are integrated into State decision-making authorities. For recommendations to be considered by States, it is important that they not only be technically sound but also resonate with political interests of the concerned States.[136]

5.2.2 Enhancing and widening cooperation

As discussed previously, many joint mechanisms have been established as international organisations with an independent legal personality. Some organisations have taken on 'a life of their own', either because States continue to refer additional tasks or widen their geographic scope, or because their mandate is designed in such a way as to allow them to enhance and expand cooperation. The pooling of resources opens up new opportunities for States to pursue activities that they would not have been able to carry out unilaterally. The 2002 decision of the ICJ in the *Case Concerning the Land and Maritime Boundary between Cameroon and Nigeria* pointed out the fact that the States participating in the Lake Chad Basin Commission (LCBC), for instance, had widened the technical scope of the Commission.[137] According to its underlying statute, the LCBC was put in charge of coordinating water uses in the basin.[138] However, following incidents between riparian States in 1983, it was given the additional task of addressing boundary and security issues, thereby widening the mechanism's tasks beyond water management.[139]

Institutional mechanisms in charge of the protection of Rhine River water quality also underwent a transformation that significantly widened their technical mandate. The 1963 *Agreement Concerning the International Commission for the Protection of the Rhine against Pollution*[140] limited the Commission's mandate to scientific research with the objective of gaining a common understanding of pollution. The institutional framework evolved since then and the ICPR, established by the 1999 *Convention on the*

136 Caflisch, '*Règles générales*', pp. 200f.

137 *Land and Maritime Boundary between Cameroon and Nigeria (Cameroon v. Nigeria: Equatorial Guinea Intervening)* (2002) ICJ Reports 332f., para. 36.

138 Annex Art. IX, 1964 *Convention and Statute Relating to the Development of the Chad Basin.*

139 *Land and Maritime Boundary between Cameroon and Nigeria (Cameroon v. Nigeria: Equatorial Guinea Intervening)* (2002) ICJ Reports, pp. 332f., para. 36.

140 29 April 1963, 994 *UNTS* 3.

Protection of the Rhine, has been charged with responsibilities concerning not only water quality and pollution but also species diversity, flow management, production of drinking water, flood prevention and protection of the North Sea from land-based pollution.

The MCR has also been transformed over time. It was initially set up in 1957 as the Committee for Coordination of Investigations of the Lower Mekong among Laos, Thailand, Vietnam and Cambodia to study and recommend potential water resources development projects. The organisation eventually – after an interlude during which Cambodia ceased cooperation for political reasons[141] – transformed into a comprehensive lower-basin planning commission. In recent years, the MRC[142] has enhanced internal cooperation processes by adopting, for example, *Procedures on Data and Information Exchange and Sharing*[143] and *Procedures on Notification, Prior Consultation and Agreement*.[144] When the Commission was established, member States had not been able to reach agreement on either set of cooperation obligations. It has been argued that the iterative process of and positive experience with cooperation within the framework of the MRC facilitated agreement on these more detailed procedures among member States.[145]

Experience with cooperation in the Lower Mekong has not only enhanced cooperative action between member States of the MRC but is also slowly showing signs of geographical expansion – namely, geographic widening of cooperation. With facilitative support from regional organisations – the ADB and the Association of Southeast Asian Nations (ASEAN) – the Lower Mekong States entered into a cooperative dialogue with the two upstream riparians, China and Myanmar.[146] Similarly, the Meeting of the Parties (MOP) of the 1992 *UNECE Water Convention* engaged in efforts to widen the *Convention*'s geographic scope. An amendment to

[141] P. Menon, 'Some Legal Aspects of the Committee for Coordination of Investigations of the Lower Mekong Basin' (1971) 18 *Netherlands International Law Review* 197–216.

[142] Established by the 1995 *Mekong Cooperation Agreement*.

[143] Approved by the MRC on 1 November 2001, www.mrcmekong.org/agreement_95/Agreement-procedures-guidelines.htm.

[144] Approved by the MRC on 30 November 2003, www.mrcmekong.org/agreement_95/Agreement-procedures-guidelines.htm.

[145] S. Schmeier, 'Regional Cooperation Efforts in the Mekong River Basin: Mitigating River-Related Security Threats and Promoting Regional Development' (2009) 2 *Austrian Journal for Southeast Asian Studies* 28–52.

[146] ADB, *The Great Mekong Subregion Economic Cooperation Program Strategic Framework (2012–2022)* (Philippines: ADB, 2011).

Articles 25 and 26 that allows non-UNECE member States to accede to the *Convention* entered into force on 6 February 2013.[147] With this, the geographic scope of activities of the MOP and its Permanent Secretariat will widen, and the geographic applicability of the obligation to establish joint management mechanisms of Article 9 of the *Convention* will expand with the new contracting parties.[148]

A different form of widening cooperation was provided for in Article 18 (6) of the 1994 *Danube River Protection Convention*. This Article called for the widening of inter-institutional cooperation. The ICPDR was authorised to decide on 'cooperation with international or national organizations or other bodies, which are engaged or interested in the protection and water management of the Danube River and of waters within this catchment area or in general questions of water protection and water management'.[149] The intention of this paragraph was to ensure that coordination is taking place among entities whose activities and responsibilities overlap in order to avoid duplication.[150]

For completeness, it must be mentioned that although joint mechanisms can be catalysts for enhanced and widening cooperation, they are not immune to political exigencies beyond water and cannot fulfill their catalytic role independent of politics. The case of the KBO highlighted the fact that the legal personality of a joint basin management organisation is more often only 'quasi-independent' and remains subject to the sovereign rights and decisions of member States. This is particularly the case when the decision making of an organisation is carried out by organs that depend on the presence of State representatives to function. In the given case, the Commission of the KBO was composed of representatives from each country, who met three times annually.[151] Their work was prepared and supported by a Permanent Secretariat but decisions were made by the Commission.[152] The KBO was established in 1977 with a far-reaching

[147] UNECE, *Progress in the Ratification Process, Including the Ratification of the Amendments to Open the Convention to Outside the Region*, 21 November 2012, MOP-6/2012/INF.1/REV.

[148] L. Boisson de Chazournes, '*Sur les rives du droit international de l'eau: entre universalité et particularismes*', in M. G. Kohen (ed.), *Promoting Justice, Human Rights and Conflict Resolution through International Law / La promotion de la justice, des droits de l'homme et du règlement des conflits par le droit international, Liber Amicorum Lucius Caflish* (Leiden: Koninlijke Brill NV, 2007), p. 699.

[149] Art. 18, 1994 *Danube Protection Convention*.

[150] Similarly, Arts. 5 and 6, 2002 *Framework Agreement on the Sava River*.

[151] Arts. 4 and 5, 1977 *Agreement for the Establishment of the Organization for the Management and Development of the Kagera River Basin*.

[152] Art. 7, *ibid*.

mandate for water resources management and economic development in the basin area and *inter alia* the capacity to approve projects and enter into financing agreements with international institutions.[153] It became defunct in the early 1990s due to political differences among its member States – Burundi, Rwanda, Tanzania and Uganda – and was formally dissolved in July 2004.[154] In most cases, however, hydrologic interdependence and institutionalised cooperation on shared waters prove resilient to political relationships. A number of joint mechanisms function despite adverse political landscapes.[155] The KBO was only formally dissolved once it was clear that another entity had been established that could assume most of the original tasks. Today, the Coordination Unit for the Nile Equatorial Lakes Subsidiary Action Program of the Nile Basin Initiative carries out parts of the original Kagera Basin Development Program.[156]

5.2.3 Ensuring compliance

Joint mechanisms also have an important role in ensuring compliance with agreed-to principles and rules for the management and utilisation of shared waters. They may accomplish this task through increase in transparency, compliance review and/or implementation support.

Enhancing compliance through data exchange

Mechanisms that increase transparency may range from *ad hoc* investigation commissions and simple mechanisms of regular information exchange to the more permanent structures of international organisations. The 1994 *Agreement between China and Mongolia on the Protection and Utilization of Transboundary Waters* refers to all lakes, rivers, streams and waters that 'straddle or rest on the boundary line' between the two States

153 Arts. 2 and 7, *ibid*.

154 R. Mbaziira, N. Senfuma and R. McDonnel, *Institutional Development in the Nile Equatorial Lakes Sub-basin: Learning from the Experience of the Kagera Basin Organisation*, IWMI Conference Paper presented at the East Africa Integrated River Basin Management Conference 7–9 March 2005, Sokoine University of Agriculture, Tanzania (2005).

155 Regarding the Mixed Commission of Israel and the Palestinian Authority, see M. Tignino, '*L'eau dans le processus de paix aux Proche-Orient: elements d'un régime juridique*', Working Paper 04/02, Graduate Institute of International Studies, Geneva (2004), p. 52; regarding the Permanent Indus Commission, see N. A. Zawahri, 'India, Pakistan and Cooperation along the Indus River System' (2009) 11 *Water Policy* 1–20.

156 The Nile Basin Initiative (NBI) was established by nine Nile riparian States in 1999 by decision of the Nile Council of Ministers of Water Affairs as a transitional institutional mechanism pending the conclusion of negotiations on the *Agreement on the Nile River Basin Cooperative Framework*.

and related ecosystems.[157] It establishes the possibility of both creating *ad hoc* investigation commissions and developing a simple mechanism of regular information exchange concerning compliance with treaty terms. Joint investigation and survey commissions can be established *inter alia* to examine the resources and quality of the boundary waters, as well as changes that affect them.[158] The use of this tool of compliance review is optional for the parties. In addition, the *Agreement* establishes a Joint Committee on Transboundary Waters, which is composed of one representative and two deputy representatives from each party.[159] This Committee convenes biannually to discuss implementation of the *Agreement.*[160] The fixation of specific intervals for meetings ensures that the parties meet regularly to assess treaty implementation and compliance with its terms.

It is worth noting that in this context, the obligation of regular data and information exchange per se already has a role with respect to compliance review. The exchange of relevant data gives the State that is receiving information insight into whether the State that is providing data is in compliance with treaty terms. The exchange of hydrometeorological and flow data, for example, permits inferences on the existence and extent of consumptive water uses.

A mandate for implementation review – similar to yet more explicit than those established for the Joint Sino-Mongolian Committee – was attributed to the MOP of the 2002 *Framework Agreement on the Sava River Basin*, a Danube River tributary. The *Framework Agreement* is a comprehensive basin management agreement governing all utilisations and aspects of water resources management, including navigation. The MOP oversees the work of the International Sava River Basin Commission and convenes at least once every two years. According to Article 14 of the *Agreement*, continuous implementation review is carried out by the MOP on the basis of the Commission's reports. Based on the outcome of this review, the MOP makes the necessary decisions to achieve the treaty's purpose.

The Conference of the Parties (COP) of the 1994 *Danube Protection Convention* equally reviews treaty compliance.[161] The COP meets only upon recommendation by the ICPDR and makes binding decisions regarding policy issues concerning implementation. With respect to decision making,

157 Art. 1 1994, *Agreement between China and Mongolia on the Protection and Utilization of Transboundary Waters.*

158 Arts. 2 and 3, *ibid.* 159 Art. 10, *ibid.* 160 Art. 11, *ibid.*

161 Art. 22, 1994 *Danube Protection Convention.*

the *Danube Protection Convention* is one of the rare treaties that include the possibility of decision making by majority.[162] If efforts to reach agreement by consensus are unsuccessful, decisions shall be made by a four-fifths majority of the parties present and voting.[163] The same decision-making rule applies to both the COP and the ICPDR.[164] This exceptional decision making by majority is due to the fact that the *Convention* has fifteen Contracting Parties: that is, fourteen riparian States and the EU.[165] In addition to compliance review on the highest political level, the 1994 *Danube Protection Convention* provides *inter alia* for the creation of data inventories of relevant point and non point source pollution and actual emissions.[166] The parties are under obligation to update these inventories periodically, the databases of which are easily accessible to all registered users via the Internet.[167] Continuous compliance monitoring by all parties is made possible using these means.

Joint mechanisms that facilitate compliance review are more common in treaties that address water quality protection. Riparian States usually have shared or aligning interests in keeping water quality at a beneficial level; cooperation is usually less controversial. Yet, compliance review mechanisms are not exclusive to agreements that govern water quality; they are also found concerning flow management. In the Columbia River Basin, compliance is monitored by a Permanent Engineering Board.[168] Cooperation between the United States and Canada on the Columbia River serves the twin objectives of flood management and valorisation of hydropower potential. These two objectives do not always align; on the Columbia River, which is characterised by a cascade of reservoirs and dams, flood prevention may require discharge of reservoir waters through overflow gates rather than storage for hydropower production. The Permanent Engineering Board assembles flow records and reports any 'substantial deviation from hydroelectric and flood control operating plans' to the Contracting Parties and it provides 'recommendations for remedial action and compensatory adjustment', when appropriate.[169]

162 Only 12 of the 219 treaties analysed in Annex allow for majority decision making in certain instances.

163 Art. 22 (4), 1994 *Danube Protection Convention*. 164 Annex IV, Article 5, *ibid*.

165 Status as of 28 February 2013. 166 Art. 8 1994, *Danube Protection Convention*.

167 http://danubis.icpdr.org/pls/danubis_sso/danubis_sso.danubis_login?

168 Art. XV, 1961 *Treaty between Canada and the United States of America Relating to Co-operative Development of the Water Resources of the Columbia River Basin*.

169 *Ibid*.

The agreements outlined herein as well as others include provisions on compliance review, yet the question is whether they also have a role in facilitating, ensuring, or even enforcing compliance. Applying Axelrod's findings on the iterative prisoner's dilemma,[170] those joint mechanisms – which are designed to ensure regular interaction and information exchange among concerned States – elicit cooperative behaviour and facilitate compliance because costs and benefits of cooperation then become transparent.[171] Commissions that convene periodically and, even more so, mechanisms that are supported by Permanent Secretariats and electronic data exchange systems facilitate compliance by 'soft means', such as transparency through data exchange and regular State interaction. Transparency has been shown to improve compliance because peers engage in continuous mutual assessment. The knowledge that a cooperative quid pro quo course of action is being pursued by others expedites a positive feedback loop of cooperation.[172]

Facilitating compliance through technical assistance

In addition to this 'soft' function of institutionalised cooperation, a number of joint mechanisms were vested with a legal mandate to reprimand States for failure to comply and, at the same time, to facilitate compliance with treaty terms. An interesting example in this context is the legal framework established by the UNECE. A compliance committee was established for implementation of the *Protocol on Water and Health to the 1992 UNECE Water Convention.*[173] According to the objective of the *Protocol* to promote human health through improved water management and ecosystem protection at the national and international levels, the instrument stipulates cooperation obligations for States including management of the transboundary impact of water-related diseases in shared hydrologic systems.[174] Based on Article 15 of the *Protocol*, the Compliance Committee is composed of nine elected individuals who serve in their personal capacity. The Committee reviews possible cases of non-compliance brought before it either (1) by parties self-submitting due to their own assessment that they will be unable to fully comply with obligations; (2) by parties that

[170] See Part I, Chapter 1, in this volume.

[171] Axelrod and Hamilton, 'The Evolution of Cooperation', pp. 1390–1396.

[172] Axelrod, 'The Evolution of Cooperation', pp. 189f.

[173] *Protocol on Water and Health to the 1992 Convention on the Protection and Use of Transboundary Watercourses and International Lakes* (*Protocol on Water and Health*), 17 June 1999, MP.WAT/2000/1, EUR/ECP/EHCO 020205/8Fin.

[174] Arts. 1 and 11–13, *ibid.*

have reservations about another party's compliance; or (3) upon referral from the Secretariat. The Committee also reviews communications from individuals.[175] The procedure is 'facilitative, non-adversarial and cooperative in nature' and is guided 'by the principles of transparency, fairness, expedition and predictability'.[176] In addition, the Committee reviews regular reports from the parties about their progress in implementing the *Protocol*.[177] The first such review exercise showed that although no non-compliance submissions, referrals, or communications were registered, compliance deficiencies nonetheless exist.[178] As a consequence, the Committee decided that it would enter into direct consultations with concerned parties with a view to offering effective and tailor-made advice to improve compliance. The MOP to the 1992 *UNECE Water Convention* equally decided to provide 'the Convention with an institutional and procedural mechanism to facilitate and support implementation and compliance on a multilateral and cooperative basis.'[179] The decision establishing an Implementation Committee 'with the aim to facilitate, promote and safeguard the implementation and application of and compliance with the Convention' was made at its sixth session held in November 2012.[180]

Experience with compliance mechanisms in the context of implementation of two other UNECE conventions, which are not entirely devoted yet are related to the management of transboundary water systems, highlights that the implementation of decisions and recommendations of compliance mechanisms is not without problems. The construction of the Danube–Black Sea Deep Water Navigation Canal (i.e., the Bystroe Channel) in the Ukrainian sector of the Danube Delta prompted *inter alia* involvement of the Implementation Committee of the 1991 *Convention on Environmental Impact Assessment in a Transboundary Context (Espoo Convention)*. In 2008, the *Convention*'s Implementation Committee found that Ukraine had not adhered to its obligations under the *Convention* during preparation and implementation of this project, which has considerable

[175] *Ibid.*

[176] MOP Decision I/2, *Report of the Meeting of the Parties to the Protocol on Water and Health to the Convention on the Protection and Use of Transboundary Watercourses and International Lakes on its First Meeting, Geneva, 17–19 January 2007*, ECE/MP.WH/2/Add.3, EUR/06/5069385/Add.3.

[177] Art. 7, 1999 *Protocol on Water and Health*.

[178] Compliance Committee, *Report of the Fifth Meeting of the Compliance Committee*, 22–23 June 2010, ECE/MP.WHC/C.1/2010/4, EUR/10/56335/16.

[179] 'Workplan for 2010–2011', in *Report of the Meeting of the Parties on its Fifth Session*, Geneva, 10–12 November 2009, ECE/MP.WAT/29/Add.1.

[180] UNECE, *Draft Decision on Support to Implementation and Compliance*, 14 September 2012, ECE/MP.WAT/2012/L.4.

transboundary impact on the sensitive ecosystem of the delta, including the Romanian sector.[181] At the fifth meeting in 2011, the MOP of the *Espoo Convention* declared the caution issued previously as effective and requested the Government of Ukraine to report on its measures to bring the project into full compliance by the end of each year.[182] Another caution was issued to Ukraine by the MOP of the *Convention on Access to Information, Public Participation in Decision-Making and Access to Justice in Environmental Matters* (hereinafter, the *Aarhus Convention*)[183] regarding its non-compliance with *Convention* obligations since 2005 within the context of the Bystroe Channel project.[184] In contrast to the caution issued under the *Espoo Convention*, which merely entails additional reporting obligations, cautions issued under the *Aarhus Convention* may result in the suspension of rights and privileges accorded to a party under the *Convention* if it remains unanswered.[185]

The Bystroe Channel case illustrates that the process of enhancing compliance and its enforcement is not easy. The role of institutionalised cooperation – in this case, through the MOPs of the *Conventions* – should not be underestimated in this respect. Institutionalised cooperation mechanisms bear an advantage with respect to compliance *vis-à-vis* the absence of such mechanisms or *vis-à-vis*, for example, third party dispute settlement mechanisms. Joint mechanisms facilitate compliance with decisions to which parties agreed to adhere because parties continue to interact within their structures and remind one another of their mutual obligations and rights. In contrast, when parties submit a dispute, for example, to a court or tribunal, in principle they have agreed to adhere to the decision that follows; however, implementation can prove to be difficult. Once the adversary proceedings are over, and unless bound by rights or obligations providing otherwise, the parties may go their separate ways. International courts and tribunals are not equipped with structures that

[181] UNECE Implementation Committee, *Findings and Recommendations Further to a Submission by Romania Regarding Ukraine (EIA/IC/S/1)*, ECE/MP.EIA/2008/6; B. Aurescu, '*L'efficacité des dispositions sur le règlement des différends dans les conventions environnementales, étude de cas: le projet ukrainien du canal navigable "Bystroe" dans le delta du Danube*', in B. Aurescu and A. Pellet (eds.), *Actualité du droit des fleuves internationaux* (Paris: Pedone, 2010), pp. 265–283.

[182] Decision V/4 adopted by the MOP of the *Espoo Convention*, in *Report of the Meeting of the Parties on its Fifth Session*, 16 August 2011, ECE/MP.EIA015, p. 19.

[183] *Convention on Access to Information, Public Participation in Decision-Making and Access to Justice in Environmental Matters (Aarhus Convention)*, 25 June 1998, 2161 *UNTS* 447.

[184] Decision IV/9h, in *Report of the Fourth Session of the Meeting of the Parties, Addendum, Chisinau Declaration and Decisions Adopted by the Meeting of the Parties, Chisinau, 29 June–1 July 2011*, ECE/MP.PP/2011/2/Add.1, pp. 57f.

[185] *Ibid.*

monitor or could facilitate subsequent compliance with their decisions; implementation is left to the States or to the good offices of a third party.

5.2.4 Preventing and settling disputes

By virtue of institutionalising State interaction and cooperation, joint mechanisms also facilitate dispute prevention and settlement.[186] This role is closely related to the institutions' role in providing platforms for dialogue. Consultations and negotiations are diplomatic means that may prevent the States' conflicting interests from leading to related disagreements or disputes. Salman affirmed that 'in practice, the line between consultations as a means of cooperation and [of] dispute avoidance, and negotiations as a dispute settlement mechanism may not be that easy to draw'.[187] The 1966 *Helsinki Rules*, for example, considered the creation of joint agencies first and foremost as a means for dispute settlement rather than as a suitable way to manage and develop shared waters equitably and reasonably.[188] It was only in 1976 that the ILA considered this second role in its *Supplementary Rules Applicable to the Administration of International Water Resources*.[189]

It is worth pointing out that the existence of a joint mechanism at the same time provides no guarantee that a legal dispute between participating States can be avoided.[190] This has been illustrated, for example, by the differences between India and Pakistan concerning the construction of the Baglihar Dam and their dispute about the construction of the Kishenganga and Neelum-Jhelum Dams. The Baglihar Dam is a hydropower project located on the Chenab River, a tributary of the Indus System that passes through Indian territory. Pakistan contested that the technical specifications of the dam did not meet the specific requirements agreed to in Annexure D of the 1960 *Indus Waters Treaty*.[191] Despite regular

186 L. Boisson de Chazournes, 'The Role of Diplomatic Means of Solving Water Disputes: A Special Emphasis on Institutional Mechanisms', in Permanent Court of Arbitration, *Resolution of International Water Disputes* (Leiden: Kluwer Law International, 2003), pp. 91–110; Salman, 'Evolution and Context', p. 89.

187 Salman, 'Evolution and Context', p. 89.

188 Arts. XXXI and XXV, 1966 *Helsinki Rules*.

189 ILA, *Report of the Fifty-Seventh Conference*, pp. 248–256.

190 'Disagreement on points of law or facts' according to the definition provided by the ICJ in *Land and Maritime Boundary between Cameroon and Nigeria, Preliminary Objections* (1998) ICJ Reports, para. 87.

191 *Baglihar Hydroelectric Plant, Expert Determination on Points of Difference Referred by the Government of Pakistan under the Provisions of the Indus Waters Treaty, Executive Summary*, 12 February 2007, http://siteresources.worldbank.org/SOUTHASIAEXT/Resources/223546-1171996340255/BagliharSummary.pdf.

meetings of the Permanent Indus Commission and its mandate to settle disputes over the application of the *Treaty*, the Commission was not able to resolve the differences; the case was eventually referred to and settled by a Neutral Expert.[192] If a dispute settlement through the Commission is unsuccessful, the *Indus Waters Treaty* provides for alternative means by decision of a Neutral Expert for any difference in the interpretation or application of the *Treaty* or the existence of any fact that might constitute a breach of the *Treaty*. In the case of a dispute according to Article IX, the Governments are called on to resolve the dispute by agreement, and they may enlist a mediator to assist in their efforts or they may submit the dispute to a Court of Arbitration.[193] A new disagreement over the application of the *Indus Waters Treaty* emerged concerning a new hydroelectric infrastructure on the Jhelum River tributaries: in May 2010, Pakistan instituted arbitral proceedings regarding India's Kishenganga Hydroelectric Project. The Court of Arbitration, constituted in accordance with the *Indus Waters Treaty* in the matter of the Kishenganga Arbitration, rendered a partial award on 18 February 2013.[194]

It is common that treaties establish an obligation to first attempt to settle any disagreement by negotiation. If a treaty also provides for the establishment of a joint mechanism, the task of dispute settlement by negotiation is frequently assigned to those institutions.[195] In cases in which negotiations through the joint institution remain unsuccessful, treaties typically stipulate that the disagreement be turned over for agreement on the political level, among the States directly, and/or that concerned States should seek dispute resolution by third party intervention. The advantage of resolving a dispute within the framework of a joint institution is that, in many cases, technical experts will be engaged. If a disagreement is of a purely technical nature, technical experts are better placed to find a pragmatic solution than legal experts, politicians, or public opinion.[196]

The International Joint Commission (IJC), which was established by the 1909 *Boundary Waters Treaty of the United States and Canada*, is an exceptional case in which the Contracting Parties vested a joint (technical) mechanism

[192] Art. IX (1), 1960 *Indus Waters Treaty*.

[193] Art. IX and Annexures F and G, 1960 *Indus Waters Treaty*.

[194] *Indus Waters Kishenganga Arbitration (Pakistan v. India)*, final award expected December 2013, www.pca-cpa.org.

[195] For example, Art. 11, *Treaty Concerning Integrated Development of the Mahakali River Including Sarada Barrage, Tanakpur Barrage and Panchewar Project*, 12 February 1996, 36 *ILM* 531.

[196] Salman, 'Evolution and Context', p. 90.

with quasi-judicial functions.[197] Differences 'involving rights, obligations, or interests of either in relation to the other or to the inhabitants of the other, along the common frontier' can be submitted by either Contracting Party for examination and report or 'may be referred for decision to the International Joint Commission by the consent of the two'.[198] The IJC is composed of six independent commissioners: three highly qualified technical experts nominated by each party,[199] who render their decisions by majority.[200] However, even though at the time of conclusion of the treaty the parties seemed confident that the IJC would be an adequate mechanism for dispute resolution, they have not yet made use of this function.[201]

5.3 Conclusion: Managing complexity through joint action

At least three factors contribute to the decision of co-riparian States to cooperate: (1) the multitude of possible water uses; (2) the complexity of interrelationships among these uses, as well as among uses and their transboundary and/or environmental impact; and (3) the inevitable interdependence established by shared hydrologic systems. States do not always regulate their collaboration through international legal instruments; when they do, they have designed their interactions in various ways. Treaty measures include the formulation of simple obligations to harmonise policy through consultation, implementation and co-ownership of joint projects and infrastructure, and establishment of joint institutions with independent decision-making power. More often, agreements are not limited to cooperation with respect to one use only. Interrelationships among uses also require coordination of interventions regarding other uses. Therefore, for example, treaties concerning ecosystem protection include rules on navigation and fisheries management; agreements to control water quality for human use include cooperation obligations with respect to ecosystem protection; and cooperation in the establishment of navigation and hydropower infrastructure results in mutual protection against natural hazards such as floods. In the majority of cases, States establish joint mechanisms to enhance management of this hydrologic complexity.

197 L. H. Legault, 'The Roles of Law and Diplomacy in Dispute Resolution: The IJC as a Possible Model' (2000) 26 *Canada–United States Law Journal* 50.

198 Arts. IX and X, 1909 *Boundary Waters Treaty*. 199 Art. VII, *ibid.*

200 Art. X, *ibid.* 201 Legault, 'The Roles of Law and Diplomacy', p. 51.

Institutionalisation of cooperation in this way is beneficial in improving continuity and frequency of dialogue and data exchange, as well as in addressing differences before they develop into disputes. Institutions can be catalysts for geographic widening of cooperation, either by serving as dialogue platforms to draw in new member States or by establishing new forms of inter-institutional cooperation. Furthermore, joint institutions that are supported by *ad hoc* or permanent expert bodies benefit from additional scientific knowledge and exchange; the level of technical information regarding sustainability of resource use and basin conditions is usually enhanced. This improvement in scientific knowledge, in particular, can lead to the expansion of cooperation into new sectors and issue areas that have been identified as requiring joint attention. This dynamic is highlighted in Chapters 6 and 7, which illustrate how States address new issue areas with which they are increasingly confronted and how institutions can be effective indeed in devising innovative responses to new water management challenges.

PART III · DEALING WITH EVOLVING CHALLENGES

"In time and with water, everything changes".

Leonardo da Vinci

States are frequently confronted with new challenges to which they must adapt. Human society, ecosystems and the water cycle are subject to constant change. Several factors precipitate change: population growth and economic development increase human demand for freshwater and anthropogenic climate change contributes to changes in precipitation patterns, increasing the likelihood of severe floods and prolonged droughts. The irregular distribution of water across regions and the increasing severity of hydrologic events require new approaches to address growing water scarcity and the harmful impacts of floods and droughts on human health, life and property, as well as on ecosystems. Part III analyses the legal response to water scarcity that aims to ensure the satisfaction of vital water needs of individuals, as well as the legal measures that were adopted to address the effects of climate change through cooperative action.

6 International protection of vital human water needs

The number of people living on this planet is steadily increasing. In 2011, world population hit the 7 billion mark and it is expected to reach close to 10 billion by 2050. At the same time, '[w]ater use has been growing much faster than population . . . '; whereas population size quadrupled during the past century, water use grew by a factor of seven.[1] The impact of population growth and economic development on water availability is much more pronounced than the impact of climate change:[2] urban sprawl in flood plains has led to ever greater flood damage;[3] industrial water uses and associated pollution have led to disputes over basic water supply for human needs at the national level;[4] construction of hydraulic infrastructure and filling of dam reservoirs, in some instances, has reduced water flows to unsustainable levels for local irrigation needs of farmers downstream;[5] and more than 880 million people in the world lack access to safe drinking water.[6]

1 The report calculated that UNDP, *Human Development Report 2006 – Beyond Scarcity: Power, Poverty and the Global Water Crisis*, (New York: Palgrave Macmillan, 2006), p. 136.

2 C. Vörösmarty *et al.*, 'Global Water Resources: Vulnerability from Climate Change and Population Growth' (2000) 289 *Science* 284–288.

3 B. E. Montz, 'The Generation of Flood Hazards and Disasters by Urban Development of Floodplains', in D. J. Parker (ed.), *Floods, Volume 1* (London: Routledge, 2000), pp. 116–128.

4 The extraction of oil from tar sands and deforestation activities, for example, threaten basic human water needs and the livelihoods of the Lubicon Lake indigenous populations in Canada. *Lubicon Lake Band v. Canada*, Communication No. 167/1984, 26 March 1990, UN Doc. CCPR/C/38/D/167/1984.

5 I. Mazlum, 'Transboundary Water Management in the Euphrates-Tigris Basin: Dynamics of Regional Cooperation, Sustainability and Governance', in C. Lipchin *et al.*, *The Jordan River and the Dead Sea: Cooperation Amid Conflict* (New York: Springer, 2009), pp. 139–164.

6 GA *Resolution 64/292 on the Human Right to Water and Sanitation*, 28 July 2010, UN Doc. A/RES/64/292.

6.1 Vital human water needs – an issue worthy of legal protection

In light of these developments, the international society of States has moved closer together in recent decades with a view to ensuring that access to sufficient and safe water can and will be provided to people in the long term. As one initiative,[7] States adopted the Millennium Development Goals (MDGs), which include specific numeric targets related to increasing water supply, and they have aligned their national and international development objectives with these goals. Goal Number Seven concerning environmental sustainability includes the target to reduce by half the proportion of people without sustainable access to safe drinking water by 2015. Other MDG targets cannot be achieved without effective water-resources management – for example, the target to reduce by half the proportion of people who suffer from hunger by 2015.[8] Additionally, both the GA and the Human Rights Council (HRC) recognised the existence of a human right to water.[9]

With recognition of 'the right to safe and clean drinking water and sanitation', the GA and the HRC added a strong affirmative voice in a protracted international debate on whether there is a self-standing human right to water.[10] Neither the International Covenant on Civil and Political Rights (ICCPR)[11] nor the International Covenant on Economic, Social and Cultural Rights (ICESCR)[12] refers to water in its text. A number of international and regional human rights instruments refer to water-related obligations but do not recognise a self-standing right to water.[13] This

[7] For others, see preamble, *ibid.*

[8] World Summit on Sustainable Development, *Johannesburg Declaration on Sustainable Development*, 26 August–4 September 2002, UN Doc. A/CONF.199/20, and the *Plan of Implementation of the World Summit of Sustainable Development* in its Annex.

[9] GA *Resolution 64/292*; GA *Resolution 54/175 on the Right to Development*, 17 December 1999, para. 12 (a), UN Doc. A/RES/51/175; HRC *Resolution 15/9 on Human Rights and Access to Safe Drinking Water and Sanitation*, 30 September 2010, UN Doc. A/HRC/RES/15/9.

[10] For example, S. McCaffrey, 'Human Right to Water: Domestic and International Implications' (1992–1993), *Georgetown International Environmental Law Review* 1–24; J. Scanlon, A. Cassar and N. Nemes, *Water as a Human Right?* (Gland: IUCN, 2004); K. Bourquain, *Freshwater Access from a Human Rights Perspective – A Challenge to International Water and Human Rights Law* (Leiden: Martinus Nijhoff Publishers, 2008); S. Salman and S. McInerney-Lankford, *The Human Right to Water – Legal and Policy Dimensions* (Washington, DC: World Bank, 2004).

[11] *International Covenant on Civil and Political Rights* (ICCPR), 16 December 1966, 999 *UNTS* 171.

[12] *International Covenant on Economic, Social and Cultural Rights* (ICESCR), 16 December 1966, 993 *UNTS* 3.

[13] Including 1979 *Convention on the Elimination of All Forms of Discrimination against Women*, 19 December 1979, 1249 *UNTS* 113; 1989 *Convention on the Rights of the Child*, 20 November

situation led to the original debate on whether such a right actually exists. The right to water was first suggested in the *Mar del del Plata Action Plan* in 1977[14] and was later recalled in Principle No. 4 of the 1992 *Dublin Statement on Water and Sustainable Development*[15] and in Agenda 21 at the 1992 *Rio Conference on the Environment and Development*.[16] In 2002, the Committee on Economic, Social and Cultural Rights (CESCR) issued General Comment 15 on the right to water,[17] in which it defines the right as entitling 'everyone to sufficient, safe, acceptable, physically accessible and affordable water for personal and domestic use'.[18] The general comments, which are published regularly by the CESCR, are not legally binding on States, yet they unfold their value in practice as sources of interpretation of the content of Covenant obligations.[19] Today, General Comment 15 represents the most comprehensive definition of human rights obligations concerning water.

In parallel to the international human rights debate on water, the protection of basic water needs of individuals has been crystallising as an issue of legal protection within the context of international water law. Cooperation obligations – such as those related to the implementation of the principle of equitable utilisation and the implementation of planned measures on transboundary water systems – seek to ensure that economic and social needs and interests of all riparian States are taken into account. Additionally, the 1997 *UN Watercourses Convention* and the 2008 ILC *Draft Articles on the Law of Transboundary Aquifers* introduced a certain priority status to the satisfaction of vital human needs *vis-à-vis* other uses; this is a novelty in the context of universal rules. Vital human water

1989, 1577 *UNTS* 3; *African Charter on the Rights and Welfare of the Child*, 11 July 1990, OAU Doc. CAB/LEG/24.9/49 (1990); 2000 *Protocol to the African Charter on Human and People's Rights on the Rights of Women in Africa*, OAU Doc. CAB/LEG/66.6 (13 September 2000); and *Convention on the Rights of Persons with Disabilities*, 13 December 2006, *UNTS* I-44910.

14 '... all peoples, whatever their stage of development and their social and economic conditions, have the right to have access to drinking water in quantities and of a quality equal to their basic needs'. Resolution II in *Report of the United Natrions Water Confernce, Mar del Plata, 14–25 March 1977* (New York: United Nations, 1977), UN Doc. E/CONF.70/29.

15 Adopted at the International Conference on Water and the Environment, 31 January 1992, Ireland, www.un-documents.net/h2o-dub.htm.

16 Chapter 18, para. 47, UN Doc. A/CONF.151/26/REV.1 (Vol. II).

17 CESCR, *General Comment No. 15 (2002) The Right to Water (Arts. 11 and 12 of the International Covenant on Economic, Social and Cultural Rights)*, UN Doc. E/C.12/2002/11.

18 *Ibid.*, para. 2.

19 M. Tignino, *Le statut de l'eau en temps de conflit armé*, Ph.D. thesis, Graduate Institute of International and Development Studies, Geneva (2008), pp. 260f.; S. Salman, 'From Marrakech (Part Two)', p. 14.

needs and the right to water are also increasingly recognised in basin treaties; to a certain extent, this trend is also supported by international jurisdiction.

This chapter discusses the process of how protection of vital human water needs emerged as a topic of regulation and analyses means by which this protection is realised and by which people can claim their right to water. It is shown herein that the concurrence of cooperation obligations of international water law and human rights law – and the complementarity of these two bodies of norms – has a key role in ensuring effective protection of vital human water needs in a transboundary context.

6.2 Emerging protection under international water law

It must be mentioned that the concept of vital human water needs, which is used in international water law, is not fully congruent with the scope and definition of the human right to water that have been presented. General Comment 15 and the HRC resolution define the right to water as 'derived from the right to an adequate standard of living and inextricably related to the right to the highest attainable standard of physical and mental health, as well as the right to life and human dignity'.[20] This definition equates the right to water – first and foremost – to economic, social and cultural rights. The content of the definition describes not merely considerations of human survival but also cultural aspects, as well as the obligation for progressive realisation and continued amelioration of living conditions, including improved accessibility to and availability of water supply.[21] In contrast, the ILC definition of 'vital human water needs' provided for in the law of transboundary water resources is more limited in scope. According to the ILC, vital human water needs are satisfied by the provision of sufficient water to 'sustain human life, including both drinking water and water required for production of food in order to prevent starvation'.[22]

[20] HRC *Res.* 15/9. [21] Art. 11, ICESCR and CESCR, General Comment 15, pp. 5f.

[22] ILC, ' Draft Articles . . . and Commentaries', p. 110; and ILC, 'Draft Articles on the Law of Transboundary Aquifers'. A similar though wider definition is offered in Art. 3 (20), 2004 *Berlin Rules*: '"Vital human needs"' means waters used for immediate human survival, including drinking, cooking and sanitary needs, as well as water needed for the immediate sustenance of a household'.

6.2.1 *Protection through the obligation not to cause significant harm and the principle of equitable utilisation*

International water law focuses primarily on State-to-State rights and obligations, but water needs of individuals are not ignored. General principles, such as the obligation not to cause significant harm and the principle of equitable and reasonable utilisation, serve to protect human water needs. With respect to the no significant harm obligation, for example, basin States are under obligation to exercise due diligence in the utilisation of the shared water resources that flow within their territory so as not to cause significant damage in the territory of other basin States.[23] The threshold of 'significant' damage is harm that can be measured by objective evidence and that causes a 'real impairment of use' by others.[24] Hence, water use in one State that pollutes shared water resources to such an extent that basin populations depending exclusively on these resources in another State can no longer use them to satisfy their vital needs constitutes such a 'significant' harm. In cases in which pollution reaches such levels that it is no longer safe for the health of the dependent population to drink the water or to eat fish from it, these basic human needs are protected through the obligation not to cause significant harm in the territory of riparian States. Some protection is also provided through related cooperation obligations that aim to prevent transboundary harm. Notification and consultation obligations related to the planning and implementation of new measures and to emergency situations further reinforce the protection of vital human needs.[25] They provide for a State whose population is potentially affected to voice its interests and rights. When a State plans to build hydraulic infrastructure that would alter the downstream flow regime, the obligation to notify potentially affected States facilitates the exchange of information between the planning State and the affected State and includes water needs of the dependent population.

The principle of equitable utilisation equally facilitates protection of basic water needs of basin populations. Factors to be considered when determining what constitutes equitable and reasonable use in individual basins include social and economic needs of concerned basin States, as well as basin populations dependent on the water system.[26] These two

[23] ILC, 'Draft Articles . . . and Commentaries', p. 103.

[24] *Ibid.*, p. 94. [25] Discussed in Part II of this volume.

[26] For example, Art. 6, 1997 *UN Watercourses Convention*; Art. 5 (1), 2008 ILC *Draft Articles*; Art. V, 1966 *Helsinki Rules*; and Art. 3 (8), *SADC Revised Protocol on Shared Watercourses*.

factors aim to ensure that the water needs of basin populations are taken into account in the processes of evaluating the distribution of uses and benefits between basin States and of deciding on new uses. Yet, the factors are weighed against other interests. They form part of a much larger, non-exhaustive list of criteria that includes geographic, hydrologic and climatic characteristics of a basin, as well as the consideration of other existing and potential water uses and the availability of alternatives. There is no predetermined hierarchy among these factors; the weight of each factor is determined by the characteristics of individual water systems.[27] A priority consideration of vital human water needs and their protection, therefore, is not guaranteed by the principle of equitable and reasonable utilisation. For example, in situations in which basic water and food needs can be met by alternative resources available nationally or through trade, the principle may not preclude uses that might impact the satisfaction of basic human needs from waters that flow in shared hydrologic systems.

Authors who analysed this issue differ somewhat in their opinion on the extent of protection provided by the obligation not to cause significant harm and the principle of equitable and reasonable use in the utilisation of shared water resources. Bourquain argued that these two fundamental rules establish a legal basis for an obligation to ensure that sufficient water quantities are left in a watercourse system for the satisfaction of vital human needs of people in other basin States.[28] According to this point of view, the concept of equity does not allow a threat to human life. Hence, read together, the two rules would contain a prohibition to threaten the livelihoods of riparian populations in other basin States through overabstraction or pollution of shared waters.[29] This argument overlooks the fact that according to the principle of equitable and reasonable utilisation, available alternatives to planned or existing uses should be considered. The availability of water must be assessed more holistically and alternative sources should be considered.

Bourquain's argument also means that there is a correlative obligation to guarantee a certain water flow in the shared system that supplies sufficient water to basin populations in other States. McCaffrey argued that an obligation of this scope does not exist because it cannot exist. In those (rare) situations in which there is not enough water for all people in the shared system, a State cannot be obliged to deprive its own population

[27] For example, Art. V (III), 1966 *Helsinki Rules*; and Art. 6 (3), 1997 *UN Watercourses Convention*.
[28] Bourquain, *Freshwater Access*, p. 39. [29] *Ibid.*

of water to satisfy the needs of others.[30] In those cases in which there is not enough water in the system, a solution based on the principle of equitable and reasonable utilisation has to be found among the riparian States.[31] Basin States are required to cooperate in finding an equitable solution that works for all. They have to identify jointly what the entitlement to an equitable share means for them in the specific context and whether there are possible alternatives to meet basic needs through other sources. Whereas Bourquain argued that States have an obligation to ensure that basic human water needs can be met in other basin States, McCaffrey proposed that such an obligation does not exist in the absolute because of possible limitations based on hydrologic conditions of individual basins. Hence, according to McCaffrey, suggesting an interpretation of norms that extends beyond the capacities of nature is not reasonable.

6.2.2 *Increasing attention in legal instruments*

The consideration of vital human needs is still uncommon in international water treaties; however, this trend is changing.

At the universal level

The texts of a small but increasing number of legal instruments accord priority to the protection of vital human needs in the context of transboundary water systems. At the universal level, both the 1997 *UN Watercourses Convention* and the 2008 ILC *Draft Articles* attribute particular attention to the satisfaction of vital human needs. The *UN Watercourses Convention* established a priority for vital human needs in the case of conflict among different utilisations. Article 10 regulates the relationship among uses; the first paragraph recalls the general rule that except for the existence of an agreement or custom to the contrary, there is no inherent priority among different types of uses. However, the second paragraph qualifies this statement and provides that in case of conflict among uses, 'it shall be resolved with reference to articles 5 to 7 [the principle of equitable utilisation and the no significant harm rule], with special regard being given to the requirements of vital human needs'.[32] When a conflict among different uses occurs, it must be resolved according to the framework established by the two rules. By referring to the principle of equitable utilisation, the

[30] McCaffrey, 'Human Right to Water', pp. 21f. [31] *Ibid.*

[32] Art. 10 of the *UN Watercourses Convention* does not resolve the question of whether uses that cause harm that significantly affects vital human needs can still be equitable.

provision re-emphasises the factors of 'social and economic needs' of concerned basin States and 'dependent basin population'.[33] In the 2008 *Draft Articles on the Law of Transboundary Aquifers*, the ILC revised this earlier approach and adopted one that can be considered even more forward-looking. Article 5 (2) of the *Draft Articles* contains the following protective clause at the end of a list of factors to be considered in the evaluation of equitable and reasonable use: 'However, in weighing different utilisations of a transboundary aquifer or aquifer system, special regard shall be given to vital human needs'. With this clause, vital human needs are already accorded a special weight in the process of determining equitable utilisation – and not only once a conflict among uses occurs. Arguably, the 2008 ILC *Draft Articles* therefore establish a somewhat stronger protection of vital human needs than the 1997 *UN Watercourses Convention*. Arguably, the due consideration of vital human needs early in the process can prevent the very occurrence of a conflict among uses (and users).

This stronger protection of vital human needs in the 2008 ILC *Draft Articles* was also affirmed by a refinement of the rules concerning emergency situations in Article 17. When an emergency occurs due to natural causes or human conduct and jeopardises vital human needs, States are allowed to take all measures during the emergency that are necessary to meet those needs, notwithstanding the obligations that might arise from the principle of equitable utilisation and the obligation not to cause significant harm. Such a strong protection clause is missing from the corresponding Article 28 of the *UN Watercourses Convention*.

Both the *UN Watercourses Convention* and the *Draft Articles on the Law of Transboundary Aquifers* remain in line with the argument that it would be unreasonable to establish legal obligations that run counter to hydrologic facts. At the same time, they require (only) a level of 'special regard' to be given to vital human needs; neither establishes an obligation 'to provide'.[34] The 2008 ILC *Draft Articles* further enhanced the protection of vital human water needs by demanding that they already be considered when determining reasonable and equitable use and by allowing States to prioritise their satisfaction in case of an emergency. In this process, the general duty to cooperate and the related obligations and rights contribute indirectly to the protection of vital human water needs through their instrumental role in implementing other general principles of international water law.

33 ILC, 'Draft Articles . . . and Commentaries', p. 110.

34 Bourquain, *Freshwater Access*, p. 41; ILC, 'Draft Articles . . . and Commentaries', p. 110, para. 4.

At the regional and basin level

At the regional level, the UNECE region and Europe lead the way; the protection of vital human needs was recognised within the legal framework of the UNECE, particularly in the 1999 *Protocol on Water and Health*[35] to the 1992 *UNECE Water Convention*. The *Protocol* applies to all water resources in a State, including those that do not flow in transboundary water systems.[36] In the first paragraph of the preamble, the parties express their conviction that 'water in quantities, and of quality, sufficient to meet basic human needs is a prerequisite both for improved health and for sustainable development'. The protection of vital human needs is 'part and parcel' of the *Protocol*'s objective to promote human health and well-being. Equally inherent in the *Protocol* is the idea of international cooperation. Read together, Articles 6 and 11 form the basis of an obligation to pursue the goal of providing safe drinking water through international actions. States are under obligation to establish national and local targets to achieve access to drinking water and provision of sanitation for everyone within a framework of integrated water management.[37] On request, States are under obligation to cooperate and assist one another in achieving those targets.[38] As described herein, in this sense, the *Protocol* reflects international cooperation obligations that are referred to in human rights covenants.

The *European Charter on Water Resources*, adopted in 2001 by the Committee of Ministers of the Council of Europe, was similar to the *Protocol on Water and Health*.[39] The *Charter* replaces the 1968 *European Water Charter* and was adopted as a recommendation to member States to apply the proclaimed principles nationally. In a formulation similar to Article 5 (2) of the 2008 ILC *Draft Articles*, the document recalls that waters are to be used according to the principle of equitable and reasonable utilisation in the public interest and that when considering all relevant factors, special regard has 'to be given to meeting vital human needs'.[40] The *Charter* further recognises a right of everyone 'to a sufficient quantity of water for his or her basic needs' and provides guidance on ecosystem protection and pollution control, evoking the principles of prevention, precaution and 'polluter pays'.[41]

[35] 17 June 1999, UN. Doc. MP.WAT/2000/1. [36] Arts. 1 and 3, *Protocol on Water and Health*.
[37] Art. 6, *ibid*. [38] Art. 11, *ibid*.
[39] Committee of Ministers, Recommendation Rec(2001)14, 17 October 2001, https://wcd.coe.int/wcd/ViewDoc.jsp?id=231615&Site=COE.
[40] Para. 2, *ibid*. [41] Paras. 5 and 8, *ibid*.

In addition, it can be said that treaties that provide for ecosystem protection, pollution prevention and water quality control make a general contribution to the protection of vital human water needs in a transboundary context.[42] When States cooperate to regulate different water uses such that water quality is maintained at a level that makes the resource safe for food production and drinking water supply, their cooperation ultimately protects basic human water needs. In some basins, States have exceeded this general protection and made their pursuit of the goal for a secure, safe water supply for individuals more explicit.

The 1909 *Boundary Waters Treaty* stands out among treaties of its time. In the early twentieth century, regional custom in Europe, for instance, still accorded priority to navigation. However, in the treaty governing the use of their boundary waters, the United Kingdom (for Canada) and the United States agreed to a hierarchy of uses that accords highest priority to the satisfaction of domestic and sanitary water needs.[43] Uses that conflict with or limit them were not permitted. Later, in 1975, the three lower riparian States of the Mekong River – Laos, Thailand and Vietnam – also declared their intention to adhere to a priority for domestic and urban water uses in their utilisation of mainstream waters.[44] This intention of the *Joint Declaration of Principles for Utilization of the Waters of the Lower Mekong Basin*, however, was not reflected in the 1995 *Mekong Agreement*; the treaty did not establish any priority among uses. In contrast, in their agreement on the utilisation and development of water resources of the Cuareim River Basin, Brazil and Uruguay agreed on a clear priority for drinking water supply to riparian populations, including an additional obligation of State organs to augment water supply within their available means.[45]

The 2002 *Senegal Water Charter* and the 2008 *Water Charter of the Niger Basin* display particular characteristics with respect to comparable treaties. These two treaties accord not only explicit priority to vital human water needs but also recognise the fundamental right to safe drinking water. The Niger Basin States consider the right to water to be a principle guiding

42 For example, Art. 4 (2) (i), 2000 *SADC Revised Protocol*; Art. 1, 2003 *Convention on the Sustainable Management of Lake Tanganyika*; Art. 16, 2002 *Senegal Water Charter*.

43 Art. VIII, 1909 *Boundary Waters Treaty*.

44 Art. XII, 1975 *Joint Declaration of Principles for Utilization of the Waters of the Lower Mekong Basin*, 31 January 1975, Mekong Committee Document E/CN 11 WRD/MKG/LK.405, p. 19.

45 Art. II, *Complementary Adjustment to the Agreement for Cooperation between Uruguay and Brazil for the Improvement of Natural Resources and the Development of the Cuareim River Basin*, 6 May 1997, http://ocid.nacse.org/tfdd/tfdddocs/575SPA.pdf.

their cooperation in managing the shared resources.[46] It is defined as the fundamental right to sufficient, physically accessible water at affordable cost and of a quality that is acceptable for personal and domestic use by everyone.[47] The *Senegal Water Charter* similarly recognises the enjoyment of the right to water, as an explicit objective of any repartition of the Senegal River's water.[48] The obligation to satisfy vital human needs has priority over technical and economic considerations of water allocation.[49] Given the likely impacts of climate change in the Sahel[50] and population-growth numbers, which are twice as high as the global average,[51] the solidarity expressed among basin States in these two treaties is encouraging. In recognising a fundamental right to drinking water, the treaties go beyond the formulas of the *UN Watercourses Convention* and the 2008 *Draft Articles* towards a guarantee of transboundary water provision for the satisfaction of vital human needs.

Recognition in jurisdiction

Recent decisions by international courts and tribunals similarly point to an affirmation that 'vital human water needs' are granted an elevated level of protection by law in the context of the utilisation of transboundary water systems. In the *Case Concerning the Dispute Regarding Navigational and Related Rights*, Costa Rica claimed and the ICJ affirmed the existence of a customary right to subsistence fishing of riparian populations of the San Juan River despite the fact that the area in question is located entirely in Nicaraguan territory as the boundary follows the riverbank on the Costa Rican side.[52] The Court recognises the right of riparian populations to draw the fruits of the transboundary watercourse for subsistence purposes as a local customary practice.

46 Preamble, 2008 *Niger Basin Water Charter*.

47 Art. 1 (11), *ibid.*: '. . . *le droit fondamental à un approvisionnement suffisant physiquement accessible et à un coût abordable, d'une eau salubre et de qualité acceptable pour les usages personnels et domestiques de chacun*'.

48 Art. 4, 2002 *Senegal Water Charter*: '. . . *Les principes directeurs de toute répartition des eaux du Fleuve visent à assurer aux populations des Etats riverains, la pleine jouissance de la ressource, dans le respect de la sécurité des personnes et des ouvrages, ainsi que du droit fondamental de l'Homme à une eau salubre, dans la perspective d'un développement durable*'.

49 Arts. 6 and 7, *ibid.*

50 B. C. Bates, Z. W. Kundzewicz, S. Wu, and J. P. Palutikof (eds.), *Climate Change and Water, IPCC Technical Paper VI* (Geneva: IPCC Secretariat, 2008), pp. 24–32.

51 World Bank, *World Bank Development Indicators* (2010), http://databank.worldbank.org.

52 *Case Concerning the Dispute Regarding Navigational and Related Rights (Costa Rica v. Nicaragua)* (2009) ICJ Reports, paras. 134–144.

The decision in the *Abyei Arbitration*[53] between the Government of Sudan (GoS) and the Sudan People's Liberation Movement/Army (SPLM/A) apparently reached a similar conclusion. In its final award, the tribunal confirmed that in the absence of an explicit agreement to the contrary, grazing and watering rights of nomadic people on transboundary waters remain unaffected by a transfer of sovereignty in the context of boundary delimitation.[54] This arbitration obtained an international dimension when South Sudan seceded from the North in July 2011.

Neither decision imposes an obligation to provide for the satisfaction of means of subsistence. Neither the GoS nor the SPLM/A is under obligation to ensure availability of water and plants for grazing. Also, the ICJ permits Nicaragua to regulate the customary fishing right, particularly regarding the protection of resources and the environment.[55] The obligation imposed is one of protection and abstention from interference with the exercise of these local rights. The discussion so far makes apparent that other than a small number of treaties, which explicitly recognise the right of individuals to water, international water law instruments do not usually stipulate an obligation to provide sufficient water to ensure their survival. What, then, are the means for individuals to claim their rights on transboundary waters, and how can States ensure that sufficient amounts of water are provided for everyone?

6.3 Enforcing satisfaction of vital human needs

This section explores the question of what constitutes the most effective legal basis or remedies for individuals and States to obtain satisfaction of vital human water needs.

6.3.1 *Enforcement of the human right to water*

The primary body of norms of international law designed to provide legal remedies for the satisfaction of vital needs of individuals is human rights law. First, human rights establish a legal relationship between an individual and the State: the individual has a claim to a certain type of behaviour

[53] *In the Matter of an Arbitration before a Tribunal Constituted in Accordance with Article 5 of the Arbitration Agreement between the Government of Sudan and the Sudan People's Liberation Movement/Army on Delimiting the Abyei Area*, Final Award, 22 July 2009, www.pca-cpa.org.

[54] *Ibid.*, p. 260, para. 753.

[55] *Case Concerning the Dispute Regarding Navigational and Related Rights (Costa Rica v. Nicaragua)* (2009) ICJ Reports, para. 141.

on the part of the State; and, second, States are under obligation to respect and provide for the enjoyment of human rights.

Increased protection through recognition as a self-standing right?

Among human rights, there are subtle differences in their protection. Civil and political rights are protected somewhat differently than economic, cultural and social rights. With respect to civil and political rights stipulated in the *ICCPR*, States are under immediate obligation 'to respect and to ensure to all individuals within [their] territory and subject to [their] jurisdiction the rights recognized in the . . . Covenant, without distinction of any kind, such as race, color, sex, language, religion, political or other opinion, national or social origin, property, birth or other status'.[56] The *ICESCR*, conversely, imposes 'promotional' or progressive obligations with respect to economic, social and cultural rights:[57] State parties agree to 'take steps, individually and through international assistance and cooperation, especially economic and technical, to the maximum of [their] available resources, with a view to achieving progressively the full realization of the rights recognized in the present Covenant by all appropriate means, including particularly the adoption of legislative measures'.

So far, human rights obligations related to water have been derived primarily from other *ICESCR* rights.[58] As economic, social and cultural rights, their implementation is partly conditioned by resources available in individual States. Links to the right to life, stipulated in the *ICCPR*, also were suggested at the universal level.[59] GA *Resolution 64/292 on the Human Right to Water and Sanitation* and the *Human Rights Council Resolution 15/9* from 2011 both establish a direct link to the right to life and recognise 'the right to safe and clean drinking water and sanitation as a human right that is essential for the full enjoyment of life and all human rights'.[60] This can be interpreted as an intention by the international community to accord an elevated level of protection to rights with respect to water and related obligations.

The right to life[61] is a non-derogable right; that is, State parties to the *ICCPR* may not derogate from their obligations concerning the right to life, even in times of public emergency that threaten the very existence

[56] Art. 2, *ICCPR*.
[57] C. Tomuschat, *Human Rights – Between Idealism and Realism*, 2nd edition (Oxford: Oxford University Press, 2008), p. 42.
[58] CESCR, General Comment 15. [59] *Ibid.*, para. 3. [60] GA *Res.* 64/292.
[61] Art. 6, *ICCPR*: '1. Every human being has the inherent right to life. This right shall be protected by law. No one shall be arbitrarily deprived of his life'.

of a State.[62] At the core of the right is the protection against arbitrary deprivation of life; however, the right and related obligations are not to be interpreted too restrictively. According to General Comment No. 6 of the Human Rights Committee, States have to adopt positive measures to promote life, including *inter alia* the reduction of infant mortality and the elimination of malnutrition.[63] This is achieved, for example, through provision of sufficient and clean drinking water. As a right of the *ICCPR*, an immediate obligation to respect and ensure enjoyment of this right is imposed on States. The following discussion considers the means available to individuals for seeking enforcement of these obligations.

Limited geographic reach of human rights enforcement mechanisms

Human rights are enforceable in national legal systems through judicial or extrajudicial remedies that allow people to assert their rights before the State under whose jurisdiction they are. Human rights claim additional protection at the universal level through a series of international mechanisms, including self-reporting procedures and special procedures (e.g., investigation of human rights situations in specific countries and on thematic issue areas) and response to individual complaints;[64] regional human rights mechanisms exist as and offer supplemental remedies.

Decisions of regional human rights commissions and courts, as well as national courts, illustrate that the water-related rights of individuals and groups have been successfully asserted so far by invoking protection of connected human rights. The Latin American experience provides examples in which the rights of indigenous groups concerning utilisation and management of natural resources, including water, were subsumed under the right to property (e.g., communal lands) and the right to life.[65]

62 Art. 4, *ICCPR*.

63 Human Rights Committee, *CCPR General Comment No. 06: The Right to Life (Art. 6)*, 30 April 1982, para. 5, www.unhchr.ch/tbs/doc.nsf/(Symbol)/84ab9690ccd81fc7c12563ed0046fae3?Opendocument.

64 The Human Rights Council, which replaced the Commission on Human Rights in 2006, continues to address specific country situations or thematic issues through special procedures that include the appointment of 'Special Rapporteurs', 'Independent Experts' or envoys with the mandate to investigate and report on human rights situations, respond to individual complaints and provide technical cooperation at the country level.

65 Decisions of the Inter-American Court of Human Rights: *Case of the Saramaka People v. Suriname* (2008); *Case of the Yakye Axa Indigenous Community v. Paraguay* (2005); *Case of the Mayagna (Sumo) Awas Tigni Community v. Nicaragua* (2001).

In at least two cases – that is, *Butan and Dragomir v. Romania*[66] and *Zander v. Sweden*[67] – claimants asserted their right to safe drinking water by invoking the right to a fair hearing before the European Court of Human Rights. In a number of countries, national courts acknowledged the right to water based on the recognition that it falls under the right to a healthy environment.[68] These decisions affirmed that human rights obligations related to water can be derived and successfully invoked through other rights; the recognition by the HRC and the GA of the right to water as self-standing gives additional support to this decision-making trend. Gleick argued that an added value in the recognition of the right to water as self-standing is primarily in encouraging renewed efforts by States to comply with their obligations and secondarily in a potentially increased focus on satisfaction of vital human needs in the context of transboundary water disputes.[69]

A challenge remains, however, with respect to geographic reach. Established human rights procedures offer protection to individuals against violations of the State in whose territory or under whose jurisdiction and control they are.[70] So far, they have not provided protection against activities that take place in the territory of and are attributable to one State, which subsequently result in situations that violate human rights of the population in another State. Hence, they do not offer an effective remedy in a situation in which the utilisation of shared water resources in one State pollutes those resources to a level that endangers the livelihoods of the riparian population in another basin State. The polluting State does not owe human rights protection to the affected population if it does not reside in territories under its jurisdiction or control.[71] In the absence of a territorial or jurisdictional link, individuals or groups lack legal means under human rights law to enforce their claims against the State in the territory in which the pollution originates.

[66] *Butan and Dragomir v. Romania* (2008), Application No. 400067/06.

[67] *Zander v. Sweden* (1993), Application No. 14282/88.

[68] For list of decisions, see Appendix II in Scanlon, Cassar and Nemes, *Water as a Human Right?*

[69] P. H. Gleick, 'The Human Right to Water' (1998) 1 *Water Policy* 489.

[70] M. Gibney, K. Tomašcevski, and J. Vedsted-Hansen, 'Transnational State Responsibility for Violations of Human Rights' (1999) 12 *Harvard Human Rights Journal* 267–295.

[71] For the exception case in which an affected population can be considered to be under effective control of the violating State, see Art. 2 (1), *ICCPR*, and *Legal Consequences of the Construction of a Wall in the Occupied Palestinian Territory* (2004) ICJ Reports, para. 111; T. Meron, 'Extraterritoriality of Human Rights Treaties' (1995) 89 *AJIL* 79.

6.3.2 *Alternative enforcement avenues*

A legal remedy to protect these rights or, rather, the core content of the rights of individuals or groups in these or similar situations was included in the 1997 *UN Watercourses Convention*. Article 32 includes an obligation for non-discrimination with regard to access to justice in the case of significant transboundary harm:

> Unless the watercourse States concerned have agreed otherwise for the protection of the interests of persons, natural or juridical, who have suffered or are under a serious threat of suffering significant transboundary harm as a result of activities related to an international watercourse, a watercourse State shall not discriminate on the basis of nationality or residence or place where the injury occurred, in granting to such persons, in accordance with its legal system, access to judicial or other procedures, or a right to claim compensation or other relief in respect of significant harm caused by such activities carried on in its territory.

This provision provides a remedy to individuals who reside in the State in whose territory significant harm or the threat of it occurs as a result of intentional or negligent conduct within the territory of another riparian State.[72] In a case in which an affected individual decided to make a claim before a court of the State that caused harm or in whose territory the harmful activity occurred, this State cannot bar this claim 'on the grounds that the harm occurred outside its jurisdiction'.[73] The provision could provide individuals with a remedy to protect their interest in the satisfaction of their vital human needs against third States once the *Convention* enters into force. However, a right of access to basic water supply cannot be derived from Article 32.[74] Although there are a few precedents for similar regulations in international water law instruments,[75] this rule

[72] The obligation not to cause significant harm is a due diligence obligation and causality must be established. ILC, 'Draft Articles . . . and Commentaries', p. 103.

[73] *Ibid.*, 133. The ILC commentary refers to jurisdiction because a change in wording between Article 32, 1994 *Draft Articles*, and Article 32, *UN Watercourse Convention*; the three last words in the 1994 *Draft Articles* state: 'under its jurisdiction'.

[74] Bourquain, *Freshwater Access*, p. 31.

[75] Article 3 of the 1974 *Nordic Environmental Protection Convention* gives 'any person who is affected or may be affected by environmentally harmful activities in another Contracting State . . . the right to bring before the appropriate Court of Administrative Authority of that State the question of the permissibility of such activities including the question of measures to prevent damage', as well as a right to appeal. Signed 19 February 1974, UNE EPL, Vol. 1, 1975–1976, p. 44; see also *OECD Recommendation of the Council for the Implementation of a Regime of Equal Right of Access and Non-Discrimination in Relation to Transfrontier Pollution*, 17 May 1977, OECD Doc. C(77)28.

does not yet constitute customary international law.[76] The 2008 *Draft Articles on Transboundary Groundwater Resources* do not include a comparable provision.

Article 32 of the *UN Watercourses Convention* has the potential to 'open a door' to extraterritorial protection of international human rights obligations. The exact nature of these obligations, which so far have been discussed more in doctrine than invoked in practice, are addressed in more detail herein. In essence, it is proposed that States are under obligation to refrain from actions that would infringe on the ability of other States to comply with their human rights obligations. This is a duty owed among States. When such an infringement is caused by infliction of significant transboundary harm, Article 32 could grant protection to individuals against the violation of the essence of the international human rights obligation. It would establish a legal basis for an extraterritorial obligation that is independent of a jurisdictional or 'effective control' link between the State inflicting the violation and the individual who is affected by it.

6.3.3 *International obligations under human rights law*

The fact that human rights are a matter not only of national implementation but also of international cooperation was already recognised in the *Charter of the UN* (UNC). The UN System was established with the objective *inter alia* to 'achieve international cooperation in . . . promoting and encouraging respect for human rights'.[77] The international society of States acts through organisations and programs of the UN System, regional human rights mechanisms and other development organisations to contribute to the progressive realisation of human rights. The adoption of the MDGs is one of many manifestations of these activities. By virtue of being a member of the UN, States assume general obligations to cooperate in order to achieve universal respect and observance of human rights and improvement of living standards.[78] Both the GA and the HRC resolutions that recognise the right to water reemphasise this obligation by calling upon 'States and international organizations

[76] McCaffrey, *The Law of International Watercourses*, p. 510.

[77] Arts. 1 (3), 55 and 56, 1945 UNC.

[78] Arts. 55 and 56, UNC; B. von Tigerstrom, *Human Security and International Law – Prospects and Problems* (Oxford/Portland, OR: Hart Publishing, 2007), p. 75.

to provide financial resources, capacity-building and technology transfer, through international assistance and cooperation . . . '[79]

Extraterritorial and international human rights obligations

In accordance with this general obligation to cooperate in achieving respect for human rights, the two international human rights covenants provide for certain extraterritorial and international human rights obligations. 'Extraterritorial obligations' are defined as obligations to individuals and groups outside a specific State's territory.[80] Obligations of States to seek assistance, to provide assistance, or to cooperate with other States in order to put themselves or others into a position to comply with human rights obligations are referred to as 'international human rights obligations'. Extraterritorial obligations are owed by States to individuals, whereas international human rights obligations concern the legal relationship among States. The definition of 'international obligations', which is adopted herein, makes a clear distinction with respect to the party to whom the obligation is owed.[81] These obligations were also referred to more generally as 'transnational' human rights obligations by authors who explored their existence.[82]

Both the *ICCPR* and the *ICESCR* comprise extraterritorial obligations. The human rights obligations described in these two *Covenants* apply not only to individuals within the territory of a State but also to those who are outside of it but nonetheless are under the jurisdiction or effective control of this State. The *ICCPR* makes this clear in the text of Article 2 (1): 'Each State Party to the present Covenant undertakes to respect and to ensure to all individuals within its territory and subject to its jurisdiction the rights recognized in the present Covenant . . . ' The ICJ confirmed the extraterritorial applicability of both *Covenants* in its advisory opinion on *Legal Consequences of the Construction of a Wall in the Occupied Palestinian*

79 Para. 2, GA *Res. 64/292*; para. 10; HRC *Res. 15/9*.

80 R. Künnemann, 'Extraterritorial Application of the International Covenant on Economic, Social and Cultural Rights', in F. Coomans and M. Kamminga (eds.), *Extraterritorial Application of Human Rights Treaties* (Antwerp: Intersentia, 2004), p. 201.

81 In contrast, the CESCR, in General Comment No. 12 on the right to food, remains vague on the question as to whom the obligations are owed that are described in paras. 36 through 41 under the heading 'international obligations'. Equally vague in this respect is the definition of 'international obligations' adopted by Künnemann; *ibid.*, p. 216.

82 Gibney, Tomaŝcevski, and Vedsted-Hansen, 'Transnational State Responsibility'; S. Skogly and M. Gibney, 'Transnational Human Rights Obligations' (2002) 24 *Human Rights Quarterly* 781–798.

Territory.[83] It stated that the word 'and' must be understood and read as 'and/or' in this context.[84] Extraterritorial obligations that go beyond jurisdiction and effective control also were derived in the specific case from the *erga omnes* character of the right to self-determination.[85] All State parties to the international human rights covenants have a duty to take positive measures to facilitate this right to self-determination 'not only in relation to their own peoples but *vis-à-vis* all peoples'.[86] Translated into the context of transboundary hydrographic systems, this means that States cannot deprive people in another riparian State of their means of subsistence – for example, by stopping or reducing the flow of a transboundary river to a level at which basic water supply is no longer provided, or by polluting water resources to such an extent that the fish population the group depends on is threatened with extinction.

International cooperation obligations

Article 2 (1) of the *ICESCR* establishes additional international human rights obligations in requiring States 'to take steps . . . through international assistance and cooperation' with a view to progressively realising *Covenant* rights.[87] On the one hand, this norm imposes a duty on States to seek assistance from other States if they lack sufficient resources to meet their own human rights obligations.[88] On the other hand, it imposes an obligation on other States to provide assistance to those States that lack the resources to ensure the satisfaction of minimum essential levels of *ICESCR* rights by their own means. The obligation to provide, however, is limited; it does not extend to a duty to meet all of the needs of the State

83 *Legal Consequences of the Construction of a Wall in the Occupied Palestinian Territory* (2004) ICJ Reports, paras. 111f.

84 *Ibid.*, para. 111; T. Meron, 'Extraterritoriality of Human Rights Treaties', 79. The *ICESCR* lacks this definition of territorial and jurisdictional applicability.

85 *Legal Consequences of the Construction of a Wall in the Occupied Palestinian Territory* (2004) ICJ Reports, para. 155; von Tigerstrom, *Human Security*, p. 76.

86 Human Rights Committee, *CCPR General Comment No. 12: The Right to Self-Determination of Peoples (Art. 1)*, 13 March 1984, para. 6, www.unhchr.ch/tbs/doc.nsf/(Symbol)/f3c99406d528f37fc12563ed004960b4?Opendocument.

87 See also Art. 4, 1989 *Convention on the Rights of the Child*: 'States Parties shall undertake all appropriate legislative, administrative, and other measures for the implementation of the rights recognized in the present Convention. With regard to economic, social and cultural rights, States Parties shall undertake such measures to the maximum extent of their available resources and, where needed, with the framework of international cooperation'.

88 CESCR, General Comment No. 3, para. 11.

that is seeking assistance.[89] Regarding the question about what type of assistance must be provided, it is important that the phrase 'international assistance and cooperation' is not restricted to the meaning of 'development assistance', as was suggested by some authors.[90] Furthermore, nothing in the text of the Article suggests that these obligations exist only between developed and developing countries. Such an interpretation of the *Covenant* would be too restrictive given and in light of the breadth of the rights it establishes. Rather, these two terms – 'international assistance' and 'cooperation'– refer to a wide range of joint activities that can include the cooperative management of shared resources.[91] International human rights obligations between co-riparian States to ensure, for example, safe water quality and sufficient water flow would therefore exist despite and regardless of differences in their economic and financial endowments. The absence of international assistance does not relieve a State from its human rights obligations.

Content of international obligations

The nature of international obligations of States to assist other States in meeting their human rights obligations can be illustrated in the three-level framework that is usually applied to human rights obligations at the national level:[92] States are under obligation to (1) respect, (2) protect and (3) fulfill human rights.[93] Applied to international obligations, the duty to respect would signify that States must refrain from interfering with the enjoyment of rights in other States. States must protect the enjoyment of rights in other States through the adoption of measures that prevent infringement of rights through third parties over which a State holds influence. Finally, the obligation to fulfill relates to international

[89] M. Craven, *The International Covenant on Economic, Social and Cultural Rights – A Perspective on its Development* (Oxford: Clarendon Press, 1995), p. 149.

[90] *Ibid.*, pp. 144–150. [91] Künnemann, 'Extraterritorial application', p. 204.

[92] Craven, *The International Covenant*, pp. 147–149; F. Coomans, 'Some Remarks on the Extraterritorial Application of the International Covenant on Economic, Social and Cultural Rights', in Coomans and Kamminga (eds.), *Extraterritorial Application of Human Rights Treaties* (Antwerp: Intersentia, 2004), pp. 192–199.

[93] This classification was adopted for international obligations in the 'Report of the Office of the United Nations High Commissioner for Human Rights on the Relationship between Climate Change and Human Rights', citing CESCR General Comments No. 12 (adequate food), No. 13 (education), No. 14 (health) and No. 15 (water) on p. 24, para. 86. Report reproduced in Office of the High Commissioner of Human Rights, *Annual Report of the United Nations High Commissioner for Human Rights and Reports of the Office of the High Commissioner and the Secretary General*, 15 January 2009, UN Doc. A/HRC/10/61.

obligations to provide assistance and enter into international agreements that promote the enjoyment of rights in other States.

Article 23 of the *ICESCR* further provides a non-exhaustive description of the range of actions with which the States asking for assistance and those providing it can fulfill their international obligations. These include 'the conclusion of conventions, the adoption of recommendations, the furnishing of technical assistance and the holding of regional meetings'. The CESCR clarified the interpretation of this norm in General Comment 14 on the right to the highest attainable standard of health: 'States parties have to respect the enjoyment of the right . . . in other countries . . . [They] should ensure that the right . . . is given due attention in international agreements and, to that end, should consider the development of further legal instruments'.[94] This obligation can be interpreted as establishing a bridge between human rights law and international water law. The activities referred to in Article 23 of the *ICESCR*, when applied to the context of the human right to water, suggest a general duty to promote this right *inter alia* through the conclusion of international water treaties. This highlights the potential for mutual supportiveness of the two bodies of law and the complementary qualities of international water law in facilitating and promoting transboundary protection of human rights.

6.4 Conclusion

As pressure on transboundary water systems increases and two out of three people may be living under conditions of 'water stress' by 2020,[95] international concern about the satisfaction of basic human needs related to water and food is on the rise. Given the extent of the challenges, it will become increasingly more difficult to satisfy vital human water needs by purely national means. Some States already currently receive temporary food aid[96] and third States provide water and water supply

[94] CESCR, *Substantive Issues Arising in the Implementation of the International Covenant on Economic, Social and Cultural Rights, General Comment No. 14 (2000), The Right to the Highest Attainable Standard of Health*, 11 August 2000, E/C.12/2000/4, para. 39.

[95] L. Lewis, 'Ecologists Warn the Planet is Running Short of Water', *TimesOnline*, 22 January 2009, www.timesonline.co.uk/tol/news/environment/article5562906.ece.

[96] 'Ethiopia Asks for Urgent Food Aid,' *BBC NEWS*, 22 October 2009, http://news.bbc.co.uk/2/hi/8319741.stm; 'UN to Resume Burma Food Flights', *BBC NEWS*, 9 May 2008, http://news.bbc.co.uk/2/hi/7392331.stm.

systems[97] in situations of humanitarian emergency. In the future, riparian States of transboundary water systems – particularly the non-recharging groundwater systems in the Sahel and Middle East – may no longer be in the position to fulfill vital human water needs of the national basin populations without engaging co-riparian or other States in international cooperation; shared waters will no longer be sufficient to supply all of the dependent basin populations.

Human rights law can have a facilitative function in this respect. The satisfaction of vital human water needs is at the core of human rights law. It contains rights and obligations that aim to do just that, as well as at the progressive improvement of living and sanitary conditions. Yet, human rights mechanisms – first and foremost – provide individuals with protection against interference in their rights by their own State. They do not leave people with adequate legal remedies to claim their rights to subsistence if they have been violated or neglected by other States. However, human rights law also provides a legal basis for international cooperation, or 'transnational' obligations. So far, States have not yet invoked these obligations; rather, they have treated them as moral or political obligations.[98] Therefore, questions remain regarding the actual enforcement (or implementation) of these obligations. This is where international water law intervenes in a complementary manner to promote the satisfaction of vital human water needs. Remedies, such as the one proposed in Article 32 of the *UN Watercourses Convention* on non-discrimination with respect to access of justice, as well as new treaties and emerging rules that employ the general duty to cooperate for the 'transboundary' protection of vital human water, will also facilitate their future satisfaction. In light of the ever-increasing water demand, international cooperation holds one of the keys to promote the use of this vital resource with a view to ensuring a sufficient supply for everyone at both the transboundary and universal levels.

97 USAID, *Haiti-Earthquake*, Fact Sheet #5, Fiscal Year (FY) 2011, 29 October 2011, www.usaid.gov/our_work/humanitarian_assistance/disaster_assistance/countries/haiti/template/fs_sr/fy2011/haiti_eq_fs05_10-29-2010.pdf.

98 Von Tigerstrom, *Human Security*, p. 83.

7 Emerging international cooperation on global water challenges

The threat of a global water crisis is ringing alarm bells.[1] Water is already becoming increasingly scarce in some regions, with around thirty-one countries currently facing chronic water shortages; a number that is expected to rise to nearly fifty in the next fifteen years, particularly in the Middle East, Sub-Saharan Africa and Southeast Asia.[2] These developments not only impact water supply for individuals but also have much wider implications.[3] Whereas some regions are affected by water shortages, others are subject to increases in the occurrence and severity of floods or must deal with land loss and inundations due to sea level rise, thus becoming the victims of 'too much' water. These events are caused by changes in the hydrologic cycle, which in turn are due in large degree to human economic activity: emissions from decades of industrialisation have accumulated in and are changing the atmosphere and the global climate. While this process carries on with continued economic development worldwide and emission levels continue to rise, States also have initiated cooperative efforts to address the causes and effects of these changes.

[1] H. Bigas (ed.), *The Global Water Crisis: Addressing an Urgent Security Issue*, Papers for the InterAction Council (Hamilton, Ontario: United Nations University–Institute for Water, Environment and Health [UNU–INIWEH], 2012).

[2] Speech of Dr. R. K. Pachauri, Chairman of the Intergovernmental Panel on Climate Change, at the World Economic Forum, Davos, 23 January 2008, www.ipcc.ch/graphics/speeches/pachauri-davos-january-2008.pdf.

[3] See discussion of rights of and remedies for individuals with respect to access to water in Part III, Chapter 6, of this volume.

7.1 Hydrologic effects of climate change

'Climate change is to a great extent water change [because] water is the primary medium through which climate change impacts [are] felt by humans and the environment'.[4] Regarding regional variability, hydrologic models predict a general increase in water availability at high latitudes and some areas of the tropics, whereas a decrease is expected in semi-arid areas at middle latitudes.[5] Higher overall temperatures are expected to result in seasonal shifts of river runoff; they will likely decrease the volume of water storage in glaciers and are expected to lead to earlier melting of snowpack.[6] Agriculture, which accounts for about 70 per cent of water use, will suffer from a decrease in water availability in the lean season; changes in temperature and seasonal water availability will necessitate crop adjustments in many regions, from water-intensive to water-efficient crops.[7] In river basins fed by glacial meltwater, river flow will first increase and then decrease with the disappearance of glaciers in the long term.[8] Melting of the fossil Andean glaciers is expected to threaten the water supply of millions of people.[9] In the Alpine and Mediterranean regions, it is expected that receding glaciers and a shift of water availability earlier in the season will impact hydropower generation and agriculture.[10]

Climate change effects will also be felt with respect to groundwater resources, although the connection is still less well understood than impacts on surface water.[11] Decrease in water availability will reduce aquifer recharge rates, which is likely to cause drastic effects in the Middle East and North Africa, where groundwater mining led to agricultural desert bloom in some areas in recent years – and recharge rates are already

4 *Water and Climate Change Towards COP-16, Statement from the Water and Climate Coalition and partners to the UNFCCC negotiations*, Bonn 2010, www.stakeholderforum.org/fileadmin/files/wccstatement.pdf.

5 B. C. Bates *et al.* (eds.), *Climate Change and Water*, Technical Paper of the Intergovernmental Panel on Climate Change (Geneva: IPCC Secretariat, 2008), p. 3.

6 *Ibid.*

7 H. Cooley *et al.*, *Understanding and Reducing the Risks of Climate Change for Transboundary Waters* (Oakland, CA: Pacific Institute, 2009), p. 14.

8 J. P. Bravard, *Impacts of Climate Change on the Management of Upland Waters: The Rhone River Case* (Banff: 5th Rosenberg International Forum, 2006).

9 M. Baraer *et al.*, 'Glacier Recession and Water Resources in Peru's Cordillera Blanca' (2012) 58 *Journal of Glaciology* 134–150.

10 G. P. Harrison, H. W. Whittington, and S. W. Gundry, 'Climate Change Impacts on Hydroelectric Power' in *Proceedings of 33rd Universities Power Engineering Conference (UPEC '98)* (Edinburgh, 1998), pp. 391–394.

11 *Ibid.*, pp. 13f.

low.[12] Higher evaporation rates are expected to increase the salinisation of groundwater, with sea level rise and saltwater intrusion expected to have the same effect on coastal aquifers.[13] Global warming will impact the quality of surface water bodies, in which higher temperatures may cause algal and bacterial bloom that results in harmful effects on fisheries, biodiversity and human health.

These impacts require adaptation by those affected; furthermore, the complexity of climate change dynamics, their unpredictability and their irreversibility[14] call for international cooperation and flexibility in the process.[15] Anthropogenic climate change renders naturally occurring hydrologic variability – which water practitioners have grown accustomed to manage over the course of history – considerably less predictable and more difficult (and costly) to address. Effects of anthropogenic climate change add to naturally occurring hydrologic variability and phenomena, such as the El Niño and La Niña events: the burning of fossil fuels and industrial activity change precipitation patterns worldwide due to mineral dust that floats in the atmosphere and impact rainfall distribution, local water availability and quality.[16] Questions about exactly where, how and when climate change impacts are most likely to occur cannot be answered with certainty; too many factors intervene.

In transboundary water systems, the involvement of multiple States adds more challenges to the implementation of effective adaptation measures. Coordination of activities becomes essential because unilateral adaptation measures of individual States may not be of adequate scale or may give rise to transboundary harm and aggravation of impacts on others (e.g., when upstream States respond to a basin-wide drought with increased water abstractions instead of demand management). By pooling technical, human and financial resources, cooperation provides States with the opportunity for innovative solutions and

[12] Saudi Aramco, 'The Greening of the Arab East' (1983) 34 *Saudi Aramco World* 5, 24–40.

[13] H. Cooley *et al.*, *Understanding and Reducing*, p. 13.

[14] Potsdam Institute for Climate Impact Research and Climate Analytics, *Turn down the Heat: Why a 4°C Warmer World Must Be Avoided* (Washington, DC: World Bank, 2012).

[15] G. Eckstein, 'Water Scarcity, Conflict and Security in a Climate Change World: Challenges and Opportunities for International Law and Policy' (2010) 27 *Wisconsin International Law Journal* 3, 432f.

[16] R. J. Moses and J. N. Corbridge, *Legal Structures for International Supervision of Weather Modification Techniques* (Washington, DC: Water for Peace Conference, Doc. P/635 1967), p. 380.

implementation of adaptation measures of a larger scale than when they act unilaterally. This chapter explores the legal frameworks and norms available to cope with impacts of climate change on the hydrologic cycle. It explores questions of whether international water agreements provide the tools that are required for flexible and coordinated responses and whether legal principles and the rules of climate change law offer new avenues for international cooperation in addressing transboundary water challenges at the universal level.

7.2 Appreciation of hydrologic variability in international water law

Hydrologic variability has long been one of the core concerns for treaty regulation of transboundary watercourses by riparian States.[17] As a consequence, the principles and rules of international water law provide a legal framework that contains elements of the flexibility necessary to address adaptation to hydrologic change.[18]

7.2.1 Flexibility in principle(s)

Climatic and hydrologic conditions of transboundary watercourses were included as a factor to be considered in the assessment of equitable and reasonable utilisation in the *UN Watercourses Convention* and the 1966 *Helsinki Rules*.[19] Although commentaries on the respective Articles do not define more specifically what the authors had in mind when they included climate, it can be assumed that this factor must be understood according to the widest meaning of the term. Arguably, both natural and anthropogenic causes for climate change must be considered with respect to the climatic conditions of a basin, even if the anthropogenic causes were not recognised when the ILA formulated the principle in the 1966 *Helsinki Rules*. A wide interpretation is supported by the fact that the instruments

[17] *Treaty on Sharing of the Ganga/Ganges Waters at Farakka*, 12 December 1996, published in Salman and Uprety, *Conflict and Cooperation*, pp. 387–391; *Agreement on Flood Warning in the Catchment Basin of the Moselle*, 1 October 1987, 1506 *UNTS* 97; 1977 *Treaty Concerning the Construction and Operation of the Gabčíkovo-Nagymaros System of Locks*; 1961 *Treaty between Canada and the United States of America Relating to Co-Operative Development of the Water Resources of the Columbia River Basin*.

[18] S. McCaffrey, 'The Need for Flexibility in Freshwater Treaty Regimes' (2003) 27 *Natural Resources Forum* 157.

[19] Art. 6 (1) (a), *UN Watercourses Convention*, and Art. V (2) (c), 1966 *Helsinki Rules*.

that stipulate the principle of equitable and reasonable utilisation clarify that the listing of factors includes examples for guidance and should not be considered exhaustive. Article 5 of the *Draft Articles on the Law of Transboundary Aquifers* does not list climatic conditions as one of the factors relevant to the assessment of equitable and reasonable utilisation; this may be because the effects were not yet well understood. However, given that this list of factors – similar to its equivalent in other legal instruments – is non-exhaustive and only indicative of elements to consider, climate should be considered equally as a relevant factor when it impacts the recharge rate or water quality and salinisation of aquifers.

Wide interpretation is argued in the ILC commentary on the provisions of Article 6 of the *UN Watercourses Convention*; it stipulates that, in general, the approach to the assessment process must remain flexible and adaptable to a change in circumstances; weight must be given to each factor according to the circumstances in specific cases.[20] The commentary clarifies that 'in order to assure that their conduct is in conformity [with the provision] watercourse States must take into account, *in an ongoing manner*, all factors that are relevant in ensuring that the equal and correlative rights of other watercourse States are respected' [emphasis added].[21] For situations in which changes in circumstances, 'such as a reduction in the quantity of water',[22] require adaptation with regard to application of the equitable utilisation principle, Paragraph 2 envisages an obligation of riparian States to 'enter into consultations in a spirit of cooperation'. Thus, the principle of equitable and reasonable utilisation gives rise to a duty to cooperate where climate change causes a significant change in basin conditions and circumstances. It is implemented, for example, through the obligation to enter into consultations with respect to adapting uses to the change in basin conditions. Climate change again emphasises the importance of understanding implementation of the principle of equitable and reasonable utilisation as a continuous and ongoing process – a process in which effectiveness becomes a function of flexibility and cooperation among States.[23]

When subjecting the provisions of Article 27 of the *UN Watercourses Convention* to a wide interpretation, it appears that its authors also provided for the possibility of harmful results of human-induced climate change. In the spirit of the principle of good neighbourliness, Article 27 stipulates that '[w]atercourse States shall, individually, and, where

[20] ILC, 'Draft Articles . . . and Commentaries', p. 101.
[21] *Ibid.* [22] *Ibid.*, p. 102. [23] McCaffrey, *The Law of International Watercourses*, pp. 401–404.

appropriate, jointly, take all appropriate measures to prevent or mitigate conditions related to an international watercourse that may be harmful to other watercourse States, whether resulting from natural causes or human conduct, such as flood or ice-conditions, water-borne diseases, siltation, erosion, salt water intrusion, drought or desertification'. This Article establishes an obligation to cooperate in the response and adaption to hydrologic change. The obligation was attenuated with the insertion of the terms 'where appropriate'; it exists only when joint action is required and useful.

7.2.2 The role of treaties in facilitating flexibility to adjust for hydrologic variability

Ongoing cooperation and coordination of activities among States allow for necessary adjustments and timely reaction to hydrologic variability. Treaties are a frequent 'tool of choice' employed by States with a view to achieving a certain level of stability and predictability in their relationships.[24] The way in which these treaties are crafted determines, to a large degree, whether the legal relationships established between States by these agreements at the same time provide the flexibility required to cope with the additional unpredictability brought about by climate change. In this sense, the use of treaties takes place in the face of competing objectives of stability and, therefore, the need for a certain level of rigidity and durability, on the one hand, and for flexibility and adaptability on the other.

States seal their legal relationships with treaties *inter alia* because they provide a stable and durable frame of reference for their interactions. Treaties are rigid in principle because they can be modified 'only pursuant to their terms or by mutual agreement' and cannot be terminated or withdrawn from except under certain conditions.[25] As a consequence, even in cases in which treaties include provisions that allow for changes in content and scope through amendments or protocols, or that envisage regular renegotiation of some of their terms,[26] there remains a considerable level of rigidity given the reality of (often lengthy) negotiations, adoption and ratification procedures required for changes in treaty provisions. At the same time, treaties unfold their value in regulating flexibility by establishing obligations of conduct (e.g., consulting and negotiating) and

[24] McCaffrey, 'The Need for Flexibility', p. 157.

[25] *Ibid.*; see also *Vienna Convention on the Law of Treaties*, 23 May 1969, 1155 *UNTS* 331.

[26] For example, 1996 *Treaty on Sharing of the Ganga/Ganges Waters at Farakka.*

more specific obligations (e.g., relative to reservoir release schedules in case of floods), which enhance predictability of the way in which concerned States will respond to hydrologic change. By specifying and establishing clear rules of conduct in written format, legal agreements increase the likelihood of coordinated adjustment.

7.2.3 Norms and mechanisms facilitating adaptation

Obligations of conduct in response to changes in circumstances that are specified in treaties can be broadly classified in three categories: (1) required conduct in reaction to certain identified changes in conditions, (2) obligations to act that can set previously undetermined processes of adaptation in motion, and (3) coordination within the framework of joint mechanisms that facilitate or implement adaptation.

Required conduct based on pre-identified changes

A number of water treaties make provisions for expected seasonal or cyclical flow changes. McCaffrey, for example, pointed to the 1944 *Treaty Relating to Utilization of Waters of the Colorado and Tijuana Rivers, and of the Rio Grande* (hereinafter, 'Rio Grande–Colorado Treaty'), which regulates State action in the case of droughts.[27] Among others, the *Treaty* allocates the waters of the Rio Grande and its tributaries between the United States and Mexico. In case of droughts or accidents that render it difficult for Mexico to make agreed runoff volumes available to the United States, flow deficiencies within one five-year cycle must be made up in the subsequent five-year cycle.[28] In the case of drought in the Colorado Basin, 'the water allotted to Mexico . . . will be reduced in the same proportion as consumptive uses in the United States are reduced'.[29] The 1996 *Treaty between India and Bangladesh on Sharing of the Ganga/Ganges Waters at Farakka*[30] equally establishes water-sharing formulas for the dry season from January to May; depending on whether the river flow is below or above a certain flow threshold, the two parties obtain different predetermined shares of the total.[31] The threshold is defined at three levels, at which India is obliged to apply different water release schedules at the Farakka Dam depending on whether the flow is less than 70,000 cubic feet per second (cusecs), between 70,000 and 75,000 cusecs, or more than 75,000 cusecs. Seasonal

[27] McCaffrey, 'The Need for Flexibility', p. 160.

[28] Art. 4, 1944 *Rio Grande–Colorado Treaty*. [29] Art. 10, 1944 *Rio Grande–Colorado Treaty*.

[30] Salman and Uprety, *Conflict and Cooperation*, pp. 387–391.

[31] Annexure I, 1996 *Treaty on Sharing of the Ganga/Ganges Waters at Farakka*.

adjustments of flow volume are also provided for in a 1959 Agreement between Norway, Finland and the Soviet Union concerning the regulation of flow in the Paatsjoki River Basin.[32] In this case, the parties agreed to water releases from Lake Inari, located in Finland, in connection with the operation of a series of downstream hydroelectric power stations located in Russia and Norway. Minimum and maximum lake levels and flow release into the Paatsjoki River are regulated for various scenarios of spring flood predictions.[33]

These treaty provisions and others, which require adjustment of State behaviour to changes in flow volume of transboundary watercourses, provide legal solutions for adaptation to changing circumstances. They accommodate some of the precipitation changes that are predicted in current climate change scenarios. However, these provisions may not be sufficient; they provide for a level of change that was within the realm of what the treaty authors were able to predict at the time of conclusion of the agreements. Some date to periods long before scientific models on climate change were developed and therefore do not consider the extent of long-term impacts on the hydrologic cycle. For example, the 1959 *Agreement on the Regulation of Lake Inari* provides for a maximum permissible water level of the lake of 118.03 metre above sea level and a maximum permissible rate of discharge of 500 cubic metre per second.[34] It remains to be seen whether the predicted long-term increase in precipitation in the northern hemisphere will require transgression of these mutually agreed-to thresholds and thus revision of the treaty. Both the maximum allowed water level and the maximum permissible discharge rate may turn out to be too low. Hence, some of the existing treaties that provide adequate flexibility for typical hydrologic variability may ultimately be too rigid relative to the level of flexibility that will be required for adaptation to future precipitation changes and will need to be renegotiated.

[32] *Agreement Concerning the regulation of Lake Inari by Means of the Kaitakoski Hydroelectric Power Station and Dam*, 29 April 1959, 346 *UNTS* 168.

[33] Annex 3, 1959 *Agreement on Lake Inari*.

[34] Annex 3, para. 1, 1959 *Agreement on Lake Inari*, stipulates that 'The flow of water from Lake Inari shall be so regulated that the water level of the lake does not rise above 118.03 meters above sea level nor fall below 115.67 meters above sea level'. Para. 2 specifies: 'If in the case of very heavy floods the water level of Lake Inari threatens to rise above 118.03 meters above sea level, the rate of discharge may, as an exceptional measure, be increased up to 500 cubic meters per second when the water level reaches 117.98 meters above sea level. This shall be the maximum permissible rate of discharge'.

Obligations of conduct that set in motion previously undetermined adaptive behaviour

Obligations of conduct that do not intend to achieve a specific result but rather set in motion a process by which States can subsequently identify the required response for a given change in circumstances are better suited to accommodate climatic challenges that lie ahead. These obligations comprise cooperation duties, which are analysed at length in Part II, such as the obligations to notify, to consult and to negotiate, as well as to regularly exchange data and information. In fact, some of the treaties that require specific adjustments in State behaviour in response to pre-identified hydrologic events also consider the possibility of unlikely events and include obligations to identify jointly the responses to those challenges. Article II of the *Treaty Concerning Releases at Farakka*, for example, established an obligation for immediate consultations between India and Bangladesh '[i]n the event flow at Farakka falls below 50,000 cusecs in any 10-day period'. The parties are called on to 'make adjustments on an emergency basis, in accordance with the principles of equity, fair play and no harm to either Party'. The 1992 *Treaty between Swaziland and South Africa on the Development of Storage Dams on the Komati River* equally provides for an obligation to negotiate and agree on further action in case '*vis major*' impairs implementation of the treaty.[35] *Vis major* is defined in broad terms in Article 1 and includes among others 'any overwhelming occurrence of nature, including extreme drought or flood, not planned for'.

Treaties that provide for regular exchange of data and information or for pooling and joint analysis of data increase the State parties' capabilities to monitor and predict long-term trends and changes, as well as to better adjust to hydrologic variability and react more swiftly to the occurrence of sudden severe events (e.g., floods). The 1989 *Agreement Concerning the Regulation Governing Lake Saimaa and the Vuoksi River* displays these normative tools: the treaty was concluded between Finland and the former Soviet Union with the objective of preventing damage to these water bodies during periods of high and low flows.[36] Discharge of Lake Saimaa is regulated by the downstream Tainionkoski Dam; both the lake and the dam are located in Finland. Water releases are regulated such that lake and river flow levels remain within normal limits; however, if

35 Art. 15, *Treaty on the Development and Utilization of Water Resources of the Komati River Basin*, 13 March 1992, FAO Legislative Study–61, p. 242.

36 26 October 1989, 1663 *UNTS* 337.

higher or lower than normal water levels are imminent, flow releases are adjusted gradually.[37] Finland is under obligation to prepare preliminary water condition appraisals for four-month periods and to collect monthly data – and more frequently when necessary. This information is transmitted to the manager of the dam as well as representatives of Russia so that necessary adjustments can be made cooperatively between the two States.[38]

Preparedness for severe and sudden precipitation events is enhanced through increased frequency of data exchange. In the case of the Moselle River, a Rhine tributary, the riparian States of Germany, France and Luxembourg agreed to the installation of an automatic information system to improve flood preparedness.[39] The system, the installation of which is carried out with coordination and cost sharing among the three States, provides remote data access digitally as well as via voice transmission in French and German at all times.[40] Each State party is responsible for monitoring the performance of the system; any malfunction must be reported immediately to the other parties.[41]

Coordination within the framework of joint mechanisms

When joint systems of data exchange and data pooling are set up, they are often accompanied by the establishment of joint management mechanisms. A Technical Committee, composed of representatives from the respective national navigation and water authorities, is in charge of managing the automatic information system on the Moselle River. The Committee has a limited mandate and is structured in a relatively loose manner; it meets at the request of any of the involved authorities, unless it decides to meet regularly. According to Article 5, the Committee is authorised to make decisions related to updating technical reports, modifying and adding equipment and moving or reconstructing water level measurement transmission stations. The Committee established by India and Bangladesh for water management at the Farakka Dam is similarly limited in its mandate; it is responsible for implementing the release schedule described in the treaty, to record flow data measured at Farakka, and to report it to the two governments.[42] These two mechanisms have

[37] Appendix para. 1, 1989 *Agreement on Lake Saimaa and the Vuoksi River*. [38] Para. 2, *ibid.*

[39] *Agreement on Flood Warning in the Catchment Basin of the Moselle*, 1 October 1987, 1506 *UNTS* 97.

[40] Arts. 2 and 3, *ibid.* [41] Art. 6, *ibid.*

[42] Arts. IV–VII, 1996 *Treaty on Sharing the Ganga/Ganges Waters at Farakka*.

specific mandates and it must be recognised that by restricting the scope of information exchange to specific objectives, States forego opportunities to create broader adaptive capacity. Article 5 of the 1987 *Moselle Agreement* 'opens a back door' that increases the value of the Technical Committee for climate change reporting because it permits the Committee to submit proposals to the Governments on issues not listed in the Article: 'in particular [concerning] the transmission of additional information from existing installations'. Joint management mechanisms that are vested with wide mandates, standing technical committees, or standing Secretariats are in a better position to respond to the variety of challenges posed by climate change than those with mandates limited to flow measurement. Changes in flow volume comprise only one of the predicted impacts; others include changes in water quality, impact on ecosystems and biodiversity, and salt intrusion in delta areas and coastal aquifers.

Joint mechanisms through which States interact on a regular or even permanent basis thus comprise a useful tool to structure cooperation for adaptation. The scope of the institutional mandate prescribed in Article 9 of the 1992 *UNECE Water Convention* provides an example that could – or probably should – serve as guidance for States planning to develop joint bodies, particularly when considering this undertaking from the perspective of adaptation to climate change. The provision envisages extensive obligations for data exchange and scientific cooperation, as well as obligations to establish early warning and alarm procedures.[43] Joint mechanisms serve as a 'forum for exchange of information on existing and planned uses of water and related installations that are likely to cause transboundary impact'.[44] Although intended to satisfy the primary objective of water quality control, the structures envisaged for joint-management mechanisms under the 1992 *UNECE Water Convention* are also appropriate with respect to climate change adaptation.

The virtues of institutional mechanisms, which have strong scientific capacities in dealing with climate change adaptation, have been evidenced in the Rhine and Danube River Basins. In both basins, multilateral water-protection agreements were concluded after adoption of the 1992 *UNECE Water Convention*. The two instruments, the 1994 *Danube Protection Convention* and the 1999 *Convention on the Protection of the Rhine*, place primary emphasis on water quality and environmental protection;

[43] Art. 9 (2), 1992 *UNECE Water Convention*. [44] *Ibid*.

they also include mandates for flood control. Over time, the institutions established under their umbrella have taken on the task of adapting basin management to climate change impacts. The ICPR included the development of response measures within the context of the *Action Plan on Floods*, which forms part of the *Rhine 2020 Strategy*; in addition to basin-wide hydrologic and meteorological data processing, former wetlands are re-naturalised to mitigate for increased quantity and severity of flood events.[45] The climate program of the ICPDR consists of enhanced data generation and sharing, joint data analysis, monitoring of impacts on ecological and chemical water stress and climate-proofing of hydraulic infrastructure, as well as the translation of scientific information for water managers.[46] New information on climate change impacts is continuously integrated into the regularly updated river basin management plan, a requirement under the EU *WFD*.[47] Since 2008, the Danube River Basin has also been covered under the European Flood Alert System (EFAS), which – similar to the automatic system of the Moselle River described previously – provides national authorities with remote access to rainfall data from more than 700 stations, 24 hours a day.[48]

In addition to comprehensive data exchange among riparian States, other factors make the Danube and Rhine Commissions adaptable and flexible to address changing requirements. In both cases, the constituting legal instruments envisage that member States can change and broaden the mandate of their joint commissions if required. Article 9 of the 1999 *Rhine Protection Convention* stipulates that the ICPR shall 'carry out any other tasks entrusted to it by the Contracting Parties'. The *Danube River Protection Convention* contains a comparable provision.[49] Other key elements are the existence of standing Secretariats and the involvement of scientific experts in technical working groups. Article 14 of the *Rhine Protection Convention*, for example, allows the ICPR to consult experts and invite them to

45 See www.iksr.org.

46 R. Mair, 'ICPDR Actions Related to Climate Change', presentation, First Meeting of the Core Group of UNECE Pilot Projects, 15–16 February 2011.

47 See www.icpdr.org/icpdr-pages/river_basin_management.htm.

48 The development of the Danube–EFAS system was effectuated based on memoranda of understanding signed by riparian States. See 'New International System for Early Flood Warning in Danube River Basin Launched', *United Nations Information Service*, 10 March 2008, www.unis.unvienna.org/unis/pressrels/2008/unisicpdr002.html.

49 Article 18 (3): 'In addition to affairs explicitly entrusted to the International Commission, it is competent to deal with all other affairs the Commission is entrusted with by mandate from the Contracting Parties in the framework of [the objectives] of this Convention'.

meetings. Annex IV of the *Danube River Protection Convention* envisages the establishment of standing and *ad hoc* expert groups. This aspect is important because international cooperation and legal development generally follow or lag behind science, particularly in the field of climate change.[50] Therefore, the integration of scientific knowledge into the institutional structure is an important factor in devising effective responses to changing circumstances.

Another factor enhancing adaptive capacity in the Danube and Rhine River Basins is that they are covered by regional legal frameworks that are progressive in promoting basin-wide cooperation. Both the ICPDR and the ICPR are coordinating bodies for the implementation of the *EU Water Framework and Flood Risk Directives*,[51] and riparian States of both basins are signatories of the 1992 *UNECE Water Convention*. The UNECE framework represents to date a singular example in promoting international cooperation on transboundary waters with respect to climate change. At its Fifth Session, the MOP to the Convention adopted the *Guidance on Water and Adaptation to Climate Change*[52] and 'invited the Parties and non-Parties to the Convention to implement the Guidance in the framework of cooperation on transboundary water management and to the extent appropriate in the national context, in particular through the development of pilot projects'.[53] The *Guidance* is not legally binding and does not supersede any prior legal obligations entered into by the Parties to the Convention; rather, it was designed to enhance implementation of international commitments as well as obligations deriving from other instruments. Chapter 2 of the *Guidance* recalls relevant obligations established by *inter alia* the *United Nations Framework Convention on Climate Change*, the *UNECE Convention on Environmental Impact Assessment in a Transboundary Context*, the *Ramsar Convention on Wetlands* and EU legislation.[54] The *Guidance* reveals its value not only for basins under the UNECE umbrella; rather, it was written to also guide adaptation in all regions of the world. Similar to the *UNECE Water Convention*, the *Guidance on Water and Adaptation to Climate Change* is comprehensive in its aspirations; recommendations include

50 P. Sands, *Principles of International Environmental Law*, pp. 344–346.

51 *Directive 2007/60/EC on the Assessment and Management of Flood Risks*, 6 November 2007, OJ 2007 L 288/27.

52 UNECE, *Guidance on Water and Adaptation to Climate Change* (Geneva: United Nations, 2009), UN Doc. ECE/MP.WAT/30.

53 *Report of the Meeting of the Parties on its Fifth Session, Geneva, 10–12 November 2009*, 15 June 2010, UN Doc. ECE/MP.WAT/29.

54 UNECE, *Guidance on Water and Climate Change*, pp. 12 and 25–33.

implementation of integrated water resources management, adoption of a cross-sectoral approach extending to the transboundary level, knowledge and information sharing and public participation, as well as a focus on water supply and sanitation.

The *Guidance* so far is the only instrument of its kind at the regional level; it is the result of State cooperation within a joint institutional framework supported by the Permanent Secretariat to the 1992 *UNECE Water Convention*. In other regions, riparian States also cooperate within joint institutional structures to prepare for climate change impacts. The MRC engaged in a formalised dialogue with Myanmar and China (i.e., the two upstream riparian States that are not its members) and, since 2004, it has received daily hydrologic data on flood forecasting from an automatic telemetry system in China.[55] As one of the rivers fed by Himalayan glacial melt, the Mekong is likely to experience significant flow changes in the long term. Regular data exchange – particularly with the most upstream riparian State, China – allows for preparatory planning of flood warning and mitigation measures and for arrangements related to long-term changes in annual runoff. The riparian States of the Senegal River equally cooperate within the framework of their joint institution, the OMVS, to develop programs in response to climate change impacts on the basin.[56]

This analysis highlights that water treaties and the principles and rules of international water law offer a variety of solutions to structure adaptive responses to climate change. Joint mechanisms that have an open mandate – or whose mandate can be widened without requiring renegotiation of constitutive treaties – give States the flexibility to adjust these structures when necessary. Although water treaties generally do not refer directly to climate change,[57] they provide a legal basis for several measures that can be used in the context of climate change adaptation. Cooperation obligations and the rights and duties related to the principle of equitable and reasonable utilisation contain elements that promote flexibility and adaptability of States to changes in hydrologic circumstances at the basin level.

55 Mekong River Commission, *MRC Work Program 2005*, December 2004, p. 27, www.mrcmekong.org.

56 World Bank, *Implementation Completion and Results Report (TF-52900) on a Grant from the Global Environmental Facility (GEF) Trust Fund to the Organisation de la Mise en Valeur due Fleuve Senegal for the Regional Project: Senegal River Basin Water and Environmental Management*, Report No. ICR0000883 (Washington, DC: World Bank, 2009), p. 11.

57 References to climate change are found in Art. 5, 2002 *Senegal Water Charter*; Art. 13, 2008 *Niger Basin Charter*; and Art. 66, 2004 ILA *Berlin Rules*.

Yet, international water law does not respond to wider water complexities and to the fact that climate change is a global phenomenon and that the transboundary nature of freshwater resources does not stop at the boundaries of individual basins. The water cycle spans the entire planet. To address root causes of climate change and its hydrologic effects, international cooperation that is limited to the transboundary water system level will not be sufficient. Climate change and adaptation to its transboundary effects require a more global response. The following discussion explores the questions of whether principles and legal instruments of climate change law can complement the facilitative role of international water law for adaptation at the basin level and whether this legal dynamic also applies vice versa with regard to global water challenges.

7.3 Appreciation of water (law) in climate change law

In contrast to international water law, which emerged as a result of State cooperation on transboundary waters at the basin level, the legal climate-change framework was established to address a global challenge. At the same time, however, its principles and rules, as well as the mechanisms created by it, have an impact at the basin level.

7.3.1 Lack of recognition?

Given that effects of climate change are felt primarily through the hydrologic cycle, it is curious that the treaties adopted in reaction to this phenomenon barely mention water. The *UNFCCC* stands almost alone in referring to water in its text, imposing an obligation on States to 'cooperate in preparing for adaptation to the impacts of climate change' by elaborating appropriate plans for water resources, particularly concerning areas affected by drought, desertification and floods.[58] Other than that, eighteen years later, the decision on the 2010 *Cancun Adaptation Framework* was the first to explicitly mention water, albeit in a footnote. Recalling the spirit of Article 4 of the *UNFCCC*, it appeals to the joint responsibility of States to cooperate in order to enhance action on adaptation, including 'in the areas of water resources; health; agriculture; and food security; infrastructure; socio-economic activities; terrestrial, freshwater

[58] Art. 4 (1) (e): 'Cooperate in preparing for adaptation to the impacts of climate change; develop and elaborate appropriate and integrated plans for coastal zone management, water resources and agriculture, and for the protection and rehabilitation of areas, particularly in Africa, affected by drought and desertification, as well as floods'.

and marine ecosystems; and coastal zones'.[59] One year later, the Seventeenth COP of the *UNFCCC* continued discussion of the topic and requested the Convention Secretariat to organise a workshop 'on water and climate change impacts and adaptation strategies'.[60] This workshop, held in July 2012, proposed water as a thematic work area under the *Nairobi Work Program* on impact, vulnerability and adaptation to climate change,[61] which would further inform scientific and technical work under the *Cancun Adaptation Framework*.[62] The decision to integrate water in the official work program of the *UNFCCC* framework will be made at the thirty-eighth session of the Subsidiary Body for Scientific and Technological Advice (SBSTA) in 2013.

The tentative and slow integration of water in the *UNFCCC* framework is deceiving: States have been well aware of the impacts of climate change on freshwater resources and have engaged in cooperation at the universal level along different avenues. This is evidenced, for example, by adoption of Agenda 21 in parallel with the *UNFCCC* at the UNCED in Rio.[63] Agenda 21 recognised the impacts of atmospheric change on freshwater ecosystems and outlined a plan of action with a view to ensuring that they 'are fully taken into account in planning and implementing policies and programs concerning terrestrial and marine resource utilization and land-use practices'.[64] Furthermore, the legal framework developed on the basis of the *UNFCCC* has evolved into a structure that enables international cooperation in response to climate-induced changes in the global water cycle.

7.3.2 The UNFCCC framework

The system of international action established by the *UNFCCC* provides an enabling framework to implement the policy objectives of Agenda 21. It is

[59] COP to the *UNFCCC*, *Decision 1/CP.16 – The Cancun Agreements: Outcome of the Work of the Ad-Hoc Working Group on Long-Term Cooperative Action under the Convention*, 15 March 2011, UN Doc. FCCC/CP/2010/7/Add.1.

[60] COP to the *UNFCCC*, *Decision 6/17 – Nairobi Work Programme on Impacts, Vulnerability and Adaptation to Climate Change*, 15 March 2012, UN Doc. FCCC/CP/2011/9/Add.2.

[61] SBSTA, *Report of the Subsidiary Body for Scientific and Technological Advice on its Twenty-Fifth Session Held at Nairobi from 6 to 14 November 2006*, 1 February 2007, UN Doc. FCCC/SBSTA/2006/11.

[62] SBSTA, *Report on the technical workshop on water and climate change impacts and adaptation strategies*, 13 September 2012, UN Doc. FCCC/SBSTA/2012/4, p. 19.

[63] *Agenda 21: Earth Summit – the United Nations Program of Action from Rio* (New York: United Nations, 1993), UN Doc. A/CONF.151/26/REV.1 (Vol. II).

[64] Section II, paras. 9.19–9.21, *ibid.*

a highly institutionalised framework in which the State parties regularly interact in joint bodies at the technical and political levels, review the latest science and engage in efforts to develop appropriate responses.

It is important to recognise that the *Convention* strictly limits its applications to climate change that can be directly or indirectly related to human activity – that is, change that is additional to natural climate variability – and to impacts that have 'significant deleterious effect on the composition, resilience or productivity of natural and managed ecosystems'.[65] The *UNFCCC* recognises that States are the main contributors to climate change and that, conversely, they are also those affected by its impacts. Furthermore, the *Convention* acknowledges that individual State contributions to the factors that cause a change in climatic conditions and the level of impact felt vary among States. Thus, the commitments for mitigation and adaptation taken by the parties to the *UNFCCC* also vary.

The concept of common but differentiated responsibilities

The concept of common but differentiated responsibilities (CBDR) permeates the *UNFCCC* framework and informs its implementation.[66] The *Convention* recalls it in the preamble: '*acknowledging* that the changes in the Earth's climate and its adverse effects are a common concern of humankind' and '*noting* that the largest share of historical and current global emissions of greenhouse gases has originated in developed countries', it calls on States to cooperate and participate 'in an effective and appropriate international response, consistent with their common but differentiated responsibilities and respective capabilities and their social and economic conditions' [emphasis in the original].[67]

Differentiating among parties to a treaty with regard to their rights and obligations is a common practice in international law and dates at least as far back as the 1919 *Treaty of Versailles*.[68] As illustrated in Chapter 6, progressive implementation obligations that take account of differences among States regarding the availability of resources and their capacities, for instance, are included in the 1966 *ICESCR*. In environmental law, the CBDR concept has been incorporated in numerous multilateral

65 Art. 1, *UNFCCC*.

66 J. Brunnée, 'Climate Change, Global Environmental Justice and International Environmental Law', in J. Ebbeson and P. Okowa (eds.), *Environmental Law and Justice in Context* (Cambridge: Cambridge University Press, 2009), pp. 324f.

67 Preamble of the *UNFCCC*.

68 C. Stone, 'Common but Differentiated Responsibilities in International Law' (2004) 98 *AJIL* 278.

agreements since the 1980s. The 1982 *UN Convention on the Law of the Sea*, for example, includes special privileges for fish-dependent and developing countries. The 1985 *Vienna Convention for the Protection of the Ozone Layer* and the 1992 *Convention on Biological Diversity* also differentiate among the parties.[69]

Already endorsed in the 1972 *Stockholm Declaration on the Human Environment*, it is the 1992 *Rio Declaration* that is lauded as the document that most comprehensively defines the CBDR concept. Recognising special needs of developing countries in Principle 6, it outlines the core content of what CBDR means with respect to the environment in Principle 7:

> States shall cooperate in a spirit of global partnership to conserve, protect and restore the health and integrity of the Earth's ecosystem. In view of the different contributions to global environmental degradation, States have common but differentiated responsibilities. The developed countries acknowledge the responsibility that they bear in the international pursuit to sustainable development in view of the pressures their societies place on the global environment and of the technologies and financial resources they command.

The question on the legal status of the concept has not yet been settled in literature and practice; most authors doubt its quality as legal principle of environmental law and a few States equally object to such qualification.[70] Similarly, the notion of 'differentiated responsibilities' must be understood in terms of referring to 'general legal and political obligations' and therefore is not comparable to the more formal legal concept of State responsibility.[71] The latter refers to legal responsibility incurred due to harm caused to others in violation of international law. The CBDR concept, in turn, is based on the idea of concerted pro-activity: it results from the recognition that resources exist that are of common concern to all, such as the global environment, the ozone layer and the global climate.[72] Therefore, all States should contribute to the protection, restoration and

[69] *Vienna Convention for the Protection of the Ozone Layer*, 22 March 1985, 1513 *UNTS* 324; see also the *United Nations Convention to Combat Desertification in Those Countries Experiencing Serious Drought and/or Desertification, Particularly in Africa*, 14 October 1994, 1954 *UNTS* 3.

[70] D. Shelton, 'Equity', in D. Bodansky, J. Brunnée and E. Hey, *The Oxford Handbook of International Environmental Law* (New York: Oxford University Press, 2007), p. 655; L. Rajamani, 'The Principle of Common but Differentiated Responsibility and the Balance of Commitments under the Climate Regime' (2000) 9 *RECIEL* 2, 124.

[71] A. Kiss and D. Shelton, *International Environmental Law*, 2nd edition (Ardsley: Transnational Publishers, Ardsley, 2000), p. 257.

[72] C. Behrmann, *Das Prinzip der angemessenen und vernünftigen Nutzung und Teilhabe nach der VN-Wassserlaufkonvention* (Berlin: Duncker & Humblot, 2008), p. 57.

sustainable management of resources that are of common concern. In articulation of the concept of equity, those States that contribute to a comparatively larger extent to the degradation of resources of common concern and that have relatively more resources available to address the adverse impacts are called on to take the lead.[73]

The way in which the CBDR is currently reflected in the climate change framework is different from its formulation in the 1992 *Rio Declaration*. Principle 7 addresses the actual impacts of societies. Worded in the present tense, the phrase 'in view of the pressures their societies *place* on the global environment' [emphasis added] does not distinguish between past and current contributions but instead treats them equally. In contrast, the *UNFCCC* places primary emphasis on historic contributions to environmental degradation, combining several considerations. First, the acceptance by all parties of general commitments acknowledges unequivocally that States share a common responsibility for anthropogenic climate change and its effects. Second, the distinction between 'developing' and 'developed' countries (Annex I) is an expression of the idea that those who contributed more historically to the anthropogenic causes of climate change bear a greater responsibility.[74] Finally, the special commitments assumed by most Organisation for Economic Cooperation and Development (OECD) countries (Annex II) reflect the consideration that those who have the financial and technical capabilities to respond should lead in the response to climate change and assist those States that lack these resources.[75]

Differentiated responsibilities for mitigation and adaptation

The primary focus of the *Convention* is mitigation and stabilisation of 'greenhouse gas concentrations in the atmosphere at a level that would prevent dangerous anthropogenic interference with the climate system'.[76] Adaptation is recognised as an additional element in the *Convention* text yet is gaining increasing attention as the findings of scientific studies increasingly converge on the point that some climate change effects will be irreversible in the medium- to possibly even the long-term.[77] Both mitigation and adaptation relate to the water sector: large reservoirs

[73] E. Louka, *International Environmental Law: Fairness, Effectiveness and World Order* (New York: Cambridge University Press, 2006), p. 54; see also Art. 3 (1), *UNFCCC*, which links the CBDR concept to the notion of inter-generational equity.

[74] L. Rajamani, 'The Principle of Common', p. 121. [75] *Ibid.* [76] Art. 2, *UNFCCC*.

[77] F. Yamin and J. Depledge, *The International Climate Change Regime: A Guide to Rules, Institutions and Procedures* (Cambridge: Cambridge University Press, 2004), p. 213.

created for water storage, for instance, cause high methane gas emissions from rotting organic matter; most adaptation measures are adopted in response to extraordinary changes in water availability. It is in this context that the climate change framework influences cooperative water resources management at the global and basin levels.

The parties to the *UNFCCC* are called on to 'formulate, implement, publish and regularly update national and, where appropriate, regional programs containing measures to mitigate climate change by addressing anthropogenic emissions by sources and removals by sinks of all greenhouse gases not controlled by the *Montreal Protocol*,[78] and measures to facilitate adequate adaptation to climate change'.[79] These provisions are also recalled in Article 10 of the *Kyoto Protocol*.[80] The *Protocol* elaborates further on the possible content and focal areas of such programs and measures, including those concerning agriculture and forest management as well as spatial planning. Agriculture and forestry are two key sectors in water resources management, in the national as well transboundary basin contexts. The agriculture sector will be most affected by climate change, and forestry management is closely related to the creation and enhancement of greenhouse gas (GHG) sinks. It also has an important function in regulating surface runoff, localised precipitation patterns and local atmospheric humidity.

Regarding mitigation, the legal framework established by the *Convention* and its *Kyoto Protocol* was designed to promote international cooperation so as to prevent or at least decelerate changes in the global hydrologic cycle.[81] All signatories of the *Convention* assumed a general commitment to work towards this goal under Article 4 (1). At the same time, Annex I States[82] assume more specialised obligations to reduce emissions and to enhance GHG sinks and reservoirs.[83] Special commitments by these States to take the lead in emission reductions were quantified in the *Kyoto*

[78] The *Montreal Protocol on Substances that Deplete the Ozone Layer*, 16 September 1987, 26 ILM 1550, regulates the phasing out of chlorofluorocarbons and hydrochlorofluorocarbons.

[79] Art. 4 (1) (b), *UNFCCC*.

[80] *Kyoto Protocol* to the *United Nations Framework Convention on Climate Change*, 11 December 1997, 37 ILM 22.

[81] Arts. 10 and 12, 1997 *Kyoto Protocol*.

[82] Annex I includes OECD member States and former Eastern Bloc States undergoing transition to a market economy at the time of adoption of the *Convention*. Annex B of the *Kyoto Protocol* quantifies emission limitations and reduction commitments for Annex I States; in contrast to Annex I *UNFCCC*, Annex B does not include Turkey in the list of States.

[83] Art. 4 (2), *UNFCCC*.

Protocol.[84] Non-Annex I States were invited to adopt the same commitments[85] and/or to commit to voluntary emission reduction targets[86] in order to contribute to the global benefit of slowing climate change. Cooperation of non-Annex I States, which include mostly developing countries but also a number of States with economies currently in transition (e.g., China and India), is encouraged through legal commitments of OECD member States (i.e., Annex II States)[87] to provide financial resources and assistance in the form of technology transfers for mitigation actions implemented by developing countries.[88]

A general obligation to cooperate also exists for all parties with respect to preparation for adaptation; this includes cooperation related to the elaboration of adaptation plans, research and data development and information exchange, as well as education and training programs for human and institutional capacity development.[89] This general duty to cooperate binds all parties and is not limited per se to assistance provided by developed to developing country parties; so-called South-South cooperation is also within its scope and encouraged. As with mitigation measures, the *Convention* and the *Kyoto Protocol* stipulate special obligations for States listed in Annex II of the *Convention*. They are obligated to provide financial resources that enable non-Annex I parties to comply with their national reporting requirements, and they are obligated to 'provide such financial resources, including for the transfer of technology, needed by developing country Parties to meet the agreed full incremental costs of implementing [their general commitments under Article 4 (1)] and that are agreed between a developing country Party and the [financial mechanism]'.[90] In addition, according to Article 4 (4) of the *UNFCCC*, developed country parties are under obligation to assist particularly vulnerable developing country parties, including least-developed countries and small island States, in meeting the costs of adaptation to adverse effects of climate change. This provision is problematic to the extent that causality links between damaging effects

84 Annex B, 1997 *Kyoto Protocol*. 85 Art. 4 (2) (g), *UNFCCC*.

86 *COP-UNFCCC*, Decision 1/CP.16, 11 March 2011, UN Doc. FCCC/CP/2010/7/Add. 1, paras. 48–50.

87 Annex II includes only OECD member States. Turkey, which was initially included in the list, was deleted according to Decision 26/CP.7, 21 January 2002, UN Doc. FCCC/CP/2001/13/Add. 4.

88 Art. 4 (3) and (7), *UNFCCC*; Arts. 10 and 12, 1997 *Kyoto Protocol*.

89 Arts. 4 (1) (e) and (g)-(i), *UNFCCC*; Art. 10 (c)-(e), 1997 *Kyoto Protocol*. 90 Art. 4 (3), *UNFCCC*.

and anthropogenic climate change are difficult to prove and quantification is difficult.[91]

The performance of developing country parties with respect to their general commitments has been linked to compliance of States listed in Annex II with their specific commitments concerning funding support and technology transfer.[92] The parties to the *Convention* accept that the ability of non-Annex I parties to adhere to general commitments is a function of international cooperation and assistance provided by developed countries.[93] Linking developing States' implementation of general commitments to the provision of assistance of developed States carries elements of both cooperation and solidarity. Although mitigation action benefits all States, adaptation benefits – first and foremost – the States implementing such measures. In this regard, payments carry an aspect of solidarity in the sense that they create benefits primarily for the receiving State. In the medium- to long-term, peripheral benefits from adaptation also may accrue to the developed countries paying for adaptation. Effective adaptation to flood and drought events in developing countries improves resilience, thereby reducing the risks of conflict or mass migration to more affluent regions.

Institutionalisation of global assistance with benefits at the basin level

Special commitments of Annex II States include the provision of technical and financial resources to assist developing countries and aim to achieve the *Convention*'s objective concerning equity. The funding obligations can be effectuated through the financial mechanism of the *UNFCCC* and through bilateral, regional and other multilateral channels.[94] The institutionalisation of assistance in that form has considerable advantages for the implementation of international cooperation because it facilitates harmonisation of activities and continuity of funding, as well as an equitable distribution of funds.

Implementation details of adaptation financing were settled largely through the *Marrakesh Accords* adopted at the Seventh Meeting of the COP in 2001.[95] The Global Environmental Facility (GEF), which was entrusted

[91] Bodansky, 'The United Nations Convention on Climate Change', p. 528.
[92] Art. 4 (7), *UNFCCC*.
[93] Yamin and Depledge, *The International Climate Change Regime*, p. 506.
[94] Art. 11 (5), *UNFCCC*; Art. 11 (3), 1997 *Kyoto Protocol*.
[95] 21 January 2002, UN Doc. FCCC/CP/2001/13/Add.1.

with the role of financial mechanism for the *UNFCCC* framework,[96] was given the responsibility to manage two funding mechanisms: (1) the Special Climate Change Fund (SCCF) for adaptation, technology transfer, energy, transport, industry, forestry and activities that support developing countries in diversifying their economies; and (2) the Least Developed Countries Fund (LDCF), which facilitates preparation and implementation of National Adaptation Programs of Action (NAPAs) in least-developed countries. It also provides Secretariat services on an interim basis to the Adaptation Fund established by the *Kyoto Protocol*.[97] This fund, which receives 2 per cent of the proceeds from certified emission reductions issued for project activities under the Clean Development Mechanism, finances adaptation projects and programs of developing country parties to the *Protocol*.[98] A new fund – which is expected to channel a significant share of new multilateral funding for mitigation and adaptation from developed to developing country parties – is the Green Climate Fund, which is in the process of being set up as a new operating entity of the financing mechanism of the *UNFCCC*.[99]

Many adaptation projects and programs funded through these mechanisms benefit national and transboundary water resources management. Approximately half of the projects financed through the LDCF and SCCF are in water-related sectors, including agriculture and forestry.[100] The same is true for projects supported by the Adaptation Fund.[101] In addition, the Strategic Priority for Adaptation (SPA) GEF funding allocation, which was established as the precursor of the LDCF and SCCF to support pilot projects on adaptation planning and assessment, included two projects concerning transboundary rivers: one on sustainable management of water resources of the La Plata Basin concerning the effects of climate variability and change, and another in the Amazon River Basin, with a similar objective.[102]

[96] Art. 21 (3), *UNFCCC*; Decision 11/CP.1, 6 June 1995, UN Doc. FCCC/CP/1995/7/Add.1; and Decision 37/CP.4, 25 January 1999, UN Doc. FCCC/CP/1998/16/Add.1.

[97] Art. 12 (3), 1997 *Kyoto Protocol*.

[98] Decision 1/CMP 3, 14 March 2008, UN Doc. FCCC/KP/CMP/2007/9/Add.1; Decision 17/CP.7, 1 February 2002, UN Doc. FCCC/CP/2001/13/Add.2.

[99] Decision 3/CP.17, 15 March 2012, UN Doc. FCCC/CP/2011/9/Add.1; Decision 1/CP.16, 15 March 2011, UN Doc. FCCC/CP/2010/7/Add.1, paras. 100–112.

[100] GEF, *Progress Report on the Least Developed Countries Fund (LDCF) and the Special Climate Change Fund (SCCF)*, 28 October 2011, GEF/LDCF.SCCF.11/Inf.02/Rev.1.

[101] See www.adaptation-fund.org/funded_projects.

[102] GEF, *Evaluation of the GEF Strategic Priority for Adaptation (SPA), Full Evaluation Report* (Washington, DC: GEF, 2010), pp. 18 and 29.

States also comply with their general commitments under Article 4 (1) of the *UNFCCC* through regional cooperation activities taking place outside of the Climate Change Convention's institutional framework.[103] Transboundary water management mechanisms at the regional and basin levels have an important role in these efforts. In 2010, the MOP of the 1992 *UNECE Water Convention* decided on a program of pilot projects that promotes implementation of the *Guidance on Water and Adaptation to Climate Change*.[104] The program aims to enhance adaptation efforts in a transboundary water context through the establishment of a platform for exchange of experiences.[105] The projects are supported by the UNECE Secretariat in partnership with other organisations, including the UNDP, the Organization for Security and Cooperation in Europe (OSCE) and the UNEP, under the Environment and Security Initiative (ENVSEC) with the objective of avoiding international security risks that might result from degradation and inequitable access to natural resources in Southeastern and Eastern Europe, Central Asia and the Caucasus.[106] Many of the projects are directly implemented by basin management mechanisms, such as the Sava River Basin Commission and the ICPR.[107] The adaptation program carried out by the ICPDR on the Danube River is also part of this pilot program.[108]

The institutional architecture facilitating technical and financial assistance for mitigation and adaptation at the universal, regional and basin levels illustrates the mutual supportiveness and interplay between the legal climate change framework and transboundary water law. The joint bodies established by the *UNFCCC* and its protocol and by water treaties work towards the common challenge of addressing climate change and anthropogenic changes to the global water cycle; basin management mechanisms assist the implementation of the *UNFCCC* objectives. Another area in which international water law may be beneficial to the climate change framework concerns the limited success of current climate change negotiations.

[103] Art. 11 (5), *UNFCCC*; Art. 11 (3) 1997 *Kyoto Protocol*.

[104] UNECE, *Report of the Meeting of the Parties on its Fifth Session*, 15 June 2010, UN Doc. ECE/MP.WAT/29, pp. 20f.

[105] UNECE Task Force on Water and Climate Change, *Overview of the Programme of Pilot Projects on Adaptation to Climate Change in Transboundary Basins*, 3 March 2010, Geneva, TFWC/2010/3, p. 1.

[106] See www.envsec.org.

[107] See www1.unece.org/ehlm/platform/display/ClimateChange/Pilot+projects.

[108] R. Mair, 'ICPDR Actions Related to Climate Change'.

7.3.3 *Differentiation as a challenge to cooperation*

Twenty years after its adoption, the *UNFCCC* framework is facing challenges concerning its effectiveness. In 2011, the Conference of the Parties noted 'with grave concern the significant gap between the aggregate effect of... mitigation pledges... and aggregate emission pathways consistent with having the likely chance of holding the increase in global average temperature below 2°C or 1.5°C above pre-industrial levels'.[109] Global emissions of GHGs continue to increase and negotiations on new quantified emission limitations and reduction commitments have not produced a new arrangement. To safeguard its positive impact, the *Kyoto Protocol* was extended for a second commitment period (i.e., 2013–2020), with the expectation that a new and more effective instrument would be ready for implementation in 2020.[110]

The need for revised differentiation

A factor that contributes to the slow speed of negotiations is the way in which the CBDR concept is currently conceived of in the legal climate change framework. Differentiated responsibilities for technical and financial assistance were included in the *UNFCCC* and its *Kyoto Protocol* partly with a view to achieving a truly global response to climate change.[111] They constitute concessions of developed States to motivate cooperation of developing States for the wider common good – that is, the protection of the atmosphere. This course of action is based on the idea that benefits to all will materialise only if those who are endowed with comparatively more substantial financial and technical capacities also contribute more comprehensively to the alleviation and mitigation of the problem.[112]

One issue that has emerged as a challenge to implementation is that the term 'developing country parties' has not been clearly defined; at times, it is used interchangeably with the term 'non-Annex I parties' even

[109] Decision 1/CP.17, 15 March 2012, UN Doc. FCCC/CP/2011/ 9/Add.1, p. 2.

[110] Conference of the Parties Serving as the Meeting of the Parties of the Kyoto Protocol, *Outcome of the Work of the Ad Hoc Working Group on Further Commitments for Annex I Parties under the Kyoto Protocol*, 8 December 2012, UN Doc. FCCC/KP/CMP/2012/L.9.

[111] J. Barret, 'The Negotiation and Drafting of the Climate Change Convention', in R. Churchill and D. Freestone (eds.), *International Law and Global Climate Change* (London: Graham & Trotmann/Martinus Nijhoff, 1991), pp. 196f. The *UNFCC* currently has 195 parties and the *Kyoto Protocol* has 191 (status as of March 2013).

[112] L. Boisson de Chazournes, 'Technical and Financial Assistance', in Bodansky, Brunnée, and Hey, *The Oxford Handbook*, p. 971.

though the latter group includes States that do not or no longer qualify as developing countries according to definitions by organisations such as the World Bank and the OECD.[113] Furthermore, the States that assumed mitigation commitments based on their historic and actual emission contributions in 1992, when the *Convention* was adopted, are no longer the biggest contributors to global GHG emissions today.[114] The more recent economic performance of some developing countries and transition economies not included in Annex I has resulted in a significant shift in relative emission contributions by individual States and, more important perhaps, in the ability to contribute to the response to climate change. The difference between developed and developing countries is not as marked as it was when the *UNFCCC* framework was designed. The distribution of obligations assumed under the special commitments is no longer accepted as equitable. Annex I currently includes the States that historically contributed most to anthropogenic causes of climate change but that no longer adequately reflect present contributions. The list of States assuming special commitments under Annex II does not include States that now have significant capacity to contribute technical and financial resources to adaptation and mitigation solutions. Thus, the conditions on which the 1992 agreement on fair burden sharing was based have been overcome by time and change in the global economic system.

The CBDR concept emerged with the aim of achieving equity in matters regulated by international environmental law.[115] As demonstrated in Chapter 3, 'equity' is a dynamic notion: with a change in circumstances, agreements that were accepted as fair by participating parties at one point can lose their validity over time and reduce the willingness to cooperate. Equity is intrinsically intertwined with the context to which it is applied. To ensure that equity can be reached and maintained over time, the respective agreements must capture its dynamic and contextual characteristics. This aspect apparently did not receive adequate attention in the drafting of the *UNFCCC*. The *UNFCCC* seeks to achieve equity by distinguishing between developed and developing countries, as well as countries otherwise particularly affected by climate change.[116] However, it does not include a list of factors to be considered in order to distinguish between developed and developing States; neither does it provide for a

113 Yamin and Depledge, *The International Climate Change Regime*, p. 272.

114 See International Energy Statistics, www.eia.gov/cfapps/ipdbproject/IEDIndex3.cfm?tid=90&*pid*=44&*aid*=8.

115 Shelton, 'Equity', p. 654.

116 For example, Art. 4 (8), *UNFCCC*.

regular review of the lists of States included in Annexes I and II. Although the possibility of amending the Annexes is provided for in Article 16 of the *UNFCCC*, the way in which the CBDR is otherwise reflected in the *Convention* does not fully appreciate the contextual and dynamic characteristics of the equity it seeks to achieve.[117] The use of more contextual norms may comprise a tool to overcome the current impasse of the framework's effectiveness.

Lessons from international water law

International water law and its general principles, which emerged from State practice in dealing with hydrologic variability, provide useful insight in the value of employing law that adapts to context. A contextual norm 'typically involves balancing multiple interests and characteristics'.[118] Providing identical treatment in principle, contextual norms additionally require or permit the consideration of factors that vary among States.[119] An inherent risk of norms that allow for differentiation of rights and obligations based on the respective circumstances of individual States is that they create uncertainty with regard to their implementation. The strength of contextual norms, which lies in the flexibility they provide to adjust to changing circumstances, is therefore also their weakness. To overcome the uncertainty in their implementation, contextual norms and principles have to be accompanied by cooperation obligations that provide for a regular assessment of the factors that are to be considered. In international water law, the principle of equitable and reasonable utilisation represents a norm that includes some of these characteristics: the consultation obligations included in Article 6 of the *UN Watercourses Convention*, for instance, seek to achieve a regular reassessment of the basin context and the factors determining equitable use. The recommendation to establish joint management mechanisms further promotes this approach.

Thus, one way to enhance the effectiveness of the *UNFCCC* framework may be in borrowing key concepts of international water law and introducing some of its flexibility into the climate change regime: more specifically, the CBDR concept could be structured in a way similar to that

[117] An initial review 'no later than 31 December 1998' is also mandated in Art. 4 (2) (f), *UNFCCC*. A revision of Annex II occurred at the Seventh COP meeting in Marrakech.

[118] D. Magraw, 'Legal Treatment of Developing Countries: Differential, Contextual, and Absolute Norms' (1990) 1 *CJIELP* 74.

[119] *Ibid.*

in which the principle of equitable and reasonable utilisation applies to transboundary water systems. More specific criteria to determine the differentiation of responsibilities among States could be established and paired with cooperation obligations to regularly reassess whether changes in underlying circumstances warrant an adjustment regarding individual commitments of States.[120] The outcomes of the Ad-Hoc Working Group on Long-Term Cooperative Action (AWG-LCA) workshop on equitable access to sustainable development indicate the criteria that are considered important in defining equity under the climate change framework.[121] These criteria should include historic and present emission contributions; national circumstances such as climate, geography, level of economic development, carbon and GHG intensity of the economy, technological capacity and availability of financial resources; and the capabilities to mitigate and adapt to climate change. A provision outlining a non-exhaustive list of criteria would need to be accompanied by an obligation to review the allocation of special commitments to States at regular intervals (e.g., the revision of Annexes I and II every seven to ten years) in order to provide for a duty to reassess whether the equity objectives of the *UNFCCC* framework are still being met. The institutional structure of the framework, which brings together the parties of the *Convention* for negotiations on a regular basis, already provides an adequate platform for dialogue and negotiation to carry out such reassessments.

7.4 Conclusion: Complementarity of systems and consideration of the hydrologic cycle as a whole

Neither water nor climate change respects international boundaries; cooperation is necessary to mitigate climate change, 'to prevent negative impacts of unilateral measures and to support coordination of adaptation measures, [in particular] at the river basin level'.[122] Climate-change law and the law of transboundary water resources complement one another in this context. They create an enabling framework for

120 A similar multi-criteria formula for burden-sharing arrangements was proposed by Norway and Iceland in the Ad-Hoc Group on the Berlin Mandate in the process of negotiating the *Kyoto Protocol*. L. Ringius *et al.*, 'Burden Sharing and Fairness Principles in International Climate Policy' (2002) 2 *International Environmental Agreements: Politics, Law and Economics* 13f.

121 AWG-LCA, *Report on the Workshop on Equitable Access to Sustainable Development*, 15 August 2012, UN Doc. FCCC/AWGLCA/2012/INF.3/Rev.1, p. 11.

122 UNECE, *Guidance on Water and Climate Change*, p. iii.

cooperation to confront regional and global challenges. State practice aimed at coping with hydrologic variability has contributed significantly to the crystallisation of principles and norms of international water law. Consequently, because the impacts of climate change are felt mostly through changes in the hydrologic cycle, the rules applicable to transboundary water systems are well suited to address the effects of climate change. They are based on legal principles that can accommodate the reality of continuously changing hydrologic circumstances, whether they are due to natural or anthropogenic causes. The general duty to cooperate and the principle of equitable and reasonable utilisation facilitate the adjustment of water utilisations and the coordination among basin States that becomes necessary due to changes in the conditions of transboundary basins. In addition, experience from a number of basins shows that climate change can induce cooperation at a multilateral level, leading to the widening of the mandates of joint mechanisms and the scope of cooperation. The framework for cooperation that was established through the *UNFCCC* at the universal level is important because it is a means for States to address their hydrologic interdependence beyond the transboundary water systems. Atmospheric change has demonstrated the fact that water resources – in fact, the global water cycle – are resources of common concern: hydrologic interconnectedness exists not only at the basin level; the water cycle links States at the universal level as well. The CBDR concept acknowledges that, in the long run, economic activities in one State can cause changes in rainfall patterns and runoff volumes in other regions of the world. The formulation of this concept in the *UNFCCC* has made near-universal cooperation possible; however, at the same time, it also poses significant challenges to implementation of the framework.

International water law can inform climate change law in this respect. The inclusion of a contextual norm modelled on the principle of equitable and reasonable utilisation offers a potential way out of the impasse currently hampering climate change negotiations. Distribution of special commitments based on the assessment of a series of objective factors relevant for each State's individual vulnerability and/or capacity to contribute to a response to climate change can provide a key to continued cooperation within the *UNFCCC* framework. The complementarities of international water law and climate change law can also become a point of departure in the response to legal questions concerning the interlinkage between surface and atmospheric water resources more generally. Geo-engineering measures, such as the use of clouds and atmospheric humidity to create rain in areas of persistent drought[123] or to cool the atmosphere,[124] are not yet sufficiently addressed in international law. Geo-engineering has been proposed as a possible solution to mitigate the effects of climate change and is likely to pose legal challenges in the near future.

123 An example is cloud-seeding by injection of chemical agents to induce rain. However, this method is not yet geographically precise. J. Ferdon, 'Federal Weather Modification Projects: Compensating the Landowner' (1984) 26 *Arizona Law Review* 683f.

124 So-called cloud-whitening was proposed to reflect more solar energy back into space to cool the Earth. This would be accomplished by thickening existing or producing new clouds through the injection of seawater droplets into the air. R. Black, 'Climate 'Technical Fix' May Yield Warming, Not Cooling', *BBC News Online*, 6 April 2011, www.bbc.co.uk/news/science-environment-12983795?print=true.

Conclusion

Water, just like the atmosphere, is one of the resources that highlight most dramatically that States do not exist as independent entities. States that share transboundary water systems are interdependent; almost invariably, they experience the impact of surface water or groundwater utilisation by their co-riparian States. Even those States that do not share any transboundary water systems with their neighbours, including island States, are interconnected through the global water cycle. They experience the effects of changes in the hydrologic cycle that occur because of water utilisation or other economic activities happening elsewhere through the changes in climatic conditions and precipitation patterns. Thus, the nature of the resource challenges the conceptualisation of the State as an independent entity and as the absolute sovereign over its territory and the resources within it. Water resides only temporarily in any State's territory before it travels to another State's territory – as surface-water or groundwater runoff, or through evaporation – flowing back into the atmosphere where it can be carried away by winds and cloud formations. These characteristics give rise to the need to coordinate utilisations of this dynamic resource. Based on principles of good neighbourliness and good faith, States restrict their uses of these shared resources to not significantly harm other States' territories, or other States' existing and future rights to equitable and reasonable utilisation of these resources. The existence of shared resources and, therefore, of factual interdependence of States – as well as the uses of those resources – resulted in a transformation of the understanding and legal meaning of sovereignty: that is, from a notion that signified absolute independence to one that recognises restrictions of freedom of action and towards a concept that acknowledges the inevitable need for cooperation in an interdependent world.

Law becomes an important tool to regulate State relationships with respect to water because treaties and joint mechanisms that are established through international agreements generally provide predictability, stability and longevity to legal relationships among States. Agreed rules of conduct and institutionalised structures establish reliable parameters for regular interaction and sustained cooperation. As a consequence, law establishes an enabling framework and provides tools for the process of international cooperation. This process is non-linear and becomes an open-ended course of iterative State interaction.[1] It is composed of a series of individual steps in which a single cooperative act alone does not necessarily create benefits for all parties involved; a single step may create benefits for only one party. A State that notifies another State and provides information does not directly benefit from this act; the State receiving information that it would not have had otherwise is the immediate beneficiary in this case. Yet, the notifying State sets an essential step within the iterative process of cooperation; it may expect that other riparian States will reciprocate with comparable measures. In a similar manner, the adoption of an international treaty equally constitutes only one step in this process. It captures an agreement among States at a given point in time; it may require adjustment at a later stage in order to continue to provide mutual benefits in the long term. It is the process of cooperation as a whole that is aimed at being mutually beneficial for those who are engaged in it.

Regarding transboundary freshwater systems, the development of international cooperation and the establishment of treaties have, over time, given rise to a legal duty to cooperate. The objective and the legal nature of this duty have been clarified in the process of codification and are described in the instruments of international water law that have been adopted at the basin, regional and universal levels. The duty's objective is twofold: it encompasses equitable and reasonable utilisation of shared waters and their adequate protection. Rather than limiting the scope solely to utilisation of water, the duty aims at the cooperative management of the resource as well as of

[1] M. Daoudy and E. Kistin, 'Beyond Water Conflict: Evaluating the Effects of International Water Cooperation', paper prepared for the 49th Annual Conference of the International Studies Association, San Francisco (2008); K. O'Neill, J. Balsiger, and S. VanDeveer, 'Actors, Norms and Impact: Recent International Cooperation Theory and the Influence of the Agent-Structure Debate' (2004) 7 *Annual Review of Political Science* 151.

the surrounding ecosystem because the quality and usability of shared water resources depend on the quality and nature of the surrounding environment.

The general duty to cooperate therefore is defined as the duty of system States to engage in a process of cooperation in order to achieve equitable and sustainable management of shared water resources. Its legal nature has distinct characteristics. On the one hand, it has the character of a legal principle. It is defined in terms that, alone, remain vague; they do not specify the individual steps and activities through which States must implement this general duty. The definition of more specific rights and obligations is required. On the other hand, the general duty is an abstract obligation that mandates States to engage in a process of cooperation in order to achieve equitable and sustainable management of transboundary bodies of water. It is a concrete duty in the sense that uncooperative behaviour that renders the achievement of this objective impossible is not permissible. The general duty to cooperate, thus, is a general rule of conduct. At the same time, this conduct is a duty in and of itself; it constitutes the result that is envisaged. The general duty to cooperate, thus, is a substantive duty; it is a duty of result as far as the cooperative conduct to achieve equitable and sustainable management of shared waters is a result in and of itself. At the same time, the general duty to cooperate contains procedural elements; it is a duty of conduct that gives rise to procedures that facilitate the achievement of mutual benefits and the implementation of other principles and rules.

The general duty to cooperate is the underlying rule that gives rise to more specific rights and obligations of cooperation.[2] Some cooperative acts are required by customary law, including the notification of emergencies and planned measures that carry the risk of inflicting (significant) transboundary harm. In addition to the customary obligations, States may decide on supplementary measures and activities to implement the general duty within the limits of international law.[3] This flexibility and freedom of choice of the means are required because conditions in individual transboundary basins vary greatly; ways to adequately and effectively manage individual

[2] McCaffrey, *The Law of International Watercourses*, 470; E. Riedel, 'International Environmental Law', p. 76.

[3] D. Caponera, 'Patterns of Cooperation in International Water Law: Principles and Institutions' (1985) 25 *Natural Resources Journal* 563.

basins therefore must vary as well. The intensity of cooperation required by riparian States differs according to the circumstances of each basin. Basins shared by two States that carry an abundant amount of water, for example, will require a different style of cooperative management than basins shared among multiple riparian States and that may provide the primary or only source of water for human consumption and economic activities in the individual basin States. Hence, the range of cooperative measures that are adopted by States varies widely: the measures may be limited to simple data exchange mechanisms and may extend to different measures of water quality control, elaborate joint infrastructure projects and cooperative flow management. In most cases, States decided to structure and institutionalise cooperation by establishing joint mechanisms, which serve as platforms for dialogue and regular interaction among riparian States. Joint mechanisms may also serve as vehicles to expand cooperation to new issue areas and to widen their geographic scope. In multi-State basins, positive results of cooperation achieved by a smaller group of basin States may induce other riparian States, which had not done so previously, to participate in joint water-management activities.

The content and the dynamics of implementing the general duty to cooperate and related obligations highlight that general principles of international water law are complementary and mutually reinforcing. Legal principles apply simultaneously and equally to all riparian States. This becomes apparent when analysing specific rights and obligations that derive more closely from individual principles. They often support the implementation not only of one but also a number of different principles. The obligation to carry out and to give notification of the results of an EIA where there is a risk of transboundary harm, for example, promotes cooperation and facilitates the prevention and mitigation of harm. The obligation to regularly exchange data and information facilitates both cooperation and equitable and reasonable utilisation. The general principles support one another directly and through the specific rights and obligations to which they give rise. In addition, in instances in which one principle seems to be at odds with another, a third may intervene to establish a balance between the former two. For example, where a utilisation planned by one State risks significant harm to another riparian State but otherwise appears to be equitable and reasonable, the general duty to cooperate intervenes by facilitating the finding of a solution that is acceptable to all States concerned. Information exchange, consultation and negotiation obligations facilitate a mutually agreeable and equitable outcome. In these cases, the general duty to cooperate builds a bridge between the other principles.

The general duty to cooperate and the related rights and obligations apply to all States sharing the same transboundary freshwater system. The principle crystallised from measures adopted by States to manage their common water challenges – such as the increase in the variety of competing utilisations and the urgency to protect aquatic ecosystems against pollution – in a state of hydrologic interdependence. Continuously increasing water demand and climate change pose new challenges to water availability as well as equitable distribution and use that must be addressed. New measures of international cooperation adopted by States to address the growing water demand, hydrologic variability and increase in the frequency of severe hydrologic events (e.g., century floods and prolonged droughts) provide insight to possible future trends concerning the development of cooperation duties in international law. They demonstrate that the scope of application of cooperation duties is expanding. Water agreements, such as the 1992 *UNECE Water Convention* and the 2000 *SADC Revised Protocol on Shared Watercourses*, and codification efforts at the universal level indicate that transboundary water resources are of concern not merely to the riparian States sharing them; there also is interest in the management of the resources by States located outside of individual basins. This is similar to what existed on European and some African rivers in the late nineteenth and early twentieth centuries with respect to the right to freedom of navigation for non-riparian States. There are signs that the hydrologic cycle as a whole – and not only the waters flowing in transboundary freshwater bodies – as well as the rights and interests of individuals in the satisfaction of their vital human water needs are starting to be considered as the object of international cooperation obligations. Recent developments indicate the emergence of a legal framework for cooperation in which specific obligations apply beyond their traditional scope of regulating the relationship among riparian States. Within the context of climate change, hydrologic interdependence has become a motivating factor for cooperation on a universal scale. Legal relationships among States are no longer determined solely by their status of 'co-riparians' of transboundary water systems. New cooperation obligations are the result of States' hydrologic interconnectedness and their relative contribution to and affectedness by the changes in the global water cycle that are due to anthropogenic climate change. In addition, cooperation between

States is expanding with an increasing focus on the rights and interests of individuals.

These developments are in an early stage. Climate change law is limited, so far, to providing the basic structures of a legal framework that enables cooperation to address the global hydrologic interdependence through cooperation. Some of the legal elements promote cooperation; the CBDR concept and the differentiation of obligations to provide technical and financial assistance represent opportunities for new partnerships among States and for international cooperation aimed at solving global problems.[4] The climate change framework recognises that cooperation for mutual benefits might not be sufficient but that solidarity is required with those who share little or no responsibility for anthropogenic climate change or who do not have the means to adapt to hydrologic changes. At this time, cooperation obligations and legal commitments to cooperate within the legal climate change framework remain general. At the same time, however, the high level of institutionalisation of the framework is promising: it promotes regular interaction, dialogue and information exchange, and it facilitates joint learning about the potential mutual benefits of cooperation in this area.

With regard to the issue of protection of legal interests of individuals in a transboundary context, it remains to be seen whether international cooperation obligations will lead to general acceptance. Whereas universal recognition of a self-standing human right to water is gaining momentum, the international cooperation obligations of human rights law remain general and are not enforced. Yet, riparian States of a small number of basins, primarily in Africa, are leading the way by giving special protection to vital human water needs and by recognising a human right to water in treaties concluded at a basin-wide level. In the manner of Article 10 of the *UN Watercourses Convention* and Article 5 (2) of the 2008 *ILC Draft Articles*, these treaties promote international cooperation with the aim of protecting vital human needs in a transboundary context. They elevate the protection of these needs from an obligation that is primarily owed by States to individuals at the national level to an obligation that is equally owed among States internationally.

These developments illustrate that international rules applicable to freshwater are in a process of diversification; the rights of individuals and global hydrologic interdependence are increasingly considered. It

4 Boisson de Chazournes, 'Technical and Financial Assistance', p. 958.

remains to be seen whether these trends will continue and ultimately expand the scope of cooperation duties towards consideration of the global water cycle as a whole, thereby expanding the applicability of the law of transboundary water resources to include atmospheric water streams that reach far beyond the watersheds of individual basins.

Annex: Consideration of cooperation in international water treaties

Methodology

A textual analysis of 219 international water treaties was carried out as background research to complement and substantiate the legal arguments presented in this book. Most of the treaties reviewed herein are available from the International Freshwater Treaties Database of the Institute for Water and Watersheds at Oregon State University.[1]

In addition to a general review of content, the treaties were subjected to a 'Yes/No' analysis in which '0' indicates the absence of a provision stipulating a cooperation obligation or a joint measure and '1' confirms its presence (Table 2). The same logic applies to the categorisation of treaties as regional documents or instruments covering boundary waters or partial or entire basins (i.e., '1' indicates that the treaty belongs to the category, '0' indicates that it does not). Similarly, regarding the use of the term 'cooperation' in a treaty text, '1' indicates that the term 'cooperation' or one of its derivatives exists in the text, '0' indicates that it does not. For those treaties that also concern issues other than water resources management (e.g., boundary treaties and regional integration treaties), the duty to cooperate was assessed only for activities related directly to water uses.

It is obvious that this type of analysis must be treated with great caution because it does not allow a detailed qualitative assessment of a document or the State action that may result from the agreement. It is, at best, indicative of legal trends in the context of State cooperation on transboundary water resources. A solely textual 'Yes/No' analysis of this type

1 See http://ocid.nacse.org/tfdd/treaties.php. Treaties analysed are those available in English, French, German, Italian, Portuguese and Spanish concluded after 1900.

cannot truly capture the spirit of individual instruments, and it can be misleading without an awareness of the underlying criteria and conditions. Some treaties are the result of considerably more State cooperation and have a more cooperative spirit than their text and obligations would indicate. A treaty outlining the terms of joint implementation of a previously agreed-to infrastructure project, for example, may not include any reference to the majority of cooperation obligations analysed here. Reference to notification, consultation and negotiation obligations, among others, may be missing because these processes would have taken place before a treaty was concluded.[2] More often, these types of treaties do not explicitly reference the term 'cooperation' or any derivatives thereof. Yet, they are the result of and provide the framework for significant cooperation among the parties to the agreements. In contrast, other treaties that frequently use the term 'cooperation' do not in fact include, or hardly include, any rule that would oblige the signatory States to set cooperative acts.[3] Hence, the question of whether use of the word 'cooperation' in the text (as in 58 per cent of the treaties analysed) indicates the existence of State cooperation cannot be responded to conclusively based on a solely textual treaty analysis.

Definition of categories

Year	Indicates year of adoption (not entry into force)
Signatories	States mentioned refer to initial signatories as indicated by the Transboundary Freshwater Treaty Database; later signatories or unilateral terminations are not considered in the same manner consistently[4]

(*cont.*)

[2] See, for example, the 1969 *Exchange of Notes between the United States and Canada Constituting Agreements for the Construction of a Temporary Cofferdam at Niagara and for the Temporary Diversion for Power Purposes of the Water Normally Flowing over the American Falls at Niagara* (America: No. 25 and No. 26 in Table 2), or the 1988 *Exchange of Notes between Brazil and Bolivia Constituting an Agreement for the Construction of a Hydroelectric Plant in Cachuela Esperanza* (America: No. 41 in Table 2).

[3] See, for example, the 1925 *Agreement to Regulate the Level of Lake of the Woods* (America: No. 3 in Table 2), or the 1938 *Convention for the Rectification of the Rio Grande in the El Paso-Juarez Valley* (America: No. 6 in Table 2).

[4] The analysis serves a merely indicative purpose to assess broad trends in State cooperation and use of legal instruments. The results represented in Table 2 should not be used in any other way because there is a risk of incoherence in interpretation otherwise.

(cont.)

Regional Agreement	Treaty covering more than one transboundary hydrologic system and involving more than two States
Boundary Waters	Treaty concerning the boundary waters between two States
Entire Basin	Treaty signed by all basin States
Partial Basin	Treaty between two or more but not all basin States
Cooperation in Text	The term 'cooperation' or any of its derivatives occurs in the text
Future Agreements Envisaged	Treaty envisages the conclusion of future legal agreements by State parties, expressing the political will of States to continue and enhance the cooperation process through formal means
Data & Information Exchange	Indicates that the treaty establishes an obligation for some type of regular exchange of data and information; 'regular' intervals range from an annual to a weekly or to even a daily exchange of data
Notification	Refers to the obligation to notify concerned parties of planned measures, emergency situations, or other information that is of legal interest; notification can take place between States directly or through joint mechanisms; the unilateral act of notification to renounce a treaty is not considered herein
Consultation	Indicates the obligation to enter into consultation with concerned parties
Negotiation	Obligation considered to be included if the treaty makes explicit reference to 'negotiation' other than with respect to dispute settlement, and/or if a treaty mandates State parties to undertake certain actions 'by agreement', or requires the parties to reach agreement on certain issues before proceeding
Prior Consent	Indicates requirement of prior consent of concerned parties for planned measures or the need for unanimous agreement before proceeding
Joint Studies or Programs	Indicates that the treaty envisages some type of study or program carried out jointly by State parties; this can be the implementation of an action plan, the establishment of a joint basin development plan, or the conducting of joint hydrologic studies by experts nominated by the State parties

Joint Mechanism	Treaty either establishes a joint mechanism or imposes specific tasks on a joint mechanism established by a previous instrument; the term 'joint mechanism' is applied in a broad sense; it refers to commissions with Permanent Secretariats as well as non-permanent commissions that meet on an annual or more frequent basis
Dispute-Settlement Procedures	Indicates that the treaty includes (a) specific provision(s) on settlement of disputes that occur during implementation of the treaty; procedures range widely including negotiation, arbitration, independent-expert commissions and referral to the International Court of Justice

Regions are categorised according to continents: Africa, America (including North and South), Asia and Europe. The Middle East, Central, and Southeast Asia, which are frequently considered as independent categories in comparable analyses, are not considered separately herein. The number of treaties available concerning the continent of Asia would not have been sufficient to derive significant indicators for distinct cooperation customs in the different regions.

Outcome of analysis

Overall, the analysis confirms the trends concerning the incidence of bilateral and multilateral treaties established by other studies on water treaties.[5] The set of treaties analysed herein included 219 agreements dating from after 1900, which consisted of 75 per cent bilateral treaties and 25 per cent multilateral treaties. Given that a majority of transboundary water systems for which treaties were established are, in fact, shared by only two States, this seems to be a logical distribution. Additionally, of these 162 bilateral treaties, 44 concern multiple basins straddling the boundaries between two States (i.e., the Boundary Waters category).

Regarding geographic scope, almost 30 per cent of the treaties analysed apply to basins as a whole. This percentage, however, is not representative of the various regional averages. Extensive treaty practice between the United States and its Northern and Southern neighbours skews the overall

[5] Brown Weiss, 'The Evolution of International Water Law', pp. 232–241; J. Hamner and A. Wolf, 'Patterns in International Water Resources Treaties: The Transboundary Freshwater Dispute Database' (1997) 9 *CJIELP* 157–177.

average, as well as the regional average on the American continent, where 60 per cent of treaties apply to whole basins. In other regions, average percentages are lower, with 24 per cent in Africa, 23 per cent in Europe and only 6 per cent in Asia. The latter is a result of the presence of numerous multi-State basins coupled with complex political dynamics in the region.

What is striking regarding individual cooperation obligations is that treaties analysed for the African continent mark a higher number of content references to those obligations than those of any other region, which is even more the case when the regional average is calculated for treaties adopted only after decolonisation.[6] For example, concerning notification obligations, the worldwide average is 36.5 per cent; in Africa for treaties concluded after 1900, the regional average is 56 per cent, and this increases to 67 per cent for treaties concluded after 1960. The percentages are similar relative to the provision for joint programs and studies: the general average is 50 per cent, whereas for Africa, it is 58.5 and 73.3 per cent for treaties concluded after 1900 and after 1960, respectively. American and Asian regional averages for notification obligations are generally below overall averages, with a significant exception for data and information exchange provisions, in which Asia, at 50 per cent, is more than 10 percentage points higher than the general average of 39 per cent. Europe is generally within a 2 to 3 per cent range of the overall average, with the exception of references made to joint studies or programs, where it is 10 percentage points lower. It must be kept in mind that these percentages are relative in nature and merely indicate broad trends; therefore, they must be interpreted with caution.

One finding that stands out, however, is that more than 70 per cent of all treaties analysed either establish some type of joint mechanism or refer specific tasks to an existing joint mechanism. On this point, all regions show similar averages (i.e., within a 2 to 3 per cent range), with a small deviation for Asia, which is 5 percentage points below the average. It is interesting that when treaty practice in Africa after 1960 is considered, the regional average increases to a striking 93 per cent of analysed treaties.

Overall, the analysis illustrates that there are no limits to the number of ways in which States can structure and, indeed, have shaped their legal relationships with respect to transboundary water resources. The multitude of possibilities for regulation is determined by the specific circumstances of individual basins and by the variety of interests of riparian States.

[6] The year 1960 has been used as cut-off year for simplified analysis.

Table 2. *Cooperation in International Water Treaties*

No.	Year	Title	Signatories	Regional Agreement	Boundary Waters	Entire Basin	Partial Basin	Cooperation in Text	Future Agreements Envisaged	Data & Information Exchange	Notification	Consultation	Negotiation	Prior Consent	Joint Studies or Programs	Joint Mechanism	Dispute Settlement Procedures
AFRICA																	
1	1902	Treaty relative to the frontiers between the Anglo-Egyptian Sudan, Ethiopia and Erythrea	Great Britain, Ethiopia	0	0	0	1	0	0	0	0	0	0	1	0	0	0
2	1906	Agreement relating to the spheres of influence of Great Britain and the Independent State of Congo in East and Central Africa	Great Britain, Belgium	0	0	0	1	0	0	0	0	0	0	1	0	0	0
3	1913	Agreement respecting the boundary between Sierra Leone and French Guinea	Great Britain, France	0	1	0	0	0	0	0	0	0	1	1	0	0	0
4	1925	Exchange of notes respecting the regulation of the utilisation of the waters of the River Gash	Great Britain, Italy	0	0	1	0	0	0	0	0	0	0	0	0	0	0

(cont.)

Table 2. *(cont.)*

No.	Year	Title	Signatories	Regional Agreement	Boundary Waters	Entire Basin	Partial Basin	Cooperation in Text	Future Agreements Envisaged	Data & Information Exchange	Notification	Consultation	Negotiation	Prior Consent	Joint Studies or Programs	Joint Mechanism	Dispute Settlement Procedures
5	1927	Convention regarding various questions of economic interest in the colonies of the Belgian Congo and Angola	Belgium, Portugal	0	0	0	1	0	0	0	1	0	0	0	0	0	0
6	1929	Exchange of notes in regard to the use of the waters of the River Nile for irrigation purposes	Egypt, Great Britain	0	0	0	1	0	0	0	0	0	1	1	1	0	0
7	1934	Agreement regarding water rights on the boundary between Tanganyika and Ruanda–Urundi	Great Britain, Belgium	0	1	0	1	0	0	0	1	0	0	0	0	0	1
8	1949	Exchange of notes regarding the construction of the Owen Falls Dam, Uganda	Great Britain, Egypt	0	0	0	1	1	0	0	0	1	1	1	0	0	1
9	1950	Exchange of notes constituting an agreement regarding cooperation in meteorological and hydrological surveys in certain areas of the Nile Basin	Great Britain, Egypt	0	0	0	1	1	0	1	0	0	0	0	0	0	0

10	1953	Exchange of Notes constituting an agreement regarding the construction of Owen Falls Dam	Great Britain, Egypt	0	0	0	1	0	0	0	0	0	0	0	0	0	0
11	1959	Agreement for the full utilization of the Nile Waters	Egypt, Sudan	0	0	0	1	1	0	0	1	0	1	1	1	1	0
12	1963	*Convention relative a l'amenagement général du bassin du Fleuve Sénegal*	Mali, Mauretania, Guinea, Senegal	0	0	1	0	1	1	0	1	1	0	1	0	1	1
13	1964	*Convention relative au statut du Fleuve Sénegal*	Mali, Mauretania, Guinea, Senegal	0	0	1	0	1	1	0	1	0	1	0	1	1	1
14	1964	Convention and Statutes relating to the development of the Chad Basin	Cameroon, Chad, Niger, Nigeria	0	0	1	0	1	0	0	1	1	0	0	0	1	1
15	1964	Agreement relating to the Central African Power Corporation	Zambia, Zimbabwe	0	0	0	1	0	0	1	1	1	0	0	0	1	0
16	1969	Agreement in regard to the first phase of development of the water resources of the Cunene River Basin	South Africa, Portugal	0	0	1	0	0	0	0	1	0	1	1	0	1	0
17	1972	*Convention portant de l'Organisation pour la mise en valeur du Fleuve Senegal*	Mali, Mauritania, Senegal	0	0	0	1	1	0	1	1	0	1	0	1	1	1
18	1972	*Convention relative au statut du Fleuve Senegal*	Mali, Mauritania, Senegal	0	0	0	1	1	1	1	1	1	0	1	0	1	1

(cont.)

Table 2. *(cont.)*

No.	Year	Title	Signatories	Regional Agreement	Boundary Waters	Entire Basin	Partial Basin	Cooperation in Text	Future Agreements Envisaged	Data & Information Exchange	Notification	Consultation	Negotiation	Prior Consent	Joint Studies or Programs	Joint Mechanism	Dispute Settlement Procedures
19	1977	Agreement for the establishment of the Organization of the Management and Development of the Kagera River Basin	Rwanda, Burundi, Tanzania	0	0	0	1	1	0	0	1	0	0	0	0	1	1
20	1978	Convention relating to the creation of the Gambia River Basin Development Organization	Guinea, Gambia, Senegal	0	0	1	0	1	0	0	0	0	0	0	1	1	1
21	1978	Convention relating to the status of the River Gambia	Guinea, Gambia, Senegal	0	0	1	0	1	1	0	1	0	0	1	1	1	1
22	1978	*Convention relative au statut juridique des ouvrages communs*	Mali, Mauritania, Senegal	0	0	0	1	1	1	0	0	0	0	0	1	1	1
23	1978	*Protocole sur l'amenagement du Fleuve Koliba-Korubal*	Guinea, Guinea-Bissau	0	0	1	0	1	0	0	0	0	0	0	1	1	1
24	1983	Agreement relative to the establishement of a tripartite permanent technical committee	South Africa, Swaziland, Mozambique	0	0	0	1	0	0	0	0	0	0	0	1	1	0

25	1986	Treaty on the Lesotho Highlands Water Project	Lesotho, South Africa	0	0	0	1	1	0	1	1	1	1	1	1	1
26	1987	Agreement on the action plan for environmentally sound management of the common Zambezi River system	Botswana, Mozambique, Tanzania, Zambia, Zimbabwe	0	0	0	1	1	1	0	0	1	0	0	1	0	0
27	1987	Agreement concerning the utilization of the Zambezi River	Zambia, Zimbabwe	0	0	0	1	1	0	1	1	1	1	0	1	1	1
28	1991	Protocol IV to the Treaty on the Lesotho Highlands Water Project – supplementary arrangements regarding Phase IA	Lesotho, South Africa	0	0	0	1	0	0	1	1	1	0	1	1	1	1
29	1992	Agreement on the establishment of a permanent water commission	South Africa, Namibia	0	0	0	1	1	0	0	0	0	1	0	1	1	0
30	1992	Treaty on the development and utilisation of the water resources of the Komati River Basin	Swaziland, South Africa	0	0	0	1	1	0	0	1	0	1	0	1	1	1
31	1992	Treaty on the establishment and functioning of the joint water commission	Swaziland, South Africa	0	0	0	1	1	0	1	0	0	0	0	1	1	1
32	1993	Framework for general cooperation	Egypt, Ethiopia	0	0	0	1	1	1	0	0	1	0	0	0	0	0
33	1994	Agreement to initiate program to strengthen regional coordination in management of resources of Lake Victoria	Kenya, Tanzania, Uganda	0	0	0	1	1	0	0	0	0	0	0	1	1	0

(cont.)

Table 2. *(cont.)*

No.	Year	Title	Signatories	Regional Agreement	Boundary Waters	Entire Basin	Partial Basin	Cooperation in Text	Future Agreements Envisaged	Data & Information Exchange	Notification	Consultation	Negotiation	Prior Consent	Joint Studies or Programs	Joint Mechanism	Dispute Settlement Procedures
34	1994	Agreement on the establishment of a Permanent Okavango River Basin Water Commission (OKACOM)	Angola, Botswana, Namibia	0	0	0	1	1	0	0	1	0	1	0	1	1	1
35	1996	Agreement on the establishment of a Joint Water Commission	South Africa, Mozambique	0	0	0	1	1	0	1	0	0	0	0	1	1	1
36	1999	Protocol VI to the Treaty on the Lesotho Highlands Water Project – supplementary arrangements regarding the system of governance for the project	Lesotho, South Africa	0	0	0	1	1	0	1	1	1	0	0	1	1	1
37	2000	SADC Revised Protocol on shared water courses	13 Signatories	1	0	0	0	1	1	1	1	1	1	0	1	1	1
38	2002	*Chartes des eaux du Fleuve Senegal*	Mali, Mauritania, Senegal Guinea	0	0	1	0	1	1	1	1	0	0	1	1	1	1
39	2002	Tripartite Interim Agreement for cooperation and protection and sustainable utilisation of the water resources of the Incomati and Maputo watercourses	South Africa, Swaziland, Mozambique	0	0	1	0	1	1	1	1	1	1	0	0	1	1

40	2003	Protocol for sustainable development of Lake Victoria Basin	Kenya, Tanzania, Uganda	0	0	0	1	1	1	1	1	1	1	0	1	1 1
41	2009	Niger Basin Water Charter	Benin, Burkina Faso, Guinea, Mali, Nigeria, Niger, Chad	0	0	0	1	1	1	1	1	1	1	0	1	1 1
AMERICA																
1	1906	Convention on equitable distribution of the waters of the Rio Grande	United States, Mexico	0	0	1	0	0	0	0	0	0	0	0	0	0 0
2	1909	Treaty relating to boundary waters and boundary questions	United States, Great Britain (Canada)	0	1	1	0	0	1	0	0	0	0	0	0	1 1
3	1925	Agreement to regulate the level of Lake of the Woods	Canada, United States	0	0	1	0	1	0	0	1	0	0	1	0	1 1
4	1929	Treaty for the settlement of the dispute regarding Tacna and Arica	Chile, Peru	0	0	0	1	0	0	0	0	0	0	0	0	0 1
5	1933	Convention regarding the determination of the legal status of the frontier between Brazil and Uruguay	Brazil, Uruguay	0	1	0	0	0	0	0	1	0	0	1	0	0 0
6	1933	Convention for the rectification of the Rio Grande in the El Paso-Juarez Valley	United States, Mexico	0	0	1	0	1	0	0	0	0	0	0	0	1 0
7	1935	Preliminary Convention for the exploitation of fisheries in Lake Titicaca	Bolivia, Peru	0	0	1	0	0	1	0	0	0	1	0	1	0 0

(*cont.*)

Table 2. *(cont.)*

No.	Year	Title	Signatories	Regional Agreement	Boundary Waters	Entire Basin	Partial Basin	Cooperation in Text	Future Agreements Envisaged	Data & Information Exchange	Notification	Consultation	Negotiation	Prior Consent	Joint Studies or Programs	Joint Mechanism	Dispute Settlement Procedures
8	1938	Convention providing for emergency regulation of the level of Rainy Lake and of the level of other boundary waters in the Rainy Lake watershed	United States, Canada	0	0	1	0	0	0	0	0	0	0	0	0	1	0
9	1940	Exchange of notes constituting an agreement regarding the development of certain portions of the Great Lakes	Canada, United States	0	0	1	0	0	0	0	0	0	0	0	1	1	0
10	1944	Treaty relating to the waters of the Colorado and Tijuana Rivers, and of the Rio Grande from Fort Quitman, Texas, to the Gulf of Mexico	United States, Mexico	0	1	1	0	1	0	1	1	1	1	1	1	1	0
11	1944	Exchange of notes constituting an agreement relating to a study to be made by the International Joint Commission with respect to the Upper Columbia River Basin	United States, Canada	0	0	1	0	0	0	0	0	0	0	0	1	1	0

12	1946	Agreement concerning the utilization of the rapids of the Uruguay River in the Salto Grande Area	Argentina, Uruguay	0	0	0	1	0	0	0	0	0	0	0	1	1	1
13	1950	Treaty relating to the uses of the waters of the Niagara River	Canada, United States	0	0	1	0	0	0	0	0	0	0	0	1	1	0
14	1952	Exchange of Notes constituting an agreement relating to the St. Lawrence Seaway Project	Canada, United States	0	0	1	0	0	0	0	0	0	0	0	0	0	0
15	1953	Exchange of Notes constituting an agreement relating to the establishment of the St. Lawrence River Joint Board of Engineers	Canada, United States	0	0	1	0	0	0	1	0	0	0	1	0	1	0
16	1955	Exchange of notes establishing a joint commission for study of the Puno-Guaqui Railway Line and joint use of the waters of Lake Titicaca	Bolivia, Peru	0	0	1	0	0	0	0	0	0	0	0	1	1	0
17	1958	Agreement concerning a study of the utilization of the water power of the Apipe Falls	Argentina, Paraguay	0	0	0	1	1	0	1	0	0	0	0	1	1	0
18	1961	Treaty relating to cooperative development of the water resources of the Columbia River Basin	Canada, United States	0	0	1	0	1	1	1	1	1	1	1	1	1	1
19	1963	Exchange of notes constituting an agreement establishing a joint commission for the development of the Mirim Lagoon	Brazil, Uruguay	0	0	0	1	0	0	0	0	0	0	0	1	1	1

(cont.)

Table 2. *(cont.)*

No.	Year	Title	Signatories	Regional Agreement	Boundary Waters	Entire Basin	Partial Basin	Cooperation in Text	Future Agreements Envisaged	Data & Information Exchange	Notification	Consultation	Negotiation	Prior Consent	Joint Studies or Programs	Joint Mechanism	Dispute Settlement Procedures
20	1964	Exchange of notes authorizing the Canadian entitlement purchase agreement provided for under the treaty relating to cooperative development of the water resources of the Columbia River Basin	Canada, United States	0	0	1	0	1	1	1	1	0	1	0	0	1	1
21	1965	Exchange of notes relating to the establishment of directions to be followed by the Permanent Engineering Board	Canada, United States	0	0	1	0	1	0	1	1	1	0	0	1	1	0
22	1966	Exchange of notes constituting an agreement concerning the loan of waters of the Colorado River for irrigation of lands in the Mexicali Valley	Mexico, United States	0	0	1	0	0	0	0	0	0	0	0	0	0	0
23	1968	Exchange of notes concerning a special operating programme for the Duncan and Arrow storages on the Columbia River System	Canada, United States	0	0	1	0	1	0	0	0	0	1	0	0	0	0

24	1969	Treaty of the River Plate Basin	Argentina, Bolivia, Brazil, Paraguay, Uruguay	0	0	1	0	1	1	0	0	1	0	0	1	1	0
25	1969	Exchange of notes constituting an Agreement for the construction of a Temporary Cofferdam at Niagara	Canada, United States	0	0	1	0	0	0	0	0	0	0	0	0	0	0
26	1969	Exchange of notes constituting an agreement for the temporary diversion for power purposes of the water normally flowing over the American Falls at Niagara	Canada, United States	0	0	1	0	0	0	0	0	0	0	0	0	0	0
27	1972	Exchange of notes constituting an agreement regarding the preservation of the quality of water in the international section of the St. John River	Canada, United States	0	0	1	0	1	0	1	0	1	0	0	1	1	0
28	1973	Treaty concerning the hydroelectric utilization of the water resources of the Paranà River owned in condominium by the two countries, from and including Salto Grande de Sete Quedas or Sal to del Guaira, to the mouth of the Iguassu River	Brazil, Paraguay	0	0	0	1	0	0	0	0	0	0	1	1	1	1
29	1975	Statute of the River Uruguay	Argentina, Uruguay	0	0	0	1	1	1	1	1	1	1	0	1	1	1

(cont.)

Table 2. *(cont.)*

No.	Year	Title	Signatories	Regional Agreement	Boundary Waters	Entire Basin	Partial Basin	Cooperation in Text	Future Agreements Envisaged	Data & Information Exchange	Notification	Consultation	Negotiation	Prior Consent	Joint Studies or Programs	Joint Mechanism	Dispute Settlement Procedures
30	1977	Treaty on cooperation for the utilization of the natural resources and the development of the Mirim Lagoon Basin	Brazil, Uruguay	0	0	1	0	1	1	0	1	0	0	0	1	1	1
31	1977	Protocol for the utilization of the water resources of the land bordering on the Jaguarao River, annexed to the Treaty on the Mirim Lagoon Basin	Brazil, Uruguay	0	0	0	1	1	0	0	0	0	0	1	0	1	0
32	1978	Treaty for Amazonian cooperation	Bolivia, Brazil, Columbia, Ecuador, Guyana, Peru, Suriname, Venezuela	0	0	1	0	1	1	1	0	0	0	0	1	1	0
33	1979	Agreement on Paraná River projects	Argentina, Brazil, Paraguay	0	0	0	1	1	0	0	1	0	1	0	0	1	0

34	1980	Agreement for water resources exploitation within the Uruguay River and its effluent the Pepiri-Guacu River	Argentina, Brazil	0	0	0	1	1	1	0	0	0	1	1	1	1	0
35	1983	Agreement on cooperation for the protection and improvement of the environment in the border area	Mexico, United States	0	1	1	0	1	1	1	1	1	1	0	1	1	0
36	1984	Agreement concerning the Cachuela Esperanza hydroelectric plant	Brazil, Bolivia	0	0	0	1	1	1	1	0	0	1	0	0	1	0
37	1984	Treaty relating to the Skagit River and Ross Lake, and the Seven Mile Reservoir on the Pend d'Oreille River	Canada, United States	0	0	1	0	1	0	0	0	0	0	0	0	0	1
38	1984	Exchange of notes constituting an agreement regarding the continued preservation and enhancement of water quality on the international section of the Saint John River	Canada, United States	0	0	1	0	1	0	1	0	1	0	0	1	1	0
39	1985	Agreement of cooperation regarding pollution of the environment along the inland international boundary by discharges of hazardous substances	Mexico, United States	0	1	1	0	1	1	1	0	0	1	0	1	1	0

(cont.)

Table 2. *(cont.)*

No.	Year	Title	Signatories	Regional Agreement	Boundary Waters	Entire Basin	Partial Basin	Cooperation in Text	Future Agreements Envisaged	Data & Information Exchange	Notification	Consultation	Negotiation	Prior Consent	Joint Studies or Programs	Joint Mechanism	Dispute Settlement Procedures
40	1987	Agreement on the protection and improvement of the environment in the border area	Mexico, Guatemala	0	1	0	0	1	1	0	0	1	1	0	1	1	0
41	1988	Exchange of notes constituting an agreement for the construction of a hydroelectric plant in Cachuela Esperanza	Brazil, Bolivia	0	0	0	1	1	0	0	0	0	0	0	0	0	0
42	1988	Exchange of notes constituting an agreement concerning the construction of a joint ring levee	United States, Canada	0	1	0	0	1	0	0	0	0	1	0	1	0	1
43	1989	Agreement for water supply and flood control in the Souris River Basin	United States, Canada	0	0	1	0	1	0	1	0	1	0	0	1	1	1
44	1991	*Acuerdo de cooperación para el aprovechamiento de los recursos nautrales y el desarollo de la cuenca del Rio Cuareim*	Brazil, Uruguay	0	0	0	1	1	1	0	0	0	0	0	0	1	0

45	1991	*Acuerdo complementario al acuerdo basico de cooperación científica y técnica sorce cooperación en el area de recursos hídricos*	Brazil, Uruguay	0	0	0	0	1	0	0	0	1	1	1	0	0	0
46	1995	*Acuerdo constitutivo de la Comisión trinacional para el desarollo de la cuenca del Rio Pilcomayo*	Argentina, Bolivia, Paraguay	0	0	0	1	1	0	0	0	0	0	0	1	1	1
47	1995	*Acuerdo para el aprovechamiento múltiple de los recursos de la alta cuenca del Río Bermejo y del Río Grande de Tarija*	Argentina, Bolivia	0	0	0	1	0	1	0	0	0	0	0	1	1	1
48	1997	*Ajuste Complementario al acuerdo de cooperación para el aprovechamiento de los recursos naturales y el desarrollo de la Cuenca del Río Cuareim*	Brazil, Uruguay	0	0	0	1	1	0	0	1	0	0	1	0	0	0
49	1998	*Tratado para la ejecución del Plan Trifinio*	Guatemala, Honduras, El Salvador	0	1	0	0	1	0	1	0	0	0	0	1	1	1
50	2010	*Acuerdo sobre el Acuifero Guarani*	Argentina, Brazil, Paraguay, Uruguay	0	0	1	0	1	1	1	1	1	1	0	1	1	1
ASIA																	
1	1903	Final Working Agreement relative to the Sirhind Canal	Great Britain, Patiala, Jind, Nabha	0	0	0	1	0	0	1	0	0	0	0	0	0	1
2	1908	Ganges-Brahmaputra-Meghna	Great Britain and the Panna State (India)	0	0	0	1	0	0	0	0	0	0	0	0	0	1

(*cont.*)

Table 2. *(cont.)*

No.	Year	Title	Signatories	Regional Agreement	Boundary Waters	Entire Basin	Partial Basin	Cooperation in Text	Future Agreements Envisaged	Data & Information Exchange	Notification	Consultation	Negotiation	Prior Consent	Joint Studies or Programs	Joint Mechanism	Dispute Settlement Procedures
3	1910	Convention regarding the water supply of Aden	Great Britain, Aden (Yemen)	0	0	0	1	0	0	0	0	0	0	0	0	0	0
4	1920	Convention on certain points connected with the mandates for Syria, Lebanon, Palestine and Mesopotamia	France, Great Britain	0	1	0	0	0	0	0	0	0	1	0	0	1	1
5	1921	Treaty for the establishment of neighborly relations	Great Britain, Afghanistan	0	1	0	0	0	0	0	0	0	0	0	0	0	0
6	1946	Treaty of friendship and neighborly relations and six annexed protocols (Protocol No. 1 relative to the regulation of the waters of the Tigris and Euphrates and of their tributaries)	Turkey, Iraq	0	0	0	1	1	0	1	1	1	1	0	1	1	1
7	1948	Inter-Dominion Agreement on the canal water dispute between East and West Punjab	India, Pakistan	0	0	0	1	0	0	0	0	0	0	0	0	0	0
8	1953	Agreement concerning the utilization of the Yarmuk waters	Syria, Jordan	0	0	0	1	1	0	1	0	0	0	0	1	1	1

9	1954	Agreement on the Kosi Project	India, Nepal	0	0	0	1	0	0	0	0	0	0	0	1	1	1
10	1956	Agreement on joint research operations to determine the natural resources of the Amur River Basin and the prospects for development of its productive potentialities and on . . .	China, Russia	0	0	0	1	1	1	1	0	0	1	0	1	1	0
11	1957	Treaty concerning the regime of the Soviet–Iranian frontier and the procedure for the settlement of frontier disputes and incidents	Soviet Union, Iran	0	1	0	0	0	1	1	0	0	1	1	1	0	0
12	1957	Agreement for the joint utilisation of the frontier parts of the river Aras and Atrak for irrigation and power generation	Soviet Union, Iran	0	1	0	0	0	0	1	0	0	1	0	1	0	0
13	1957	Statute of the Committee for Coordination of Investigations of the Lower Mekong Basin	Cambodia, Laos, Thailand, Vietnam	0	0	0	1	1	0	1	0	0	0	0	0	1	0
14	1959	Agreement on the Gandak Irrigation and Power Project	India, Nepal	0	0	0	1	0	0	0	0	0	0	0	1	0	1
15	1960	Indus Waters Treaty	India, Pakistan, International Bank for Reconstruction and Development	0	0	0	1	1	0	1	1	1	1	0	1	1	1

(cont.)

Table 2. *(cont.)*

No.	Year	Title	Signatories	Regional Agreement	Boundary Waters	Entire Basin	Partial Basin	Cooperation in Text	Future Agreements Envisaged	Data & Information Exchange	Notification	Consultation	Negotiation	Prior Consent	Joint Studies or Programs	Joint Mechanism	Dispute Settlement Procedures
16	1966	Agreement concerning the Kosi Project	India, Nepal	0	0	0	1	1	0	1	0	0	0	0	1	1	1
17	1972	Statute of the Indo-Bangladesh Joint Rivers Commission	India, Bangladesh	0	1	0	0	1	0	0	0	0	0	0	1	1	1
18	1973	Agreement concerning administrative arrangements as to the border between Papua New Guinea and Indonesia	Australia, Indonesia, Papua New Guinea	0	1	0	0	1	0	1	0	1	1	0	0	0	0
19	1975	Treaty concerning the State frontier and neighborly relations between Iran and Iraq and Protocol	Iran, Iraq	0	1	0	0	1	0	0	0	0	1	0	0	1	1
20	1977	Agreement on sharing of the Ganges waters at Farakka and on augmenting its flows	India, Bangladesh	0	0	0	1	1	0	0	0	0	0	0	1	1	1
21	1982	Indo-Bangladesh memorandum of understanding on the sharing of Ganga waters at Farakka	India, Bangladesh	0	0	0	1	0	0	0	0	1	0	0	1	1	0

22	1983	Agreement on *ad-hoc* sharing of the Teesta waters	India, Bangladesh	0	0	0	1	0	0	0	0	0	0	0	1	1	0
23	1993	Agreement on joint activities in addressing the Aral Sea and the zone around the Sea crisis, improving the environment and ensuring the social and economic development of the Aral Sea region	Kazakhstan, Kyrgyzstan, Tajikistan, Turkmenistan, Uzbekistan	0	0	0	1	1	1	1	0	0	1	0	0	1	0
24	1994	Agreement on the protection and utilization of transboundary waters	China, Mongolia	0	1	0	0	1	0	1	0	1	0	0	1	1	1
25	1994	Treaty of Peace	Israel, Jordan	0	0	0	1	1	0	1	1	0	0	1	1	1	0
26	1995	The Israeli-Palestinian Interim Agreement on the West Bank and Gaza Strip	Israel, Palestine Liberation Organization	0	1	0	0	1	0	1	0	0	0	1	0	1	1
27	1995	Agreement on the cooperation for the sustainable development of the Mekong River Basin	Cambodia, Laos, Thailand, Vietnam	0	0	0	1	1	1	1	1	1	1	0	1	1	1
28	1996	Treaty concerning the integrated development of the Mahakali River including Sarada Barrage, Tanakpur Barrage, and Pancheshwar Project	India, Nepal	0	0	0	1	1	1	0	0	0	1	0	1	1	1
29	1996	Treaty on sharing the Ganga/Ganges Waters at Farakka	India, Bangladesh	0	0	0	1	1	0	0	0	1	0	0	0	1	1

(*cont.*)

Table 2. *(cont.)*

No.	Year	Title	Signatories	Regional Agreement	Boundary Waters	Entire Basin	Partial Basin	Cooperation in Text	Future Agreements Envisaged	Data & Information Exchange	Notification	Consultation	Negotiation	Prior Consent	Joint Studies or Programs	Joint Mechanism	Dispute Settlement Procedures
30	1998	Agreement on cooperation in the area of environment and rational nature use	Kazakhstan, Kyrgyzstan, Uzbekistan	0	1	0	0	1	0	1	1	0	0	0	1	0	1
31	1998/1999	Agreement and Protocol on the use of water and energy resources of the Syr Darya Basin	Kazakhstan, Kyrgyzstan, Uzbekistan, Tajikistan	0	0	1	0	1	1	0	0	0	1	0	0	1	1
32	2000	Agreement on utilization of the water facilities of interstate use on the Chu and Talas Rivers	Kazakhstan, Kyrgyzstan	0	0	1	0	1	0	0	1	0	0	0	1	0	1
EUROPE																	
1	1913	*Übereinkunft betreffend der Gewinnung der Wasserkräfte der Rhone zwischen dem projektierten Kraftwerk von La Plaine und einem noch zu bestimmenden Punkt oberhalb der Brücke von Pougny-Chancy*	Switzerland, France	0	0	1	0	1	1	0	0	0	1	1	0	0	0
2	1920	Peace Treaty and Protocol	Russia, Lithuania	0	1	0	0	0	1	0	0	0	1	0	0	0	0
3	1920	Treaty of Peace	Russia, Latvia	0	1	0	0	0	1	0	0	0	1	0	0	0	0
4	1921	Treaty of Peace	Poland, Russia, Ukraine	0	1	0	0	0	0	0	0	0	1	0	0	0	0

5	1921	Convention concerning the delimitation of the frontier and various questions connected therewith	Austria, Czechoslovakia	0	0	0	1	0	0	0	0	0	0	0	0	0	1
6	1921	Convention instituting the definitive statute of the Danube	Austria, Belgium, Bulgaria, Czechoslovakia, France, Germany, Greece, Hungary, Italy, Romania, Serb-Croat-Slovene State	0	0	1	0	0	0	1	1	0	1	0	0	1	1
7	1922	Convention instituting the statute of navigation of the Elbe	Belgium, Czechoslovakia, France, Germany, Great Britain, Italy	0	0	1	0	0	0	0	1	0	1	1	0	1	1
8	1922	Convention concerning the maintenance of river channels and the regulation of fishing on watercourses forming part of the frontier between Finland and Russia	Finland, Russia	0	1	0	0	0	0	0	0	0	0	0	0	0	0
9	1923	Convention supplementary to the statute of navigation of the Elbe	Belgium, Czechoslovakia, France, Germany, Great Britain, Italy	0	0	1	0	0	0	0	0	0	0	0	0	1	1

(cont.)

Table 2. *(cont.)*

No.	Year	Title	Signatories	Regional Agreement	Boundary Waters	Entire Basin	Partial Basin	Cooperation in Text	Future Agreements Envisaged	Data & Information Exchange	Notification	Consultation	Negotiation	Prior Consent	Joint Studies or Programs	Joint Mechanism	Dispute Settlement Procedures
10	1924	*Traite pour le redressement du Rhin des l'embouchure de l'Ill jusqu'au Lac de Constance*	Switzerland, Austria	0	0	0	1	1	1	0	0	0	1	0	1	1	1
11	1925	Convention concerning the international legal regime of the waters of the Pasvik and the Jakobslev	Finland, Norway	0	0	0	1	0	0	0	1	0	1	1	0	0	1
12	1926	Treaty for the settlement of frontier questions	Germany, Poland	0	1	0	0	0	0	0	1	0	0	0	0	0	0
13	1926	*Convention pour le règlement des rapports au sujet de certaines clauses du regime juridique de la future derivation de Kembs*	Switzerland, France	0	0	0	1	0	0	1	1	1	1	1	1	1	1
14	1928	Convention regarding the maintenance and administration of the frontier waterways	German Reich, Lithuania	0	1	0	0	0	0	1	1	0	1	1	0	0	1
15	1930	*Übereinkommen über die Verleihung der Wasserkräfte des Doubs bei Chatelot*	Switzerland, France	0	0	1	0	0	0	0	0	1	1	1	0	1	1

16	1948	Convention regarding the regime of navigation on the Danube	Bulgaria, Czechoslovakia, Hungary, Romania, Ukraine, Soviet Union Yugoslavia	0	0	0	1	0	1	1	1	1	0	0	0	1	1
17	1949	*Accord au sujet de la concession de forces hydrauliques du Reno di Lei*	Switzerland, Italy	0	0	0	1	0	0	0	0	0	1	1	1	1	1
18	1950	Protocol to establish a tripartite standing committee on polluted waters	Belgium, France, Luxemburg	0	0	0	1	0	0	1	0	0	0	0	1	1	0
19	1950	Agreement concerning the diversion of water in the Rissbach, Durrach and Walchen Districts	Austria, Bavaria	0	0	0	1	0	0	0	0	0	1	0	0	0	0
20	1951	Exchange of notes amending Article 14 (2) of the Convention of 11 August 1951	Portugal, Spain	0	0	1	0	0	0	0	0	0	0	0	0	1	0
21	1951	Agreement on the transfer from the course of the Näätämo (Neiden) River to the course of the Gandvik River of water from the Garsjöen, Kjerringvatn and Förstevannene Lakes	Finland, Norway	0	1	0	0	0	1	0	0	0	1	0	0	0	0
22	1951	*Convention concernant la correction de la Roggia Molinara*	Switzerland, Italy	0	0	0	1	0	0	0	0	0	1	1	1	0	0
23	1952	Agreement concerning the *Donaukraftwerk-Jochenstein-Aktiengesellschaft*	Austria, Germany, Bavaria	0	0	0	1	1	1	0	0	1	1	0	1	1	1

(cont.)

Table 2. *(cont.)*

No.	Year	Title	Signatories	Regional Agreement	Boundary Waters	Entire Basin	Partial Basin	Cooperation in Text	Future Agreements Envisaged	Data & Information Exchange	Notification	Consultation	Negotiation	Prior Consent	Joint Studies or Programs	Joint Mechanism	Dispute Settlement Procedures
24	1952	Convention concerning measures to prevent floods and to regulate the water regime of the River Prut	Soviet Union, Romania	0	0	0	1	0	0	1	1	0	1	0	0	0	0
25	1952	Agreement concerning navigation in frontier waters and the use and maintenance of frontier waters	German Democratic Republic, Poland	0	1	0	0	1	1	1	1	1	1	1	0	1	1
26	1954	Convention concerning water economy questions relating to the Drava	Austria, Yugoslavia	0	0	0	1	1	0	1	1	1	1	0	0	1	1
27	1954	Agreement concerning water economy questions in respect of the frontier sector of the Mura	Austria, Yugoslavia	0	0	0	1	0	0	0	1	0	0	0	0	1	1
28	1954	*Traité pour la regularisation du Rhin de l'embouchure de l'Ill au Lac de Constance*	Switzerland, Austria	0	0	0	1	0	1	1	0	1	1	1	1	`1	1
29	1955	*Convention au sujet de la regularisation du Lac de Lugano*	Switzerland, Italy	0	0	0	1	0	0	0	1	1	1	0	0	1	1
30	1956	Treaty concerning the regulation of water economy questions	Austria, Hungary	0	1	0	0	1	0	1	1	1	1	1	0	1	1

31	1956	Protocol concerning amendments to the Regulation of 24 April 1947 for the regulation of Lake Inari in connexion with the use of the Niskakoski Dam and to the Protocol of 19 April 1954 concerning amendments to paragraph 2 of the said regulation	Finland, Norway, Soviet Union	0	0	1	0	0	0	1	1	0	0	0	0	0	0
32	1957	*Convention au sujet de l'utilisation de la force hydraulique du Spoel*	Switzerland, Italy	0	0	0	1	0	0	0	1	1	1	1	1	1	1
33	1958	Agreement concerning water economy questions	Bulgaria, Yugoslavia	0	1	0	0	1	1	0	1	0	1	0	0	1	1
34	1958	Agreement relating to Lake Lanoux	Spain, France	0	0	1	0	0	0	0	1	0	0	0	0	1	1
35	1958	Agreement concerning the use of water resources in frontier water	Czechoslovakia, Poland	0	1	0	0	1	0	1	1	1	1	1	1	1	0
36	1958	State treaty concerning the construction of hydroelectric power installations on the Our	Germany (Rhineland), Luxembourg	0	0	0	1	0	0	1	1	0	1	0	1	1	1
37	1959	Agreement concerning hydro-economic questions	Greece, former Yugoslavia	0	1	0	0	1	0	0	1	0	0	0	1	1	1
38	1959	*Abkommen über die Korrektion der Hermance*	Switzerland, France	0	0	1	0	0	0	0	0	0	0	0	0	0	0
39	1959	*Abkommen über die Korrektion des Baches 'Le Boiron'*	Switzerland, France	0	0	1	0	0	0	0	0	0	0	0	0	0	0

(cont.)

Table 2. *(cont.)*

No.	Year	Title	Signatories	Regional Agreement	Boundary Waters	Entire Basin	Partial Basin	Cooperation in Text	Future Agreements Envisaged	Data & Information Exchange	Notification	Consultation	Negotiation	Prior Consent	Joint Studies or Programs	Joint Mechanism	Dispute Settlement Procedures
40	1960	Treaty concerning the course of the common frontier, the boundary waters, real property situated near the frontier, traffic crossing the frontier on land and via inland waters and on further questions	Germany, Netherlands	0	1	0	0	1	1	0	1	1	1	0	0	1	1
41	1961	Exchange of notes constituting an agreement concerning the treaty of 12 May 1863 to regulate the diversion of water from the River Meuse and the convention of 11 January 1873 amending that Treaty	Belgium, Netherlands	0	0	0	1	0	0	0	0	0	1	0	0	0	0
42	1961	Protocol concerning the establishment of an international commission to protect the Saar against pollution	Germany, France	0	0	0	1	1	0	0	0	0	0	0	0	1	1

43	1961	Protocol concerning the establishment of an international commission to protect the Moselle against pollution	Germany, France, Luxembourg	0	0	0	1	1	0	0	0	0	0	0	0	1	1
44	1962	Convention concerning protection of the waters of Lake Geneva against pollution	Switzerland France	0	0	1	0	0	1	0	0	0	0	0	1	1	0
45	1963	*Abkommen über den Ausbau der Wasserkräfte bei Emosson*	Switzerland, France	0	0	1	0	0	0	0	0	1	1	1	0	1	1
46	1963	Final Act relating to the establishment and operation of the Iron Gates Water Power and Navigation System	Yugoslavia, Romania	0	0	0	1	1	0	0	0	0	0	0	1	1	0
47	1963	Agreement concerning the construction and operation of the Iron Gates Water Power and Navigation System	Yugoslavia, Romania	0	0	0	1	1	0	0	0	1	1	1	1	1	1
48	1963	Convention concerning compensation for damage caused by the construction of the Iron Gates WP &N System	Yugoslavia, Romania	0	0	0	1	0	0	0	0	0	0	0	1	0	0
49	1963	Convention concerning the execution of works for the Iron Gates WP&N System	Yugoslavia, Romania	0	0	0	1	1	0	1	1	1	0	1	1	1	1
50	1963	Convention concerning the operation of the Iron Gates WP&N System	Yugoslavia, Romania	0	0	0	1	0	0	1	0	0	0	0	1	1	0

(cont.)

Table 2. *(cont.)*

No.	Year	Title	Signatories	Regional Agreement	Boundary Waters	Entire Basin	Partial Basin	Cooperation in Text	Future Agreements Envisaged	Data & Information Exchange	Notification	Consultation	Negotiation	Prior Consent	Joint Studies or Programs	Joint Mechanism	Dispute Settlement Procedures
51	1963	Convention concerning the preparation of designs for the construction of the Iron Gates WP&N System	Yugoslavia, Romania	0	0	0	1	0	0	0	0	0	0	1	1	1	0
52	1963	Convention concerning the determination of the value of investments and mutual accounting in connection with the construction of the Iron Gates WP&N System	Yugoslavia, Romania	0	0	0	1	0	0	1	0	0	0	0	0	1	0
53	1963	Agreement on the International Commission for the protection of the Rhine against pollution (and Protocol)	Germany, France, Luxembourg, Netherlands, Switzerland	0	0	0	1	1	0	0	0	0	0	0	0	1	0
54	1964	Agreement concerning the use of water resources in frontier waters	Poland, Soviet Union	0	1	0	0	1	1	1	1	1	1	1	1	1	0
55	1966	Agreement relating to the withdrawal of water from Lake Constance	Austria, Germany, Switzerland	0	0	0	1	0	0	0	1	1	1	0	0	0	1

56	1967	Treaty concerning regulation of water management questions relating to frontier waters	Austria, Czechoslovakia	0	1	0	0	1	0	0	1	0	0	1	0	1	1
57	1968	Agreement concerning the establishment of a river administration in the Rajka-Gonyiu sector of the Danube	Czechoslovakia, Hungary	0	0	0	1	0	0	1	0	0	0	0	1	1	0
58	1968	Agreement regulating the use and development of the water resources of the international reaches of the rivers Mino, Limia, Tajo, Guadiana and Chanza and of their tributaries	Spain, Portugal	0	1	0	0	1	1	0	1	0	0	1	1	1	1
59	1968	Agreement concerning cooperation in the use of waters of rivers flowing through the territory of both countries	Bulgaria, Turkey	0	1	0	0	1	1	1	1	0	1	0	0	1	1
60	1970	Exchange of letters amending the arrangement of 12 July 1958 relating to Lake Lanoux	Spain, France	0	0	1	0	0	0	0	0	0	0	0	0	1	1
61	1971	Agreement on the joint construction of the Stanca-Costesti hydraulic engineering scheme on the river Prut and the establishment of the conditions for its operation	Romania, Soviet Union	0	0	0	1	0	0	0	0	0	0	0	1	1	1

(cont.)

Table 2. *(cont.)*

No.	Year	Title	Signatories	Regional Agreement	Boundary Waters	Entire Basin	Partial Basin	Cooperation in Text	Future Agreements Envisaged	Data & Information Exchange	Notification	Consultation	Negotiation	Prior Consent	Joint Studies or Programs	Joint Mechanism	Dispute Settlement Procedures
62	1971	Exchange of letters constituting an agreement on the implementation of the Convention of 16 November 1962 concerning protection of the waters of Lake Geneva against pollution	Switzerland, France	0	0	1	0	0	0	0	0	0	0	0	1	1	0
63	1972	Convention concerning the protection of Italo-Swiss waters against pollution	Italy, Switzerland	0	1	0	0	1	0	0	0	0	0	0	1	1	0
64	1972	*Convention relative a l'amenagement hydraulique du torrent Breggia a la frontière italo-suisse*	Switzerland, Italy	0	0	0	1	0	0	0	0	0	0	0	0	0	0
65	1972	Agreement concerning the production of electric power in the part of the Vuoksi River bounded by the Imatra and Svetogorsk hydroelectric stations	Finland, Soviet Union	0	0	1	0	0	0	0	0	0	0	0	0	1	1
66	1975	Agreement on the development of economic cooperation	Italy, Yugoslavia	0	1	0	0	1	1	0	0	1	1	0	1	1	0

67	1976	Agreement concerning the maintenance, restoration and operation of the section of the Moselle common to the two States	Germany, Luxembourg	0	0	0	1	0	0	0	1	1	1	0	0	0 1
68	1976	Convention on the protection of the Rhine against chemical pollution	Switzerland, Germany, France, Luxembourg, Netherlands, EC	0	0	0	1	0	0	1	1	1	0	0	0	1 1
69	1976	Convention on the protection of the Rhine against pollution by chlorides	Switzerland, Germany, France, Luxembourg, Netherlands	0	0	0	1	1	0	1	1	1	1	0	0	1 1
70	1977	Treaty concerning the construction and operation of the Gabcikovo–Nagymaros systems of locks	Hungary, Czechoslovakia	0	0	0	1	1	0	0	0	0	1	0	1	1 1
71	1977	Agreement concerning the activities of agencies for the control of accidental water pollution by hydrocarbons or other substances capable of contaminating water and recognised as such under the 1962 Convention concerning protection of the water of Lake Geneva against pollution	Switzerland, France	0	0	1	0	1	0	0	1	0	0	0	1	1 0

(cont.)

Table 2. *(cont.)*

No.	Year	Title	Signatories	Regional Agreement	Boundary Waters	Entire Basin	Partial Basin	Cooperation in Text	Future Agreements Envisaged	Data & Information Exchange	Notification	Consultation	Negotiation	Prior Consent	Joint Studies or Programs	Joint Mechanism	Dispute Settlement Procedures
72	1980	Convention on the subject of the water of the Sure	Belgium, Luxembourg	0	0	0	1	0	0	0	0	1	0	1	0	1	1
73	1980	*Accord sur la dephosphatation des eaux du Lac Leman*	Switzerland, France	0	0	1	0	0	0	1	0	0	0	0	0	0	0
74	1982	*Convention modifiant et completant la Convention additionelle du 16 juillet 1975 a la Convention du 4 juillet 1969 au sujet de l'amenagement du Rhin entre Strasbourg et Lauterbourg-Neuburgweier*	France, Germany	0	0	0	1	0	0	1	1	1	1	0	0	1	1
75	1982	Convention concerning improvement of the common rivers Lys between Deulemont et Menin	France, Belgium	0	0	0	1	0	0	0	0	1	1	1	1	1	1
76	1986	*Convention pour la realisation et l'exploitation de certaines implantations industrielles sur la Moselle*	France, Luxembourg	0	0	0	1	0	0	1	0	1	0	0	0	1	1

77	1986	Exchange of notes constituting an agreement concerning the execution of improvement works on the River Gander at Mondorff and at Mondorf-les-Bains	France, Luxembourg	0	0	0	1	0	0	0	0	0	0	0	0	0	0
78	1987	Agreement on flood warning for the catchment basin of the Moselle	Germany, France, Luxembourg	0	0	0	1	1	0	1	0	0	1	0	0	1	0
79	1987	Agreement on cooperation on management of water resources in the Danube Basin	Austria, Germany, EC	0	0	0	1	1	0	1	1	1	0	0	1	1	1
80	1989	Agreement concerning regulations governing Lake Saimaa and the Vuoksi River	Finland, Soviet Union	0	0	1	0	1	0	0	0	0	0	0	0	1	1
81	1990	Convention on the International Commission for the Protection of the Elbe	Germany, Czech Republic, EC	0	0	1	0	1	1	1	0	0	0	0	0	1	0
82	1991	Additional Protocol to the Convention on the protection of the Rhine against chlorides	Switzerland, Germany, France, Luxembourg, Netherlands	0	0	0	1	0	0	1	0	0	0	0	0	0	0
83	1992	Treaty on mutual relations and principles of cooperation in agriculture, food industry, forestry and water economy under the conditions of the customs union	Czech Republic, Slovakia	0	1	0	0	1	1	0	0	0	0	0	1	0	0

(cont.)

Table 2. *(cont.)*

No.	Year	Title	Signatories	Regional Agreement	Boundary Waters	Entire Basin	Partial Basin	Cooperation in Text	Future Agreements Envisaged	Data & Information Exchange	Notification	Consultation	Negotiation	Prior Consent	Joint Studies or Programs	Joint Mechanism	Dispute Settlement Procedures
84	1992	Convention on the Protection and Use of Transboundary Watercourses and International Lakes	UNECE (including Central Asia – 37 Parties, including EU)	1	0	0	0	1	1	1	1	1	1	0	1	1	1
85	1994	Convention on cooperation for the protection and sustainable use of the River Danube	Austria, Bulgaria, Croatia, Czech Republic, EC, Germany, Hungary, Moldova, Romania, Slovakia, Slovenia, Ukraine	0	0	1	0	1	1	1	0	1	0	0	1	1	1
86	1994	Agreement on joint boundary waters management and protection	Ukraine, Moldova	0	1	0	0	1	0	1	1	1	0	0	0	0	0
87	1994	Agreement on the protection of the (River) Meuse	France, Netherlands, Belgium	0	0	0	1	1	0	1	0	0	0	0	0	1	0

88	1994	Agreement on the protection of the River Scheldt	France, Netherlands, Belgium	0	0	1	0	1	1	1	0	0	0	0	0	1	0
89	1995	*Vertrag über die Zusammenarbeit auf dem Gebiet der Wasserwirtschaft an den Grenzgewässern*	Germany, Czech Republic	0	1	0	0	1	0	1	1	1	1	0	1	1	0
90	1996	*Vertrag über die Internationale Kommission zum Schutz der Oder gegen Verunreinigung*	Germany, Poland, Czech Republic, EU	0	0	1	0	1	0	0	0	0	0	0	1	1	0
91	1996	*Convention relative à la collecte, au depot et la reception des dechets survenant en navigation Rhenane et interieure*	Germany, Belgium, France, Luxembourg, Netherlands, Switzerland	0	0	0	1	0	1	0	0	0	1	0	0	1	0
92	1998	*Convenção sobre cooperação para a protecção e o aproveitamento sustentável das aguas das bacias hidrográficas luso-espanholas*	Spain, Portugal	0	1	0	0	1	1	1	1	1	1	1	1	1	1
93	1999	Convention on the protection of the Rhine	Germany, France, Luxembourg, Netherlands, Switzerland, EC	0	0	0	1	1	1	1	1	1	0	0	1	1	1
94	2002	Framework Agreement on the Sava River Basin	Bosnia & Herzegovina, Croatia, Slovenia, Yugoslavia	0	0	0	1	1	1	1	0	0	0	0	1	1	1

(cont.)

Table 2. *(cont.)*

No.	Year	Title	Signatories	Regional Agreement	Boundary Waters	Entire Basin	Partial Basin	Cooperation in Text	Future Agreements Envisaged	Data & Information Exchange	Notification	Consultation	Negotiation	Prior Consent	Joint Studies or Programs	Joint Mechanism	Dispute Settlement Procedures
95	2002	*Internationales Massübereinkommen*	Netherlands, Belgium, Luxembourg, Germany	0	0	0	1	1	1	1	1	1	0	0	1	1	1
96	2003	Agreement on the establishment of the Permanent Greek-Albanian Commission on Transboundary Freshwater Issues	Greece, Albania	0	1	0	0	1	0	1	0	1	0	0	1	1	1

Bibliography

Abi-Saab, G., '*Cours Général de Droit International Public*' (1987) 207 *RCADI* 9–464.

Ad-Hoc Working Group on Long-Term Cooperative Action, *Compilation of Information on Nationally Appropriate Mitigation Actions to be Implemented by Parties not Included in Annex I to the Convention*, 18 March 2011, UN Doc. FCCC/AWGLCA/2011/INF.1.

Advisory Committee of Jurists, *Procés-Verbaux of the Proceedings of the Committee June 16th–July 24th 1920* (The Hague: Van Langenhuysen Brothers, 1920).

African Union, *The AU/NEPAD African Action Plan 2010–2015: Advancing Regional and Continental Integration in Africa* (African Union, 2009).

Akanda, D., 'International Organizations', in M. D. Evans, *International Law* (New York: Oxford University Press, 2003), pp. 269–297.

Alam, U., 'Questioning the Water Wars Rationale: A Case Study of the Indus Water Treaty' (2002) 168 *The Geographical Journal* 4, 341–353.

Alam, U., 'Water Rationality: Mediating the Indus Water Treaty', unpublished Ph.D. Thesis, University of Durham, England (1998).

Anand, R. P., 'Sovereign Equality of States in International Law' (1986) 197 *RCADI* 1–228.

Andrassy, J., '*Les relations internationales de voisinage*' (1951) 79 *RCADI* 73–182.

Arechaga, J., 'International Law in the Past Third of a Century' (1978) 159 *RCADI* 1–343.

Asian Development Bank, *The Great Mekong Subregion Economic Cooperation Program Strategic Framework (2012–2022)* (Philippines: ADB, 2011).

Aurescu, B., '*L'efficacité des dispositions sur le règlement des différends dans les conventions environnementales, étude de cas: le projet ukrainien du canal navigable "Bystroe" dans le delta du Danube*', in B. Aurescu and A. Pellet (eds.), *Actualité du droit des fleuves internationaux* (Paris: Pedone, 2010), pp. 265–283.

Aurescu, B., and A. Pellet (eds.), *Actualité du droit des fleuves internationaux* (Paris: Pedone, 2010).

Axelrod, R., *The Evolution of Cooperation*, revised edition (New York: Basic Books, 2006).

Axelrod, R., and W. D. Hamilton, 'The Evolution of Cooperation' (1981) 211 *Science* 1390–1396.

Bangkok Post, 'Reshaping the Mekong', 23 August 2010, https://groups.google.com/forum/?fromgroups#!topic/lancang-mekong/0_q5YagXBs4.

Baraer, M., B. G. Mark, J. M. McKenzie, T. Condom, J. Bury, K.-I. Huh, C. Portocarrero, J. Gomez, and S. Rathay, 'Glacier Recession and Water Resources in Peru's Cordillera Blanca' (2012) 58 *Journal of Glaciology* 134–150.

Barberis, J., *Los Recursos Naturales Compartidos entre Estados y el Derecho Internacional* (Madrid: Editorial Tecnos, 1979).

Barret, J., 'The Negotiation and Drafting of the Climate Change Convention', in R. Churchill and D. Freestone (eds.), *International Law and Global Climate Change* (London: Graham & Trotmann/Martinus Nijhoff, 1991), pp. 183–200.

Bates, B. C., Z. W. Kundzewicz, S. Wu, and J. P. Palutikof (eds.), *Climate Change and Water, IPCC Technical Paper VI* (Geneva: Intergovernmental Panel on Climate Change Secretariat, 2008).

Baxter, A. (ed.), *International Navigable Waterways – Financial and Legal Aspects of their Improvement and Maintenance*, Study No. 6 (New York: United Nations Institute for Training and Research, 1975), UN Doc. UNITAR/ /74/ST/6.

Baylis, J., and S. Smith (eds.), *The Globalization of World Politics*, 3rd edition (New York: Oxford University Press, 2005).

Behrmann, C., *Das Prinzip der angemessenen und vernünftigen Nutzung und Teilhabe nach der VN-Wassserlaufkonvention* (Berlin: Duncker & Humblot, 2008).

Bernad y Alvarez de Eulate, M., '*La coopération transfrontalière régionale et locale*' (1993) 243 *RCADI* 293–417.

Bernauer, T., 'Explaining Success and Failure in International River Management' (2002) 64 *Aquatic Science* 1–19.

Bernhardt, R. (ed.), *Encyclopedia of Public International Law, Installment 7* (Amsterdam: North Holland Publishing, 1984).

Biermann, F., *Internationale Meeresumweltpolitik – Auf dem Weg zu einem Umweltregime für die Ozeane* (Berlin: Peter Land, 1994).

Bigas, H. (ed.), *The Global Water Crisis: Addressing an Urgent Security Issue*, Papers for the InterAction Council (Hamilton, Ontario: United Nations University–Institute for Water, Environment and Health, 2012).

Birnie, P., and A. Boyle, *International Law and the Environment*, 3rd edition (New York: Oxford University Press, 2009).

Black, R., 'Climate "Technical Fix" May Yield Warming, Not Cooling', *BBC News Online*, 6 April 2011, www.bbc.co.uk/news/science-environment-12983795?print=*true*.

Blake, G. H., W. Hildesley, M. Pratt, R. Ridley, and C. Schofield (eds.), *The Peaceful Management of Transboundary Resources* (London: Graham & Trotman/Martinus Nijhoff, 1995).

Bodansky, D., 'The United Nations Convention on Climate Change: A Commentary' (1993) 18 *Yale Journal of International Law* 453–554.

Bodansky, D., J. Brunnée, and E. Hey (eds.), *The Oxford Handbook of International Environmental Law* (New York: Oxford University Press, 2007).

Boehmer-Christiansen, S., 'Environmental Quality Objectives versus Uniform Emission Standards', in D. Freeston and T. Ijlstra, *The North Sea: Perspectives on Regional Environmental Cooperation* (London: Graham and Trotman, 1990), pp. 139–149.

Boisson de Chazournes, L., *Freshwater and International Law: The Interplay between Universal, Regional and Basin Perspectives* (Paris: United Nations Educational, Scientific and Cultural Organization, 2009).

Boisson de Chazournes, L., '*Sur les Rives du Droit International de l'Eau: entre Universalité et Particularismes*', in M. G. Kohen (ed.), *Promoting Justice, Human Rights and Conflict Resolution through International Law / La promotion de la justice, des droits de l'homme et du règlement des conflits par le droit international, Liber Amicorum Lucius Caflish* (Leiden: Koninlijke Brill NV, 2007), pp. 685–700.

Boisson de Chazournes, L., 'Precaution in International Law: Reflection on its Composite Nature', in T. M. Ndiaye and R. Wolfrum (eds.), *Law of the Sea, Environmental Law and Settlement of Disputes – Liber Amicorum Judge Thomas A. Mensah* (Leiden/Boston: Martinus Nijhoff Publishers, 2007), pp. 21–34.

Boisson de Chazournes, L., 'Technical and Financial Assistance', in D. Bodansky, J. Brunnée, and E. Hey (eds.), *The Oxford Handbook of International Environmental Law* (New York: Oxford University Press, 2007), pp. 948–972.

Boisson de Chazournes, L., '*Eaux internationales et droit international: vers l'idée de gestion commune*', in L. Boisson de Chazournes and S. Salman (eds.), *Les ressources en eau et le droit international – Water Resources and International Law*, The Hague Academy of International Law (Leiden/Boston: Martinus Nijhoff Publishers, 2005), pp. 3–43.

Boisson de Chazournes, L., 'The Global Environmental Facility (GEF): A Unique and Crucial Institution' (2005) 14 *RECIEL* 1993–201.

Boisson de Chazournes, L., 'The Role of Diplomatic Means of Solving Water Disputes: A Special Emphasis on Institutional Mechanisms', in Permanent Court of Arbitration, *Resolution of International Water Disputes* (Leiden: Kluwer Law International, 2003), pp. 91–110.

Boisson de Chazournes, L., 'Changing Perspective in the Management of International Watercourses: An International Law Perspective', in A. Turton, P. Ashton, and E. Cloete (eds.), *Transboundary Rivers, Sovereignty and Development: Hydropolitical Drivers in the Okavango Basin* (Pretoria: African Water Issues Research Unit and Green Cross International, 2003), pp. 213–228.

Boisson de Chazournes, L., R. Desgagné, M. Mbengue, and C. Romano, *Protection internationale de l'environnement*, 2nd edition (Paris: Pedone, 2005).

Boisson de Chazournes, L., and M. Tignino, '*Droit international et eau douce*' (2007) Fasc. 2900 *Jurisclasseur – Environnement* 1–32.

Boisson de Chazournes, L., and S. Salman (eds.), *Les ressources en eau et le droit international – Water Resources and International Law*, The Hague Academy of International Law (Leiden/Boston: Martinus Nijhoff Publishers, 2005).

Bourne, C., 'Procedures in the Development of International Drainage Basins: Notice and Exchange of Information' (1972) 22 *University of Toronto Law Journal* 172–206.

Bourne, C., 'Procedure in the Development of International Drainage Basins: The Duty to Consult and to Negotiate' (1972) 10 *Canadian Yearbook of International Law* 212–234.

Bourne, C., 'Mediation, Conciliation and Adjudication in the Settlement of International Drainage Basin Disputes' (1971) 9 *Canadian Yearbook of International Law* 114–158.

Bourquain, K., *Freshwater Access from a Human Rights Perspective – A Challenge to International Water and Human Rights Law* (Leiden: Martinus Nijhoff Publishers, 2008).

Brandenburg, E. C. (ed.), *Official Opinions of the Attorney-General of the United States Advising the President and Heads of Departments in Relation to their Official Duties*, Vol. 21 (Washington, DC: Government Printing Office, 1898).

Bravard, J. P., *Impacts of Climate Change on the Management of Upland Waters: The Rhone River Case* (Banff: 5th Rosenberg International Forum, 2006).

Brown, C., and J. King, *Environmental Flows: Concepts and Methods*, Water Resources and Environment: Technical Note C.1 (Washington, DC: World Bank, 2003).

Brown Weiss, E., 'The Evolution of International Water Law' (2007) 331 *RCADI* 167–404.

Brown Weiss, E., 'Opening the Door to the Environment and to Future Generations', in L. Boisson de Chazournes and P. Sands, *International Law, the International Court of Justice and Nuclear Weapons* (Cambridge: Cambridge University Press, 1999), pp. 338–353.

Brown Weiss, E., 'Our Rights and Obligations to Future Generations for the Environment' (1990) 84 *AJIL* 198–207.

Brown Weiss, E., L. Boisson de Chazournes, and N. Bernasconi-Osterwalder (eds.), *Fresh Water and International Economic Law* (New York: Oxford University Press, 2005).

Bruhács, J., *The Law of Non-Navigational Uses of International Watercourses* (Dordrecht/Boston: M. Nijhoff, 1993).

Brunnée, J., 'Climate Change, Global Environmental Justice and International Environmental Law', in J. Ebbeson and P. Okowa (eds.), *Environmental Law and Justice in Context* (Cambridge: Cambridge University Press, 2009), pp. 316–330.

Brunnée, J., '"Common Interest" – Echoes from an Empty Shell' (1989) 49 *Zeitschrift für ausländisches öffentliches Recht und Völkerrecht* 791–808.

Brunnée, J., and S. J. Toope, 'Environmental Security and Freshwater Resources: Ecosystem Regime Building' (1997) 91 *AJIL* 1, 26–59.

Burchi, S., and K. Mechlem, *Groundwater in International Law – Compilation of Treaties and Other Legal Instruments*, FAO Legislative Study 86 (Rome: FAO/UNESCO, 2005).

Caflisch, L., 'Regulation of the Uses of International Watercourses', in S. Salman and L. Boisson de Chazournes (eds.), *International Watercourses: Enhancing*

Cooperation and Managing Conflict: Proceedings of a World Bank Seminar (Washington, DC: World Bank, 1998), pp. 3–17.

Caflisch, L., '*Règles Générales du droit des cours d'eau internationaux*' (1989) 219 *RCADI* 13–225.

Caponera, D., 'Patterns of Cooperation in International Water Law: Principles and Institutions' (1985) 25 *Natural Resources Journal* 563–587.

Cheng, B., *General Principles of Law as Applied by International Courts and Tribunals*, reprint (Cambridge: Grotius Publications Unlimited, 1987).

Churchill, R., and D. Freestone (eds.), *International Law and Global Climate Change* (London: Graham & Trotmann/Martinus Nijhoff, 1991).

Clausen, T. J., and C. Bjerg, *The Blue Revolution: Adapting to Climate Change*, Thought Leadership Series No. 6 (Copenhagen: Copenhagen Climate Council, 2009).

Committee on Economic, Social and Cultural Rights, *General Comment No. 15 (2002), The Right to Water (Arts. 11 and 12 of the International Covenant on Economic, Social and Cultural Rights)*, 20 January 2003, UN Doc. E/C.12/2002/11.

Committee on Economic, Social and Cultural Rights, *Substantive Issues Arising in the Implementation of the International Covenant on Economic, Social and Cultural Rights, General Comment No. 14 (2000), The Right to the Highest Attainable Standard of Health*, 11 August 2000, E/C.12/2000/4.

Committee on Economic, Social and Cultural Rights, *General Comment No. 3: The Nature of States Parties Obligations (Art. 2, Par. 1)*, 14 December 1990, UN Doc. E/1991/23.

Committee on Public Health, *Statement on Preservation of New York City's Drinking Water Quality* (New York: The New York Academy of Medicine, 1989).

Compliance Committee of the 1999 Protocol on Water and Health, *Report of the Fifth Meeting of the Compliance Committee*, 22–23 June 2010, ECE/MP.WHC/C.1/2010/4, EUR/10/56335/16.

Compliance Committee of the 1999 Protocol on Water and Health, *Report of the Meeting of the Parties to the Protocol on Water and Health to the Convention on the Protection and Use of Transboundary Watercourses and International Lakes on its First Meeting*, 17–19 January 2007, ECE/MP.WH/2/Add.3, EUR/06/5069385/Add.3.

Conca, K., F. Wu, and J. Neukirchen, *Is There a Global Rivers Regime? Trends in the Principled Content of International River Agreements*, Harrison Program Research Report, University of Maryland (2003).

Conference of Rhine Ministers, *Rhine 2020: Program on the Sustainable Development of the Rhine* (Koblenz: ICPR, 2001).

Conference of the Parties Serving as the Meeting of the Parties of the Kyoto Protocol, *Outcome of the Work of the Ad Hoc Working Group on Further Commitments for Annex I Parties under the Kyoto Protocol*, 8 December 2012, UN Doc. FCCC/KP/CMP/2012/L.9.

Conference of the Parties of the UNFCCC, *Report of the Conference of the Parties on its Seventeenth Session, Held in Durban from 28 November to 11 December 2011*, 15 March 2012, UN Doc. FCCC/CP/2011/9/Add.1.

Conference of the Parties of the UNFCCC, *Report of the Conference of the Parties on its Seventeenth Session, Held in Durban from 28 November to 11 December 2011*, 15 March 2012, UN Doc. FCCC/CP/2011/9/Add.2.

Conference of the Parties of the UNFCCC, *Report of the Conference of the Parties on its Sixteenth Session, Held in Cancun from 29 November to 10 December 2010*, 15 March 2011, UN Doc. FCCC/CP/2010/7/Add.1.

Conference of the Parties of the UNFCCC, *Report of the Conference of the Parties on its Fifteenth Session, Held in Copenhagen from 7 to 19 December 2009*, 30 March 2010, UN Doc. FCCC/CP/2009/11/Add.1.

Conference of the Parties of the UNFCCC, *Report of the Conference of the Parties on its Thirteenth Session, Held in Bali from 3 to 15 December 2007*, 14 March 2008, UN Doc. FCCC/CP/2007/Add.1.

Conference of the Parties of the UNFCCC, *Report of the Conference of the Parties on its Seventh Session, Held at Marrakech from 29 October to 10 November 2001*, 21 January 2002, UN Doc. FCCC/CP/2001/13/Add.1, Add.2, and Add.4.

Conference of the Parties of the UNFCCC, *Report of the Conference of the Parties on its Fourth Session, Held at Buenos Aires from 2 to 14 November 1998*, 25 January 1999, UN Doc. FCCC/CP/1998/16/Add.1.

Conference of the Parties of the UNFCCC, *Report of the Conference of the Parties on its First Session, Held at Berlin from 28 March to 7 April 1995*, 6 June 1995, UN Doc. FCCC/CP/1995/7/Add.1.

Cooley, H., J. Christian-Smith, P. Gleick, L. Allen, and M. Cohen, *Understanding and Reducing the Risks of Climate Change for Transboundary Waters* (Oakland, CA: Pacific Institute, 2009).

Coomans, F., 'Some Remarks on the Extraterritorial Application of the International Covenant on Economic, Social and Cultural Rights', in F. Coomans and M. Kamminga (eds.), *Extraterritorial Application of Human Rights Treaties* (Antwerp: Intersentia, 2004), pp. 183–199.

Craven, M., *The International Covenant on Economic, Social and Cultural Rights – A Perspective on its Development* (Oxford: Clarendon Press, 1995).

Davidson, B., *The History of West Africa to the Nineteenth Century* (New York: Anchor Books, 1966).

Daoudy, M., and E. Kistin, 'Beyond Water Conflict: Evaluating the Effects of International Water Cooperation', paper prepared for the 49th Annual Conference of the International Studies Association, San Francisco (2008).

De los Cobos, G., 'The Transboundary Aquifer of the Geneva Region (Switzerland and France): Successfully Managed for 30 Years by the State of Geneva and French Border Communities', International Shared Aquifer Resource Management (ISARM) 2010 International Conference on Transboundary Aquifers, Paris, 6–8 December 2010.

Delbrück, J. (ed.), *International Law of Cooperation and State Sovereignty – Proceedings of an International Symposium of the Kiel Walther-Schücking-Institute of International Law, May 23–26, 2001* (Berlin: Duncker & Humblot, 2002).

Delbrück, J. (ed.), *New Trends in International Lawmaking – International 'Legislation' in the Public Interest* (Berlin: Duncker & Humblot, 1997).

Dellapenna, J. W., 'The Customary International Law of Transboundary Fresh Waters' (2001) 1 *International Journal Global Environmental Issues* 264–305.
Dellapenna, J. W., 'Adapting the Law of Water Management to Global Climate Change and Other Hydropolitical Stresses' (1999) 35 *JAWRA* 6, 1301–1326.
Dickstein, H., 'International Lake and River Pollution Control: Questions of Method' (1973) 12 *Columbia Journal of Transnational Law* 487–519.
Drafting Committee of the Sixth Committee, *Report of the Drafting Committee*, 22 October 1996, UN Doc. A/C.6/NUW/WG/L.1.
Dunne, T., and B. C. Schmidt, 'Realism', in J. Baylis and S. Smith (eds.), *The Globalization of World Politics*, 3rd edition (New York: Oxford University Press, 2005), pp. 141–160.
Ebbeson, J., and P. Okowa (eds.), *Environmental Law and Justice in Context* (Cambridge: Cambridge University Press, 2009).
Eckstein, G., 'Water Scarcity, Conflict and Security in a Climate Change World: Challenges and Opportunities for International Law and Policy' (2010) 27 *Wisconsin International Law Journal* 3, 409–461.
Eckstein, G., 'A Hydrological Perspective of the Status of Ground Water Resources under the UN Watercourses Convention' (2005) 30 *Columbia Journal of Environmental Law* 3, 525–564.
Ellis, J., 'Sustainable Development as a Legal Principle: A Rhetorical Analysis', Working Paper Series, McGill University (2008).
Elver, H., *Peaceful Uses of International Rivers – The Euphrates and Tigris Rivers Dispute* (Ardsley, NY: Transnational Publishers Inc., 2002).
Evans, M. D., *International Law* (New York: Oxford University Press, 2003).
Farrajota, M., 'Notification and Consultation in the Law Applicable to International Watercourses', in L. Boisson de Chazournes and S. Salman (eds.), *Les ressources en eau et le droit international – Water Resources and International Law*, The Hague Academy of International Law (Leiden: Martinus Nijhoff Publishers, 2005), pp. 281–339.
Faruqui, N. I., A. K. Biswas, and M. J. Bino, (eds.), *Water Management in Islam* (Tokyo: United Nations University Press, 2001).
Fawcett, J., 'The Legal Character of International Agreements' (1953) 30 *BYIL* 381–400.
Ferdon, J., 'Federal Weather Modification Projects: Compensating the Landowner' (1984) 26 *Arizona Law Review* 681–698.
Finnish Branch of the ILA, *The Work of the International Law Association on the Law of International Water Resources* (Helsinki: Finnish Branch of the ILA, 1988).
Fitzmaurice, G., 'The General Principles of International Law' (1957) 92 *RCADI* 1–227.
Fleishman, M., 'Reciprocity Unmasked: The Role of the Mexican Government in Defense of its Foreign Nationals in United States Death Penalty Cases' (2003) 20 *Arizona Journal of International and Comparative Law* 2, 359–407.
Food and Agricultural Organization (FAO), *Treaties Concerning the Non-navigational Uses of International Watercourses: Africa – Traités concernant l'utilisation des cours*

d'eau internationaux à des fins autres que la navigation: Afrique, FAO Legislative Study – 61 (Rome: FAO, 1997).
Food and Agricultural Organization (FAO), *Code of Conduct for Responsible Fishers* (Rome: FAO, 1959).
Fowler, M. R., and J. M. Bunck, *Law, Power, and the Sovereign State – The Evolution and Application of the Concept of Sovereignty* (University Park: Pennsylvania State University Press, 1995).
Freeston, D., and T. Ijlstra, *The North Sea: Perspectives on Regional Environmental Cooperation* (London: Graham and Trotman, 1990).
Friedman, W., 'General Course in Public International Law' (1969) 127 *RCADI* 39–224.
Garcia, B., *The Amazon from an International Law Perspective* (Cambridge: Cambridge University Press, 2011).
Garner, B. (ed.), *Black's Law Dictionary*, 8th edition (St. Paul, MN: Thomson West, 2004).
Garretson, A. H., R. D. Hayton, and C. J. Olmstead (eds.), *The Law of International Drainage Basins* (Dobbs Ferry, NY: Oceana Publications, 1967).
General Assembly, *Resolution 64/292 on the Human Right to Water and Sanitation*, 28 July 2010, UN Doc. A/RES/64/292.
General Assembly, *Resolution 63/124 on the Law of Transboundary Aquifers*, 11 December 2008, UN Doc. A/RES/63/124.
General Assembly, *Official Records of the General Assembly, Fifty-Seventh Session, Supplement No. 10 and Corrigendum*, 30 September 2002, UN Doc. A/57/10 and Corr.1, para. 518.
General Assembly, *Resolution 54/175 on the Right to Development*, 17 December 1999, UN Doc. A/RES/51/175.
General Assembly, *Convention on the Law of Non-navigational Uses of International Watercourses, Report of the Secretary-General*, 6 August 1996, UN Doc. A/51/275.
General Assembly, *Resolution 3201 (S-VI), Declaration on the Establishment of a New International Economic Order*, 1 May 1974, UN Doc. A/RES/S-6/3201.
General Assembly, *Resolution 2995 (XXVII) on Cooperation between States in the Field of Environment*, 15 December 1972, UN Doc. A/RES/2995(XXVII).
General Assembly, *Resolution 2669 (XXV) on Progressive Development and Codification of the Rules of International Law Relating to International Watercourses*, 8 December 1970, UN Doc. A/RES/2669(XXV).
General Assembly, *Resolution 2625 (XXV), Declaration on Principles of International Law Concerning Friendly Relations and Cooperation among States in Accordance with the Charter of the United Nations*, 24 October 1970, UN Doc. A/8082.
General Assembly, *Resolution 1401 (XIV) on Preliminary Studies on the Legal Problems Relating to the Utilization and Use of International Rivers*, 21 November 1959, UN Doc. A/RES/1401 (XIV).
General Assembly, *Resolution 626 (VII) on the Right to Exploit Freely Natural Wealth and Resources*, 21 December 1952, UN Doc. A/RES/626(VII).
General Assembly, *Resolution 375 (IV), Draft Declaration on Rights and Duties of States*, 6 December 1949, UN Doc. A/RES/375 (IV).

Gibney, M., K. Tomašcevski, and J. Vedsted-Hansen, 'Transnational State Responsibility for Violations of Human Rights' (1999) 12 *Harvard Human Rights Journal* 267–295.
Giordano, M., and A. Wolf, 'Sharing Waters: Post-Rio International Water Management' (2003) 27 *Natural Resources Forum* 163–171.
Gleick, P., 'The Human Right to Water' (1998) 1 *Water Policy* 487–503.
Gleick, P., 'Water Resources: A Long-Range Global Evaluation' (1993) 20 *Ecology Law Quarterly* 141–149.
Gleick, P., M. Palaniappan, M. Morikawa, J. Morrison, and H. Cooley, *World's Water 2008–2009* (Washington, DC: Island Press, 2009).
Global Environmental Facility, *Progress Report on the Least Developed Countries Fund (LDCF) and the Special Climate Change Fund (SCCF)*, 28 October 2011, GEF/LDCF.SCCF.11/Inf.02/Rev.1.
Global Environmental Facility, *Evaluation of the GEF Strategic Priority for Adaptation (SPA), Full Evaluation Report* (Washington, DC: GEF, 2010).
Global Environmental Facility, *Programming to Implement the Guidance for the Special Climate Change Fund Adopted by the Conference of the Parties to the United Nations Framework Convention on Climate Change at its Ninth Session* (Washington, DC: GEF, 2004).
Godana, B. A., *Africa's Shared Water Resources – Legal and Institutional Aspects of the Nile, Niger and Senegal River Systems* (London: Frances Pinter [Publishers] Limited, 1985).
Gore, P. J. W., 'Floods and Flood Control', in C. W. Allin (ed.), *Encyclopedia of Global Resources* (Pasadena, CA: Salem Press, 2010), pp. 441–445.
Grotius, H., *The Rights of War and Peace*, reprint (New York: Cosimo Classics, 2007).
Hafner, G., 'The Optimum Utilization Principle and the Non-Navigational Uses of Drainage Basins' (1993) 45 *Austrian Journal of Public and International Law* 113–146.
Hamner, J., and A. Wolf, 'Patterns in International Water Resources Treaties: The Transboundary Freshwater Dispute Database' (1997) 9 *Colorado Journal of International Environmental Law and Policy* 157–177.
Harris, P. G., 'Collective Action on Climate Change: The Logic of Regime Failure' (2007) 47 *Natural Resources Journal* 195–224.
Harrison, G. P., H. W. Whittington, and S. W. Gundry, 'Climate Change Impacts on Hydroelectric Power', in *Proceedings of 33rd Universities Power Engineering Conference (UPEC '98)* (Edinburgh, 1998), pp. 391–394.
Hayward, K., 'Supporting Basin-Wide Reforms with an Independent Assessment Applying International Water Law: Case Study of the Dnieper River' (2007) 18 *CJIEL* 3, 633–664.
He, Y., 'On the Performance of the Principle of Equitable and Reasonable Utilization in the Practice of Utilization of Transboundary Water Resources' (2009) 1 *IBusiness* 40–45.
Hettne, B., and F. Söderbaum, *Regional Cooperation* (Göteborg: International Task Force on Public Goods, 2004).
Higgins, R., 'General Course on Public International Law' (1991) 230 *RCADI* 9–342.

Higgins, R., 'The United Nations and Lawmaking: The Political Organs' (1970) 64 *AJIL* 37–48.

Hinds, C., *Umweltrechtliche Einschränkungen der Souveränität – Völkerrechtliche Präventionspflichten zur Verhinderung von Umweltschäden* (Frankfurt am Main: Peter Lang Europäischer Verlag der Wissenschaften, 1997).

Hirji, R., and R. Davis, *Environmental Flows in Water Resources Policies, Plans and Projects* (Washington, DC: World Bank, 2009).

Hirsch, A. J., 'From the Indus to the Jordan: Characteristics of Middle East International River Disputes' (1956) 71 *Political Science Quarterly* 2, 203–222.

Hoffmeister, F., 'Outsider or Frontrunner? Recent Developments under International and European Law on the Status of the European Union in International Organizations and Treaty Bodies' (2007) 44 *Common Market Law Review* 41–68.

Homer-Dixon, T. F., *Environmental Scarcity and Global Security*, Headline Series No. 300 (Ephrata, PA: Foreign Policy Association, 1993).

Human Rights Committee, *CCPR General Comment No. 12: The Right to Self-determination of Peoples (Art. 1)*, 13 March 1984.

Human Rights Committee, *CCPR General Comment No. 06: The Right to Life (Art. 6)*, 30 April 1982.

Human Rights Council, *Resolution 15/9 on Human Rights and Access to Safe Drinking Water and Sanitation*, 30 September 2010, UN Doc. A/HRC/RES/15/9.

Hussein, I., and O. Al-Jayousi, 'Management of Shared Waters: A Comparison of International and Islamic Law', in N. Faruqui, A. Biswas, and M. Bino (eds.), *Water Management in Islam* (Tokyo: United Nations University Press, 2001), pp. 128–135.

Institute of International Law, *Annuaire de l'Institut de droit international, Session d'Athènes*, 58(I) (Paris: S. Karger, 1979).

Institute of International Law, *Annuaire de l'Institut de droit international, Session d'Athènes 1979*, 58(II) (Paris: S. Karger, 1979).

Institute of International Law, *Annuaire de l'Institut de Droit International – Session de Salzbourg, September 1961*, 49(II) (Basle: Verlag für Recht und Gesellschaft AG, 1961).

Inter-American Bar Association, *Proceedings of the Tenth Conference, Buenos Aires, 14–21 November 1957*, Vol. I, Buenos Aires (1958).

International Commission of the Protection of the Danube River (ICPDR), *Shared Waters – Joint Responsibilities, ICPDR Annual Report 2011* (Vienna: ICPDR, 2011).

International Commission for the Protection of the Rhine (ICPR), *Internationally Coordinated Management Plan for the International River Basin District of the Rhine, Part A* (Koblenz: ICPR, 2009).

International Commission for the Protection of the Rhine (ICPR), *Conference of Rhine Ministers 2001: Rhine 2020 – Program on the Sustainable Development of the Rhine* (Koblenz: ICPR, 2001).

International Law Association, *Report of the Seventy-First Conference (Berlin)* (London: ILA, 2004).

International Law Association, *Report of the Sixty-First Conference Held at Paris, August 26th to September 1st, 1984* (London: ILA, 1985).
International Law Association, *Report of the Sixtieth Conference Montreal 1982* (London: ILA, 1982).
International Law Association, *Report of the Fifty-Ninth Conference, Belgrade, August 17th, 1980, to August 23rd, 1980* (London: ILA, 1980).
International Law Association, *Report of the Fifty-Seventh Conference Held at Madrid, August 30th, 1976, to September 4th, 1976* (London: ILA, 1978).
International Law Association, *Report of the Fifty-Sixth Conference Held at New Delhi, December 29th, 1974, to January 4th, 1975* (London: ILA, 1976).
International Law Association, *Report of the Fifty-Fifth Conference, New York 1972* (London: ILA, 1974).
International Law Association, *Report of the Fifty-Second Conference Held at Helsinki, August 14th to August 20th, 1966* (London: ILA, 1967).
International Law Commission, 'Draft Articles on the Law of Transboundary Aquifers, with Commentaries', in ILC, *Report on the Work of its Sixtieth Session (5 May to 6 June and 7 July to 8 August* 2008), UN Doc. A/63/10.
International Law Commission, *Fifth Report on Shared Natural Resources: Transboundary Aquifers by Chusei Yamada, Special Rapporteur*, 21 February 2008, UN Doc. A/CN.4/591.
International Law Commission, *Third Report by the Special Rapporteur Chusai Yamada on Shared Natural Resources: Transboundary Groundwaters*, 11 February 2005, UN Doc. A/CN.4/551.
International Law Commission, *Second Report on Shared Natural Resources: Transboundary Groundwaters by Mr. Chusei Yamada, Special Rapporteur*, 9 March 2004, UN Doc. A/CN.4/539.
International Law Commission, 'Draft Articles on the Law of the Non-navigational Uses of International Watercourses and Commentaries thereto and Resolution on Transboundary Confined Groundwater (1994) II(2) *YBILC* 89–135.
International Law Commission, 'Sixth Report on the Law of the Non-navigational Uses of International Watercourses, by Mr. Stephen C. McCaffrey, Special Rapporteur' (1990) II(1) *YBILC* 42–82.
International Law Commission, 'Draft Articles on the Law of the Non-navigational Uses of International Watercourses. Titles and Texts Adopted by the Drafting Committee: Titles of Parts II and III of the Draft Articles; Articles 8 to 21 – Reproduced in A/CN.4/SR.2070 to SR.2073' (1988) I *YBILC* 179–207.
International Law Commission, 'Fourth Report on the Law of the Non-navigational Uses of International Watercourses, by Mr. Stephen C. McCaffrey, Special Rapporteur' (1988) II(1) *YBILC* 205–250.
International Law Commission, 'Third Report on the Law of the Non-navigational Uses of International Watercourses, by Mr. Stephen McCaffrey, Special Rapporteur' (1987) II(1) *YBILC* 15–45.

International Law Commission, 'Summary Records of the Meetings of the Thirty-Ninth Session, 4 May–17 July 1987' (1987) I *YBILC* 70–95.

International Law Commission, 'Second Report on the Law of the Non-navigational Uses of International Watercourses, by Mr. Jens Evensen, Special Rapporteur' (1984) II(1) *YBILC* 101–127.

International Law Commission, 'First Report on the Law of the Non-navigational Uses of International Watercourses, by Mr. J. Evensen, Special Rapporteur' (1983) II(1) *YBILC* 155–195.

International Law Commission, 'Second Report on the Law of the Non-navigational Uses of International Watercourses, by Mr. Stephen Schwebel, Special Rapporteur' (1980) II(1) *YBILC* 159–198.

International Law Commission, 'Third Report on the Law of the Non-navigational Uses of International Watercourses, by Mr. Stephen M. Schwebel, Special Rapporteur' (1982) II(1) *YBILC* 65–191.

International Law Commission, 'First Report on the Law of the Non-navigational Uses of International Watercourses, by Mr. Stephen Schwebel, Special Rapporteur' (1979) II(1) *YBILC* 143–177.

International Rivers Network, *Trading Away the Future – the Mekong Power Grid* (Berkeley, CA: International Rivers Network, 2006).

Kahler, M., 'Economic Security in an Era of Globalization: Definition and Provision' (2004) 17 *The Pacific Review* 4, 485–502.

Kandel, R., *Water from Heaven* (New York: Columbia University Press, 1998).

Kaya, I., *Equitable Utilization: The Law of Non-Navigational Uses of International Watercourses* (Aldershot: Ashgate, 2003).

Kelsen, H., '*Théorie du Droit International Public*' (1953) 84 *RCADI* 5–200.

Keohane, R. O., *After Hegemony – Cooperation and Discord in the World Political Economy*, 1st edition (Princeton, NJ: Princeton University Press, 1984/2005).

Kiss, A., 'The Protection of the Rhine against Pollution' (1985) 25 *Natural Resources Journal* 613–638.

Kiss, A., and D. Shelton, *International Environmental Law*, 2nd edition (Ardsley, NY: Transnational Publishers, 2000).

Klabbers, J., *An Introduction to International Institutional Law* (Cambridge: Cambridge University Press, 2002).

Kliot, N., *Water Resources and Conflict in the Middle East* (New York: Routledge, 1994).

Kohen, M. G. (ed.), *Promoting Justice, Human Rights and Conflict Resolution through International Law / La promotion de la justice, des droits de l'homme et du règlement des conflits par le droit international, Liber Amicorum Lucius Caflish* (Leiden: Koninlijke Brill NV, 2007).

Kohen, M., 'Is the Notion of Territorial Sovereignty Obsolete?', in M. Pratt and J. A. Brown (eds.), *Borderlands under Stress* (London: Kluwer Law International, 2000), pp. 35–47.

Kolb, R., *La bonne foi en droit international public – Contribution à l'étude des principes généraux de droit*, Ph.D. Thesis No. 593, Université de Genève (1999).

Künnemann, R. 'Extraterritorial Application of the International Covenant on Economic, Social and Cultural Rights', in F. Coomans and M. Kamminga

(eds.), *Extraterritorial Application of Human Rights Treaties* (Antwerp: Intersentia, 2004), pp. 201–231.
Lammers, J. G., *Pollution of International Watercourses – A Search for Substantive Rules and Principles of Law* (The Hague: Martinus Nijhoff Publishers, 1984).
Landheer, M. B., '*Les théories de la sociologie contemporaine et le droit international*' (1957) 92 *RCADI* 519–627.
Landman, T., *Studying Human Rights* (New York: Routledge, 2006).
Lauterpacht, H., *Private Law Sources and Analogies of International Law* (London: Longmann, Green and Co. Ltd., 1927).
Lautze, J., M. Giordano, and M. Borghese, 'Driving Forces behind African Transboundary Water Law: Internal, External, and Implications', Conference Paper, *African Water Laws: Plural Legislative Frameworks for Rural Water Management in Africa*, Johannesburg, South Africa, 26–28 January 2005.
Legault, L. H., 'The Roles of Law and Diplomacy in Dispute Resolution: The IJC as a Possible Model' (2000) 26 *Canada–United States Law Journal* 47–55.
Lejeune, Y., *Receuil des accords internationaux conclus par les Cantons suisses: en vigeur 1er janvier 1980* (Bern: P. Lang, 1982).
Le Marquand, D. G., *International Rivers: The Politics of Cooperation* (Vancouver: University of British Columbia, Westwater Research Centre, 1977).
Lewis, L., 'Ecologists Warn the Planet is Running Short of Water', *TimesOnline*, 22 January 2009, www.timesonline.co.uk/tol/news/environment/article5562906.ece.
Lipper, J., 'Equitable Utilization', in A. H. Garretson, R. D. Hayton, and C. J. Olmstead (eds.), *The Law of International Drainage Basins* (Dobbs Ferry, NY: Oceana Publications, 1967), pp. 15–88.
Louka, E., *International Environmental Law: Fairness, Effectiveness and World Order* (New York: Cambridge University Press, 2006).
Magee, D., 'Powershed Politics: Yunnan Hydropower under Great Western Development' (2006) 185 *The China Quarterly* 23–41.
Magraw, D., 'Legal Treatment of Developing Countries: Differential, Contextual, and Absolute Norms' (1990) 1 *CJIEL* 69–99.
Mair, R., 'ICPDR Actions Related to Climate Change', Presentation, First Meeting of the Core Group of UNECE Pilot Projects, 15–16 February 2011.
Mazlum, I., 'Transboundary Water Management in the Euphrates–Tigris Basin: Dynamics of Regional Cooperation, Sustainability and Governance', in C. Lipchin *et al.*, *The Jordan River and the Dead Sea: Cooperation Amid Conflict*, NATO Science for Peace and Security Series C: Environmental Security (New York: Springer, 2009), pp. 139–164.
Mbaziira, R., N. Senfuma, and R. McDonnell, *Institutional Development in the Nile Equatorial Lakes Sub-Basin: Learning from the Experience of the Kagera Basin Organisation*, International Water Management Institute Conference Paper, East Africa Integrated River Basin Management Conference, Sokoine University of Agriculture, Morogoro, Tanzania, 7–9 March 2005.
McCaffrey, S., 'International Law Commission Adopts Draft Articles on Transboundary Aquifers' (2009) 103 *AJIL* 272–293.

McCaffrey, S., *The Law of International Watercourses: Non-Navigational Uses*, 2nd edition (New York: Oxford University Press, 2007).

McCaffrey, S., 'Some Developments in the Law of International Watercourses', in M. Kohen (ed.), *Promoting Justice, Human Rights and Conflict Resolution through International Law – Liber Amicorum Lucius Caflisch* (Leiden: Martinus Nijhoff, 2007), pp. 785–791.

McCaffrey, S., 'The Need for Flexibility in Freshwater Treaty Regimes' (2003) 27 *Natural Resources Forum* 156–162.

McCaffrey, S., 'The UN Convention on the Law of the Non-Navigational Uses of International Watercourses: Prospects and Pitfalls', in S. Salman and L. Boisson de Chazournes, *International Watercourses – Enhancing Cooperation and Managing Conflict*, World Bank Technical Paper No. 414 (Washington, DC: World Bank, 1998), pp. 17–28.

McCaffrey, S., 'Human Right to Water: Domestic and International Implications' (1992–1993), 5 *Georgetown International Environmental Law Review* 1–24.

McCaffrey, S., 'International Organizations and the Holistic Approach to Water Problems' (1991) 31 *Natural Resources Journal* 139–165.

McDonnell, R., 'Challenges for Integrated Water Resources Management: How Do We Provide the Knowledge to Support Truly Integrated Thinking?' (2008) 24 *International Journal of Water Resources Management* 1, 131–143.

McIntyre, O., *Environmental Protection of International Watercourses under International Law* (Aldershot: Ashgate, 2007).

Mechlem, K., 'International Law Commission Adopts Draft Articles of a Transboundary Aquifers Convention' (2008) 12 *ASIL Insights* 18.

Menon, P., 'Some Legal Aspects of the Committee for Coordination of Investigations of the Lower Mekong Basin' (1971) 18 *Netherlands International Law Review* 197–216.

Meron, T., 'Extraterritoriality of Human Rights Treaties' (1995) 89 *AJIL* 78–82.

Molle, F., T. Foran, and M. Käkönen (eds.), *Contested Waterscapes in the Mekong Region – Hydropower, Livelihoods and Governance* (London: Earthscan, 2009).

Moller, L., 'Transboundary Water Conflicts over Hydropower and Irrigation: Can Multilateral Development Banks Help?', CREDIT Research Paper No. 05/09, University of Nottingham (2009).

Montz, B. E., 'The Generation of Flood Hazards and Disasters by Urban Development of Floodplains', in D. J. Parker (ed.), *Floods, Volume 1* (London: Routledge, 2000), pp. 116–128.

Moses, R. J., and J. N. Corbridge, *Legal Structures for International Supervision of Weather Modification Techniques* (Washington, DC: Water for Peace Conference), Doc. P/635, 1967.

Mosler, H., 'The International Society as a Legal Community' (1974) 140 *RCADI* 1–282.

Nollkaemper, A., *The Legal Regime for Transboundary Water Pollution: Between Discretion and Constraint* (Dordrecht: Martinus Nijhoff/Graham & Trotman, 1993).

O'Connor, J. F., *Good Faith in International Law* (Aldershot: Dartmouth Publishing Company Limited, 1991).
Office of the High Commissioner of Human Rights, *Annual Report of the United Nations High Commissioner for Human Rights and Reports of the Office of the High Commissioner and the Secretary General*, 15 January 2009, UN Doc. A/HRC/10/61.
Okowa, P., 'Procedural Obligations in International Environmental Agreements' (1996) 67 *BYIL* 275–336.
Olson, M., *The Logic of Collective Action: Public Goods and the Theory of Groups*, 2nd edition (New York: Schocken Books, 1968).
O'Neill, K., J. Balsiger, and S. VanDeveer, 'Actors, Norms and Impact: Recent International Cooperation Theory and the Influence of the Agent-Structure Debate' (2004) 7 *Annual Review of Political Science* 149–175.
Pakenham, T., *The Scramble for Africa – White Man's Conquest of the Dark Continent from 1876 to 1912* (New York: Avon Books, 1991).
Paisley, R., 'Adversaries into Partners: International Water Law and the Equitable Sharing of Downstream Benefits' (2002) 3 *Melbourne Journal of International Law* 280–300.
Palaniappan, M., and Gleick, P., 'Peak Water', in P. Gleick *et al.*, *World's Water 2008–2009* (Washington, DC: Island Press, 2009), pp. 1–15.
Pan American Union, *Seventh International Conference of American States, Plenary Sessions, Minutes and Antecedents* (Montevideo: Pan American Union, 1933).
Permanent Court of Arbitration, *Resolution of International Water Disputes* (Leiden: Kluwer Law International, 2003).
Perrez, F. X., *Cooperative Sovereignty – From Independence to Interdependence in the Structure of International Environmental Law* (The Hague: Kluwer Law International, 2000).
Petri, M., 'Water Quality of Lake Constance', in T. Knepper, *The Rhine – Handbook of Environmental Chemistry* (SpringerLink, 2006), pp. 127–138.
Phillips, D., M. Daoudy, S. McCaffrey, J. Öjendal, and A. Turton, *Trans-boundary Water Co-operation as a Tool for Conflict Prevention and Broader Benefit Sharing* (Stockholm: Ministry of Foreign Affairs, 2006).
PONS Fachwörterbuch Recht English-Deutsch/Deutsch-Englisch (München: Ernst Klett Verlag, 1997).
Potsdam Institute for Climate Impact Research and Climate Analytics, *Turn down the Heat: Why a 4°C Warmer World Must Be Avoided* (Washington, DC: World Bank, 2012).
Pyatt, F. B., 'Acid Rain in Scandinavia – Some Current Data' (1987) 7 *The Environmentalist* 3, 197–200.
Raison, J.-P., and G. Magrin (eds.), *Des fleuves entre conflits et compromis* (Paris: Karthala, 2009).
Rajamani, L., 'The Principle of Common but Differentiated Responsibility and the Balance of Commitments under the Climate Regime' (2000) 9 *Review of European Community & International Environmental Law* 2, 120–131.
Riedel, E., 'International Environmental Law – A Law to Serve the Public Interest? An Analysis of the Scope of the Binding Effect of Basic Principles (Public

Interest Norms)', in J. Delbrück (ed.), *New Trends in International Lawmaking – International 'Legislation' in the Public Interest* (Berlin: Duncker & Humblot, 1997), pp. 61–98.

Ringius, L., A. Torvanger, and A. Underdal, 'Burden Sharing and Fairness Principles in International Climate Policy' (2002) 2 *International Environmental Agreements: Politics, Law and Economics* 1–22.

Rub, A., *Hans Kelsens Völkerrechtslehre – Versuch einer Würdigung* (Zürich: Schulthess Polygraphischer Verlag, 1995).

Sadoff, C., T. Greiber, M. Smith, and G. Berkamp, *SHARE – Managing Water across Boundaries* (Gland, Switzerland: International Union for the Conservation of Nature, 2008).

Sadoff, C., and D. Grey, 'Cooperation on International Rivers: A Continuum for Securing and Sharing Benefits' (2005) 30 *Water International* 4, 420–427.

Sadoff, C., and D. Grey, 'Beyond the River: The Benefits of Cooperation on International Rivers' (2002) 4 *Water Policy* 5, 389–403.

Sahovič, M., 'Codification of the Law of Friendly Relations between States' (1972) 137 *RCADI* 243–310.

Salman, S., 'Downstream Riparians Can also Harm Upstream Riparians: The Concept of Foreclosure of Future Uses' (2010) 34 *Water International* 4, 350–364.

Salman, S., 'The United Nations Watercourses Convention Ten Years Later: Why Has Entry into Force Proven Difficult?' (2007) 32 *Water International* 1, 1–15.

Salman, S., 'Evolution and Context of International Water Resources Law', in L. Boisson de Chazournes and S. Salman (eds.), *Les resources en eau et le droit international – Water Resources and International Law*, The Hague Academy of International Law (Leiden: Martinus Nijhoff Publishers, 2005), pp. 45–94.

Salman, S., 'From Marrakech through The Hague to Kyoto: Has the Global Debate on Water Reached a Dead End? Part Two' (2004) 29 *Water International* 1, 11–19.

Salman, S., 'From Marrakech through The Hague to Kyoto: Has the Global Debate on Water Reached a Dead End? Part One' (2003) 28 *Water International* 4, 491–500.

Salman, S., and L. Boisson de Chazournes (eds.), *International Watercourses – Enhancing Cooperation and Managing Conflict*, World Bank Technical Paper No. 414 (Washington, DC: World Bank, 1998).

Salman, S., and S. McInerney-Lankford, *The Human Right to Water – Legal and Policy Dimensions* (Washington, DC: World Bank, 2004).

Salman, S., and K. Uprety, *Conflict and Cooperation on South Asia's International Rivers* (The Hague: Kluwer Law International, 2002).

Samboly Ba, A., and M. Mbengue, '*Le régime juridique du fleuve Sénégal: aspects du droit des cours d'eau dans un context regional*' (2004) 12 *African Yearbook of International Law* 309–347.

Sands, P., *Principles of International Environmental Law*, 2nd edition (Cambridge: Cambridge University Press, 2003).

Saudi Aramco, 'The Greening of the Arab East' (1983) 34 *Saudi Aramco World* 5, 24–40.
Sauser-Hall, G., '*L'utilisation industrielle des fleuves internationaux*' (1953) 83 *RCADI* 465–582.
Scanlon, J., A. Cassar, and N. Nemes, *Water as a Human Right?* (Gland: IUCN, 2004).
Schmeier, S., 'Regional Cooperation Efforts in the Mekong River Basin: Mitigating River-Related Security Threats and Promoting Regional Development' (2009) 2 *Austrian Journal for Southeast Asian Studies* 2, 28–52.
Schwarzenberger, G., 'The Fundamental Principles of International Law' (1955) 87 *RCADI* 193–383.
Seck, S. M., A. Lericollais, and G. Magrin, '*L'aménagement de la vallée du Sénégal. Logiques nationales, crises et coopération entre les Etats riverains*', in J.-P. Raison and G. Magrin (eds.), *Des fleuves entre conflits et compromis* (Paris: Karthala, 2009), pp. 31–76.
Shelton, D., 'Equity' in D. Bodansky, J. Brunnée, and E. Hey (eds.), *The Oxford Handbook of International Environmental Law* (New York: Oxford University Press, 2007), pp. 640–663.
Shrijver, N., *Sovereignty over Natural Resources: Balancing Rights and Duties* (Cambridge: Cambridge University Press, 1977).
Simma, B., 'Reciprocity', in R. Bernhardt (ed.), *Encyclopedia of Public International Law, Volume Four* (Amsterdam: North-Holland/Elsevier Science, 2000), pp. 29–32.
Sixth Committee, *Summary Record of the 24th Meeting of the Sixth Committee*, held on 25 October 1996, UN Doc. A/C.6/51/SR.24.
Skogly, S., and M. Gibney, 'Transnational Human Rights Obligations' (2002) 24 *Human Rights Quarterly* 781–798.
Slaughter Burley, A.-M., 'International Law and International Relations Theory: A Dual Agenda' (2003) 87 *American Journal of International Law (AJIL)* 205–238.
Smakhtin, V., C. Revenga, and P. Doell, 'A Pilot Global Assessment of Environmental Water Requirements and Scarcity' (2004) 29 *Water International* 3, 307–317.
Smith, H., *The Economic Uses of International Rivers* (London: P. S. King, 1931).
Sohnle, J., *Le droit international des ressources en eau douce: solidarité contre souveraineté* (Paris: La documentation Française, 2002).
Solomon, S., G.-K. Plattner, R. Knutti, and P. Friedlingstein, 'Irreversible Climate Change due to Carbon Dioxide Emissions' (2009) 106 *Proceedings of the National Academy of Sciences of the United States of America* 6, 1704–1709.
Stone, C., 'Common but Differentiated Responsibilities in International Law' (2004) 98 *AJIL* 276–301.
Streck, C., 'The Global Environmental Facility – A Role Model for International Governance?' (2001) 1 *Global Environmental Politics* 2, 71–94.
Subsidiary Body for Scientific and Technological Advice, *Report on the Technical Workshop on Water and Climate Change Impacts and Adaptation Strategies*, 13 September 2012, UN Doc. FCCC/SBSTA/2012/4.

Subsidiary Body for Scientific and Technological Advice, *Report of the Subsidiary Body for Scientific and Technological Advice on its Twenty-Fifth Session Held at Nairobi from 6 to 14 November 2006*, 1 February 2007, UN Doc. FCCC/SBSTA/2006/11.

Tanzi, A., and M. Arcari, *The United Nations Convention on the Law of International Watercourses* (The Hague: Kluwer Law International, 2001).

Tarlock, A. D., and P. Wouters, 'Are Shared Benefits of International Waters an Equitable Apportionment?' (2007) 18 *CJIELP* 3, 523–536.

Teclaff, L. A., *The River Basin in History and Law* (The Hague: Martinus Nijhoff, 1967).

Thomson, D., *Europe since Napoleon* (London: Penguin Books, 1990).

Tignino, M., *'Les contours du principe de la participation publique et la protection des ressources en eau transfrontières'* (2010) 7 *Vertigo – La Revue électronique en sciences de l'environnement*.

Tignino, M., *Le statut de l'eau en temps de conflit armé*, Ph.D. thesis, Graduate Institute of International and Development Studies, Geneva (2008).

Tignino, M., *L'eau dans le processus de paix aux Proche-Orient: elements d'un régime juridique*, Working Paper 04/02, Graduate Institute of International Studies, Geneva (2004).

Tomuschat, C., *Human Rights – Between Idealism and Realism*, 2nd edition (Oxford: Oxford University Press, 2008).

United Nations, *Guidance on Water and Adaptation to Climate Change* (Geneva: United Nations, 2009), UN Doc. ECE/MP.WAT/30.

United Nations, *Agenda 21: Earth Summit – the United Nations Program of Action from Rio* (New York: United Nations, 1993), UN Doc. A/CONF.151/26/REV.1 (Vol. II).

United Nations, Press Release, 21 May 1997, UN Doc. GA/9248.

United Nations, *Report of the United Nations Water Conference, Mar del Plata, 14–25 March 1977* (New York: United Nations, 1977), UN Doc. E/CONF.70/29.

United Nations, *Report of the United Nations Conference on the Human Environment, Stockholm, 5–16 June 1972* (New York: United Nations, 1973).

United Nations, *Legal Problems in Relation to the Non-Navigational Uses of International Watercourses*, 15 April 1963, UN Doc. A/5409.

United Nations, *Legislative Texts and Treaty Provisions Concerning the Utilization of International Rivers for Other Purposes than Navigation*, 1963, UN Doc. ST/LEG/SER.B/12.

United Nations Conference on Environment and Development, *Rio Declaration on Environment and Development*, 12 August 1992, UN Doc. A/CONF.151/26 (Vol. I).

United Nations Conference on the Human Environment, 'Stockholm Declaration on the Human Environment', in *Report of the United Nations Conference on the Human Environment, Stockholm, 5–16 June 1972* (New York: United Nations, 1972).

United Nations Development Program (UNDP), *Human Development Report 2006 – Beyond Scarcity: Power, Poverty and the Global Water Crisis* (New York: Palgrave McMillan, 2006).

United Nations Economic and Social Commission for Western Asia, *Assessment of Legal Aspects of the Management of Shared Water Resources in the ESCWA Region* (New York: United Nations, 2002), UN Doc. GENERAL E/ESCWA/ENR/2001/3.

United Nations Economic Commission for Europe, *Progress in the Ratification Process, Including the Ratification of the Amendments to Open the Convention to Outside the Region*, 21 November 2012, MOP-6/2012/INF.1/REV.

United Nations Economic Commission for Europe, *Report of the Fourth Session of the Meeting of the Parties, Addendum, Chisinau Declaration and Decisions Adopted by the Meeting of the Parties, Chisinau, 29 June–1 July 2011*, UN Doc. ECE/MP.PP/2011/2/Add.1.

United Nations Economic Commission for Europe, *Report of the Meeting of the Parties on its Fifth Session*, 15 June 2010 (Geneva: UNECE, 2010), UN Doc. ECE/MP.WAT/29.

United Nations Economic Commission for Europe, *Overview of the Programme of Pilot Projects on Adaptation to Climate Change in Transboundary Basins*, 3 March 2010 (Geneva: UNECE Task Force on Water and Climate Change, 2010), TFWC/2010/3.

United Nations Economic Commission for Europe, *Legal Aspects of Hydroelectric Development of Rivers and Lakes of Common Interest* (Geneva: UNECE, 1952).

United Nations Environmental Program, *UNEP Goals and Principles of Environmental Impact Assessment*, UNEP Governing Council Decision 14/25, 16 January 1987, UN Doc. UNEP/GC/Dec/14/25.

United Nations Population Fund, *State of World Population 2001* (New York: United Nations Population Fund, 2001).

Utton, A. E., and J. Utton, 'The International Law of Minimum Stream Flows' (1999) 10 *CJIELP* 7–38.

Van Lier, I. H., *Acid Rain and International Law* (Alphen aan den Rijn: Sijthoff & Noordhoff, 1980).

Verdross, A., and B. Simma, *Universelles Völkerrecht*, 3rd edition (Berlin: Duncker & Humblot, 1984).

Verzijl, J., *International Law in Historical Perspective: Part III State Territory* (Leyden: A. W. Swijthoff, 1970).

Villiger, M. E., *Customary International Law and Treaties – A Manual on the Theory and Practice of the Interrelation of Sources*, 2nd edition (The Hague: Kluwer Law International, 1997).

Virally, M., '*La Réciprocité dans le droit international*' (1967) 122 *RCADI* 1–105.

Vitanyi, B., '*Les positions doctrinales concernant le sens de la notion de principes généraux de droit reconnus par les nations civilisées*' (1982) 86 *RGDIP* 48–116.

Vitànyi, B., *The International Regime of River Navigation* (Alphen aan den Rijn: Sijthoff & Noordhoff, 1979).

Voigt, C., *Sustainable Development as a Principle of International Law: Resolving Conflicts between Climate Measures and WTO Law* (Leiden: Martinus Nijhoff Publishers, 2009).

Von Tigerstrom, B., *Human Security and International Law – Prospects and Problems* (Oxford/Portland, OR: Hart Publishing, 2007).

Vörösmarty, C., P. Green, J. Salisbury, and R. Lammers, 'Global Water Resources: Vulnerability from Climate Change and Population Growth' (2000) 289 *Science* 284–288.

Warner, J., 'Contested Hydrohegemony: Hydraulic Control and Security in Turkey' (2008) 1 *Water Alternatives* 2, 271–288.

Waterbury, J., 'Between Unilateralism and Comprehensive Accords: Modest Steps toward Cooperation in International River Basins' (1997) 13 *Water Resources Development* 3, 279–289.

Wehberg, H., *Tableau Général des Résolutions (1873–1956)* (Basel: Verlag für Recht und Gesellschaft AG, 1957).

Wescoat, J., 'Beyond the River Basin: The Changing Geography of International Water Problems in International Watercourse Law' (1992) 3 *CJIELP* 301–330.

West's Law and Commercial Dictionary in Five Languages (St. Paul, MN: West Publishing Company, 1985).

Wolfers, A., 'National Security as an Ambitious Symbol', in A. Wolfers (ed.), *Discord and Collaboration: Essays on International Politics* (Baltimore, MD: Johns Hopkins University Press, 1962), pp. 147–165.

World Bank, *Implementation Completion and Results Report (TF-52900) on a Grant from the Global Environmental Facility (GEF) Trust Fund to the Organisation de la Mise en Valeur due Fleuve Senegal for the Regional Project: Senegal River Basin Water and Environmental Management* (Washington, DC: World Bank, 2009).

World Commission on Environment and Development, *Report of the World Commission on Environment and Development: Our Common Future*, 4 August 1987, UN Doc. A/42/427.

World Meteorological Organization (WMO), *Legal and Institutional Aspects of Integrated Flood Management* (Geneva: WMO, 2006).

World Summit on Sustainable Development, *Johannesburg Declaration on Sustainable Development*, 4 September 2002, UN Doc. A/CONF.199/20.

Wouters, P., 'The Legal Response to International Water Conflicts: The UN Watercourses Convention and Beyond' (1999) 42 *German Yearbook of International Law* 293–336.

Wouters, P., S. Salman, and P. Jones, 'The Legal Response for the World's Water Crisis: What Legacy from The Hague? What Future in Kyoto?' (2001) 4 *University of Denver Water Law Review* 2, 418–425.

Wouters, P., S. Vinogradov, A. Allan, P. Jones, and A. Rieu-Clarke, *Sharing Transboundary Waters – An Integrated Assessment of Equitable Entitlement: The Legal Assessment Model*, IHP-VI, Technical Documents in Hydrology, No. 74 (Paris: UNESCO, 2005).

Yamin, F., and J. Depledge, *The International Climate Change Regime: A Guide to Rules, Institutions and Procedures* (Cambridge: Cambridge University Press, 2004).

Yu, W., *Benefit Sharing in International Rivers: Findings from the Senegal River Basin, the Columbia River Basin, and the Lesotho Highlands Water Project*, Africa Region Water Resources Unit Working Paper 1 (Washington, DC: World Bank, 2008).

Zacklin, R., and L. Caflisch (eds.), *The Legal Regime of International Rivers and Lakes – Le régime juridique des fleuves et des lacs internationaux* (The Hague/Boston/London: Martinus Nijhoff Publishers, 1981).

Zartman, I. W., 'Concept: Cooperation' (2008) *PIN Points* 30, 5–7.

Zawahri, N. A., 'India, Pakistan and Cooperation along the Indus River System' (2009) 11 *Water Policy* 1–20.
Zeitoun, M., and N. Mirumachi, 'Transboundary Water Interaction I: Reconsidering Conflict and Cooperation' (2008) 8 *International Environmental Agreements* 4, 297–316.
Zeitoun, M., and J. Warner, 'Hydro-hegemony – A Framework for Analysis of Trans-boundary Water Conflicts' (2008) 8 *Water Policy* 435–460.
Zekster, I. S., and L. G. Everett, *Groundwater Resources of the World and their Use*, IHP-VI, Series on Groundwater No. 6 (Paris: UNESCO, 2004).
Zemanek, K., 'Unilateral Legal Acts Revisited', in K. Wellens (ed.), *International Law: Theory and Practice – Essays in Honour of Eric Suy* (The Hague: Martinus Nijhoff Publishers, 1998), pp. 209–221.
Zimmermann, A., C. Tomuschat, and K. Oellers-Frahm, (eds.), *The Statute of the International Court of Justice – A Commentary* (New York: Oxford University Press, 2006).

INTERNET SOURCES

Adaptation Fund: www.adaptation-fund.org.
Commission for the Hydrology of the Rhine Basin: www.chr-khr.org.
Department of Ecology, State of Washington: www.ecy.wa.gov/programs/wr/rules/images/pdf/pol1020.pdf.
Global Environmental Facility: www.thegef.org.
Global Water Partnership: www.gwpforum.org.
International Commission for the Protection of the Danube River: www.icpdr.org.
International Commission for the Protection of the Rhine: www.iksr.org.
International Court of Justice: www.icj-cij.org.
International Freshwater Treaties Database, Institute for Water and Watersheds, Oregon State University: http://ocid.nacse.org/tfdd/treaties.php.
International River Basin Registry, Oregon State University: www.transboundarywaters.orst.edu/publications/register/tables/IRB_southamerica.html.
Millennium Development Goals: www.un.org/millenniumgoals/goals.html.
Permanent Court of Arbitration: www.pca-cpa.org.
Permanent Court of International Justice: www.icj-cij.org/pcij/index.php?p1=9.
Second World Water Forum: www.waternunc.com/gb/secwwf12.htm.
Southern African Development Community: www.sadc.int.
Tonle Sap Cambodia: http://tonlesap.net.
United Nations Audiovisual Library of International Law: http://untreaty.un.org/cod/avl/ha/ccc/ccc.html.
United Nations Framework Convention on Climate Change: www.unfccc.int.

U.S. Census Bureau, International Data Base, December 2008 Update: www.census.gov/ipc/www/idb/worldpopgraph.html.
U.S. Energy Information Administration: www.eia.gov.
World Bank Development Indicators, 2010: http://databank.worldbank.org.

Index

CAMBRIDGE STUDIES IN INTERNATIONAL AND COMPARATIVE LAW

Books in the series

Cooperation in the Law of Transboundary Water Resources
Christina Leb

Underwater Cultural Heritage and International Law
Sarah Dromgoole

State Responsibility: The General Part
James Crawford

The Origins of International Investment Law
Kate Miles

The Crime of Aggression under the Rome Statute of the International Criminal Court
Carrie McDougall

Crimes against Peace and International Law
Kirsten Sellars

The Non-Legal in International Law
Fleur Johns

Armed Conflict and Displacement: The Protection of Refugees and Displaced Persons under International Humanitarian Law
Mélanie Jacques

Foreign Investment and the Environment in International Law
Jorge Viñuales

The Human Rights Treaty Obligations of Peacekeepers
Kjetil Larsen

Cyberwarfare and the Laws of War
Heather Harrison Dinniss

The Right to Reparation in International Law for Victims of Armed Conflict
Christine Evans

Global Public Interest in International Investment Law
Andreas Kulick

State Immunity in International Law
Xiaodong Yang

Reparations and Victim Support in the International Criminal Court
Conor McCarthy

Reducing Genocide to Law: Definition, Meaning, and the Ultimate Crime
Payam Akhavan

Decolonizing International Law: Development, Economic Growth and the Politics of Universality
Sundhya Pahuja

Complicity and the Law of State Responsibility
Helmut Philipp Aust

State Control over Private Military and Security Companies in Armed Conflict
Hannah Tonkin

'Fair and Equitable Treatment' in International Investment Law
Roland Kläger

The UN and Human Rights: Who Guards the Guardians?
Guglielmo Verdirame

Sovereign Defaults before International Courts and Tribunals
Michael Waibel

Making the Law of the Sea: A Study in the Development of International Law
James Harrison

Science and the Precautionary Principle in International Courts and Tribunals: Expert Evidence, Burden of Proof and Finality
Caroline E. Foster

Transition from Illegal Regimes in International Law
Yaël Ronen

Access to Asylum: International Refugee Law and the Globalisation of Migration Control
Thomas Gammeltoft-Hansen

Trading Fish, Saving Fish: The Interaction between Regimes in International Law
Margaret Young

The Individual in the International Legal System: Continuity and Change in International Law
Kate Parlett

The Participation of States in International Organisations: The Role of Human Rights and Democracy
Alison Duxbury

'Armed Attack' and Article 51 of the UN Charter: Evolutions in Customary Law and Practice
Tom Ruys

Science and Risk Regulation in International Law
Jacqueline Peel

Theatre of the Rule of Law: Transnational Legal Intervention in Theory and Practice
Stephen Humphreys

The Public International Law Theory of Hans Kelsen: Believing in Universal Law
Jochen von Bernstorff

Vicarious Liability in Tort: A Comparative Perspective
Paula Giliker

Legal Personality in International Law
Roland Portmann

Legitimacy and Legality in International Law: An Interactional Account
Jutta Brunnée and
Stephen J. Toope

The Concept of Non-International Armed Conflict in International Humanitarian Law
Anthony Cullen

The Challenge of Child Labour in International Law
Franziska Humbert

Shipping Interdiction and the Law of the Sea
Douglas Guilfoyle

International Courts and Environmental Protection
Tim Stephens

Legal Principles in WTO Disputes
Andrew D. Mitchell

War Crimes in Internal Armed Conflicts
Eve La Haye

Humanitarian Occupation
Gregory H. Fox

The International Law of Environmental Impact Assessment: Process, Substance and Integration
Neil Craik

The Law and Practice of International Territorial Administration: Versailles to Iraq and Beyond
Carsten Stahn

Cultural Products and the World Trade Organization
Tania Voon

United Nations Sanctions and the Rule of Law
Jeremy Farrall

National Law in WTO Law: Effectiveness and Good Governance in the World Trading System
Sharif Bhuiyan

The Threat of Force in International Law
Nikolas Stürchler

Indigenous Rights and United Nations Standards
Alexandra Xanthaki

International Refugee Law and Socio-Economic Rights
Michelle Foster

The Protection of Cultural Property in Armed Conflict
Roger O'Keefe

Interpretation and Revision of International Boundary Decisions
Kaiyan Homi Kaikobad

Multinationals and Corporate Social Responsibility: Limitations and Opportunities in International Law
Jennifer A. Zerk

Judiciaries within Europe: A Comparative Review
John Bell

Law in Times of Crisis: Emergency Powers in Theory and Practice
Oren Gross and
Fionnuala Ní Aoláin

Vessel-Source Marine Pollution: The Law and Politics of International Regulation
Alan Tan

Enforcing Obligations Erga Omnes *in International Law*
Christian J. Tams

Non-Governmental Organisations in International Law
Anna-Karin Lindblom

Democracy, Minorities and International Law
Steven Wheatley

Prosecuting International Crimes: Selectivity and the International Law Regime
Robert Cryer

Compensation for Personal Injury in English, German and Italian Law: A Comparative Outline
Basil Markesinis,
Michael Coester,
Guido Alpa and
Augustus Ullstein

Dispute Settlement in the UN Convention on the Law of the Sea
Natalie Klein

The International Protection of Internally Displaced Persons
Catherine Phuong

Imperialism, Sovereignty and the Making of International Law
Antony Anghie

Necessity, Proportionality and the Use of Force by States
Judith Gardam

International Legal Argument in the Permanent Court of International Justice: The Rise of the International Judiciary
Ole Spiermann

Great Powers and Outlaw States: Unequal Sovereigns in the International Legal Order
Gerry Simpson

Local Remedies in International Law
C. F. Amerasinghe

Reading Humanitarian Intervention: Human Rights and the Use of Force in International Law
Anne Orford

Conflict of Norms in Public International Law: How WTO Law Relates to Other Rules of International Law
Joost Pauwelyn

Transboundary Damage in International Law
Hanqin Xue

European Criminal Procedures Edited by
Mireille Delmas-Marty and
John Spencer

The Accountability of Armed Opposition Groups in International Law
Liesbeth Zegveld

Sharing Transboundary Resources: International Law and Optimal Resource Use
Eyal Benvenisti

International Human Rights and Humanitarian Law
René Provost

Remedies against International Organisations
Karel Wellens

Diversity and Self-Determination in International Law
Karen Knop

The Law of Internal Armed Conflict
Lindsay Moir

International Commercial Arbitration and African States: Practice, Participation and Institutional Development
Amazu A. Asouzu

The Enforceability of Promises in European Contract Law
James Gordley

International Law in Antiquity
David J. Bederman

Money Laundering: A New International Law Enforcement Model
Guy Stessens

Good Faith in European Contract Law
Reinhard Zimmermann and
Simon Whittaker

On Civil Procedure
J. A. Jolowicz

Trusts: A Comparative Study
Maurizio Lupoi

The Right to Property in Commonwealth Constitutions
Tom Allen

International Organizations before National Courts
August Reinisch

The Changing International Law of High Seas Fisheries
Francisco Orrego Vicuña

Trade and the Environment: A Comparative Study of EC and US Law
Damien Geradin

Unjust Enrichment: A Study of Private Law and Public Values
Hanoch Dagan

Religious Liberty and International Law in Europe
Malcolm D. Evans

Ethics and Authority in International Law
Alfred P. Rubin

Sovereignty over Natural Resources: Balancing Rights and Duties
Nico Schrijver

The Polar Regions and the Development of International Law
Donald R. Rothwell

Fragmentation and the International Relations of Micro-States: Self-Determination and Statehood
Jorri Duursma

Principles of the Institutional Law of International Organizations
C. F. Amerasinghe